T0135777

Augsburger Schriften zur Mathematik, Physik und Informatik
Band 14

herausgegeben von:
Professor Dr. F. Pukelsheim
Professor Dr. W. Reif
Professor Dr. D. Vollhardt

Bibliografische Information der Deutschen Nationalbibliothek

Die Deutsche Nationalbibliothek verzeichnet diese Publikation in der
Deutschen Nationalbibliografie; detaillierte bibliografische Daten sind
im Internet über http://dnb.d-nb.de abrufbar.

ISBN 978-3-8325-2796-9
ISSN 1611-4256

Logos Verlag Berlin GmbH
Comeniushof, Gubener Str. 47,
10243 Berlin
Tel.: +49 030 42 85 10 90
Fax: +49 030 42 85 10 92
INTERNET: http://www.logos-verlag.de

Harmonic Limits
of
Dynamical and Control Systems

Dissertation von

Tobias Wichtrey

Institut für Mathematik
Universität Augsburg
86135 Augsburg

Universität
Augsburg
University

Mündliche Prüfung abgelegt am 21. September 2010 an der Universität Augsburg vor Prof. Dr. Fritz Colonius, Prof. Dr. Dirk Blömker und Prof. Dr. Malte Peter.

Contents

Introduction

In a linear differential equation $\dot{x} = Ax$ in $\mathbb{C}^n$, $A \in \mathbb{C}^{n \times n}$, the real parts of the eigenvalues of the matrix A describe how fast trajecories grow or decay. These real parts can be generalized for nonlinear systems to Lyapunov exponents, which describe how fast two trajectories starting close to each other separate over time. On the other hand, the imaginary parts of the eigenvalues of A describe the rotational behaviour of the solutions of the linear differential equation. There have been several attempts to generalize them to nonlinear systems, and to describe the rotational behaviour of dynamical systems in general. See, e. g., [Fra92; Fra03; KH06; Poi85; Rob99] for different notions of rotation numbers for discrete-time systems, and [JM82; Ruf97; San88; Ste09] for rotation numbers in continuous time.

In this thesis, we will analyze a different approach to describe the rotational behaviour, namely the concept of rotational factor maps, which was introduced in [MB04, Section 3.1] and further pursued, e. g., in [LM08; Row09]. The general idea is to find a complex-valued map F on the state space that maps the dynamics onto a rotation around the origin in the complex plane. More formally, for a discrete-time system given by the iteration of a map $T : X \to X$ on a metric space X, we will look for a map $F : X \to \mathbb{C}$, $F \not\equiv 0$, with $F \circ T = \mathrm{e}^{\mathrm{i}\omega} \cdot F$ for some angle $\omega \in [0, 2\pi)$. We will call such a map a *rotational factor map to the angle* ω. For systems in continuous time, e. g., semi-flows Φ_t on a metric space X, we similarly look for a map $F : X \to \mathbb{C}$, $F \not\equiv 0$, with $F \circ \Phi_t = \mathrm{e}^{\mathrm{i}\omega t} \cdot F$ for some $\omega \in \mathbb{R}$ and all $t \geq 0$, which we will call a *rotational factor map to frequency* $\omega/2\pi$.

This concept of rotational factor maps is closely connected to harmonic limits, which are defined for a map $f : X \to \mathbb{C}$, an angle $\omega \in [0, 2\pi)$, and a point $x \in X$ by

$$ f_\omega^*(x) := \lim_{n \to \infty} \frac{1}{n} \sum_{j=0}^{n-1} \mathrm{e}^{\mathrm{i}j\omega} f(T^j x) $$

in the discrete-time case, and for $f : X \to \mathbb{C}$, $\omega \in \mathbb{R}$, $x \in X$ by

$$ f_\omega^*(x) := \lim_{T \to \infty} \frac{1}{T} \int_0^T \mathrm{e}^{\mathrm{i}\omega t} f(\Phi_t x) \mathrm{d}t $$

in the continuous-time case, if the limits exist. It turns out that there is a rotational factor map to some angle ω or some frequency $\omega/2\pi$ if and only if there is a map $f : X \to \mathbb{C}$ such that the harmonic limit f_ω^* is not constant zero. Because of this equivalence, harmonic limits are the key object of our analysis.

1

In [MB04], Mezić and Banaszuk use harmonic limits as a tool to compare dynamical systems, and to determine model parameters. We are interested in the properties of the harmonic limits themselves instead. We will, e. g., investigate existence of the harmonic limit under the presence of an invariant measure, and show how periodicity properties of the system affect the angles or frequencies that can occur.

This thesis is split into three parts: Chapter 1 deals with the harmonic analysis of discrete-time systems, while in Chapter 2, continuous-time systems are analyzed. In Chapter 3, we will look at the harmonic analysis of continuous-time control systems.

The first two chapters share a common structure: After some preliminary results and definitions in Section 1.1, we first introduce rotational factor maps and harmonic limits and show their relation in Section 1.2 and Section 2.2. In Section 1.3 and Section 2.3, we discuss properties of the harmonic limit, including boundedness and existence, and what effect certain dynamical properties—like periodicity or asymptotical behaviour—have on the harmonic limit. Then we discuss the relation between rotational factor maps and the Koopman operator, and, in the discrete-time case, also the Perron-Frobenius operator, see Section 1.4 and Section 2.4. Finally, in both chapters, we apply these concepts to a special class of systems, namely h-partitioned systems in the discrete-time case, see Section 1.5, and solutions of linear ordinary differential equations in the continuous-time case, see Section 2.5.

In the chapter about control systems (Chapter 3), we first perform a pointwise analysis, i. e., we fix a control and analyze the resulting system, see Section 3.2. In Section 3.3, we interpret the system as a semi-flow on the product of the state space and the set of control functions, and apply the results from Chapter 2. Finally, we look at convergent systems as a special class of control systems.

In this thesis, we have used the ideas from [MB04] to propose harmonic limits as a tool to analyze the rotational behaviour of dynamical systems in the sense of rotational factor maps. To this end, we stated and rigorously proved numerous properties of the harmonic limit, including, e. g., what angles or frequencies can occur at periodic and quasi-periodic orbits. We also transferred the concept of rotational factor maps and harmonic limits to continuous-time systems. For these systems, we provide a full proof of the Wiener-Wintner Ergodic Theorem, which gives a strong existence result in the case of ergodicity. We apply these results to h-partitioned systems in the discrete-time case, and to linear ordinary differential equations in the continuous-time case, and also transfer them to control systems, particularly to convergent systems.

For a list of the notation and symbols we use, see page 197. In this thesis, the sign $\Box$ marks the end of a proof, and the sign $\lrcorner$ marks the end of a definition, an example, or a remark. All numerical computations have been performed with the help of Matlab R2007a[1]. Furthermore,Maple 13[2] has been used for some analytical

---

[1]http://www.mathworks.com/products/matlab/
[2]http://www.maplesoft.com/products/maple/

computations.

I wish to express my sincere thanks to my advisor Prof. Dr. Fritz Colonius for his advice and help, my friends and colleagues Ralph Lettau and Gregory Pitl for proof-reading, my family for their constant support, Anna for her patience, and Jasmin for being there in the right moment.

Augsburg, May 2010   Tobias Wichtrey

Chapter 1

Harmonic analysis for discrete-time dynamical systems

In this chapter, we consider discrete-time dynamical systems given by the iteration of a map $T : X \to X$ on a metric space X. We want to analyze the rotational behaviour of such dynamical systems. More precisely, we investigate if the system is semi-conjugate to a rotation in the complex plane, i.e., if there are an angle $\omega \in [0, 2\pi)$ and a map $F : X \to \mathbb{C}$ such that $F \circ T = e^{i\omega} \cdot F$. See Subsection 1.2.1 for a discussion of these so-called *rotational factor maps* F.

A commonly considered object is the average of a map $f : X \to \mathbb{C}$ along a trajectory $x, Tx, T^2x, \ldots$, i.e., $\lim_{n \to \infty} 1/n \cdot \sum_{j=0}^{n-1} f(T^j x)$, if the limit exists, compare, e.g., [KH06, Section 4.1; Tak08]. This kind of average is called *time average*, and is the object of interest, e.g., in the well-known Birkhoff Ergodic Theorem [KH06, Theorem 4.1.2] and other ergodic theorems. One can generalize these averages to so-called *harmonic limits* $f_\omega^*(x) := \lim_{n \to \infty} 1/n \cdot \sum_{j=0}^{n-1} e^{ij\omega} f(T^j x)$ for $\omega \in [0, 2\pi)$. For $\omega = 0$, the harmonic limit equals the usual average. See Subsection 1.2.2 for the definition of these harmonic limits.

Harmonic limits turn out to be useful in the analysis of the rotational behaviour of dynamical systems. In fact, for $\omega \in [0, 2\pi)$, there is a map $f : X \to \mathbb{C}$, such that $f_\omega^* \not\equiv 0$ if and only if the system admits a rotational factor map by angle ω. See Subsection 1.2.3 for a discussion of this important relation.

These concepts were introduced in [MB04]. Mezić and Banaszuk use them in order to compare dynamical systems and to determine model parameters. In this thesis, we want to have a closer look at the properties of rotational factor maps and harmonic limits themselves. Therefore we first prove two simple properties of the harmonic limit in Subsection 1.3.1, and then show that, for given $x \in X$ and $f : X \to \mathbb{C}$ under a weak condition, there can only be countably many $\omega \in [0, 2\pi)$ such that $f_\omega^*(x) \neq 0$, i.e., only countably many angles can occur in the system.

In Subsection 1.3.2, we show some existence results under the presence of an invariant measure. The harmonic limit f_ω^* exists μ-almost everywhere, if there is an invariant measure μ, and if f is μ-integrable, see Theorem 1.3.10. Note that the null set of points, where the harmonic limit does not exist, can depend on the angle ω. For ergodic systems, there is a stronger result, known as the Wiener-Wintner Ergodic Theorem, see Theorem 1.3.14. It shows that the set of points, where the harmonic limit does not exist, can be chosen independently of the angle. Finally, for uniquely ergodic systems, the harmonic limit exists everywhere, see Proposition 1.3.11.

In the following two sections, we show what impact certain dynamical properties have on the properties of harmonic limits. Particularly, we have a look at asymptotic behaviour in Subsection 1.3.3, and at periodic behaviour in Subsection 1.3.4. If, e. g., two trajectories $x, Tx, \ldots$ and $y, Ty, \ldots$ asymptotically approach each other, then the harmonic limits $f_\omega^*(x)$ and $f_\omega^*(y)$ coincide for every $\omega \in [0, 2\pi)$ and all continuous functions $f : X \to \mathbb{C}$, see Corollary 1.3.16. One important result regarding periodicity is that, along almost periodic trajectories, harmonic limits always exist, see Lemma 1.3.20. Furthermore, one can characterize the set of angles that can possibly occur along quasi-periodic and periodic trajectories, see Proposition 1.3.22 and Proposition 1.3.24. In the periodic case, for each of these frequencies, one can actually find a continuous function $f : X \to \mathbb{C}$ whose harmonic limit is not constant zero and hence proves the existence of a rotational factor map.

There is a close connection between rotational factor maps and eigenfunctions of the Koopman operator $f \mapsto f \circ T$. This connection and its consequence for the map $f \mapsto f_\omega^*$, which maps a function onto its harmonic limit, will be treated in Section 1.4. Furthermore, we will see that the harmonic limit $f_\omega^*(x)$ can be given by $\int f \mathrm{d}\mu$ for some complex measure μ on X. This measure is an eigenmeasure, which is connected to eigenfunctions of the Perron-Frobenius operator. In fact, these measures have eigenfunctions of the Perron-Frobenius operator as densities.

Finally, we will discuss so-called h-partitioned systems in Section 1.5. The state space of this class of systems is divided into finitely many regions such that the system, loosely speaking, "jumps" through these partition elements in a given order. We will discuss the connection to rotational factor maps, and how the partition can be reconstructed using harmonic limits.

1.1 Preliminaries

Let us first collect some definitions and results, which we will use later in this chapter.

The concept of almost periodic functions dates back to Bohr [Boh47], and was later generalized (see, e. g., [Zha03]). In this work, we use almost periodic functions on $\mathbb{R}$ with values in metric spaces.

Definition 1.1.1 (Almost periodicity). Let X be a metric space with metric d. A function $f : \mathbb{R} \to X$ is called *almost periodic*, if it is continuous, and if for all $\varepsilon > 0$, there is an interval length $L(\varepsilon) > 0$ such that, in every interval $I \subset \mathbb{R}$ of length $L(\varepsilon)$, there is a so-called *translation number* $\tau \in I$ such that $d\big(f(t + \tau), f(t)\big) \leq \varepsilon$ holds for all $t \in \mathbb{R}$. ⌟

Lemma 1.1.2. *Let X, Y be metric spaces. Let $f : \mathbb{R} \to X$ be an almost periodic function, and $\phi : X \to Y$ continuous. Then $\phi \circ f : \mathbb{R} \to Y$ is almost periodic.*

Proof. This follows from [Maa67, Satz 6 in Section 7 and Satz 3 in Section 24]. □

Almost periodic functions can also be characterized as trigonometric series, i. e., as a sum of countably many periodic functions. A combination of *finitely* many periodic functions is called *quasi-periodic*. In this regard, compare also [Far94, Definition 6.5.1].

Definition 1.1.3 (Quasi-periodicity). Let X be a topological space. We call a function $f : \mathbb{R} \to X$ *quasi-periodic*, if there are $n \in \mathbb{N}$ and a function $F : \mathbb{R}^n \to X$, which is τ_j-periodic in its j-th argument for some $\tau_j > 0$, $j = 1, \ldots, n$, such that $F(t, \ldots, t) = f(t)$ for all $t \in \mathbb{R}$. The map F is called a *generating function* for f, and the numbers $\tau_1, \ldots, \tau_n$ are called *periods* of f. ⌟

As a last preliminary lemma, we compute antiderivatives of $\mathrm{e}^{\pm \mathrm{i}\omega t} \sin \omega t$ and of $\mathrm{e}^{\pm \mathrm{i}\omega t} \cos \omega t$.

Lemma 1.1.4. *Let $\omega \in \mathbb{R} \setminus \{0\}$. An antiderivative of $\mathrm{e}^{\pm \mathrm{i}\omega t} \sin \omega t$ with respect to t is given by $\mp \mathrm{i}/(2\omega) \cdot \mathrm{e}^{\pm \mathrm{i}\omega t} \sin \omega t \pm \mathrm{i}/2 \cdot t$. An antiderivative of $\mathrm{e}^{\pm \mathrm{i}\omega t} \cos \omega t$ with respect to t is given by $\mp \mathrm{i}/(2\omega) \cdot \mathrm{e}^{\pm \mathrm{i}\omega t} \cos \omega t + 1/2 \cdot t$.*

Proof. It holds that

$$
\begin{aligned}
\frac{\mathrm{d}}{\mathrm{d}t}\left[\mp \frac{\mathrm{i}}{2\omega} \mathrm{e}^{\pm \mathrm{i}\omega t} \sin \omega t \pm \frac{\mathrm{i}}{2} t \right] &= \mp \frac{\mathrm{i}}{2\omega}\left(\pm \mathrm{i}\omega \mathrm{e}^{\pm \mathrm{i}\omega t} \sin \omega t + \omega \mathrm{e}^{\pm \mathrm{i}\omega t} \cos \omega t \right) \pm \frac{\mathrm{i}}{2} \\
&= \frac{1}{2} \mathrm{e}^{\pm \mathrm{i}\omega t} \sin \omega t \mp \frac{\mathrm{i}}{2} \mathrm{e}^{\pm \mathrm{i}\omega t} \cos \omega t \pm \frac{\mathrm{i}}{2} \\
&= \frac{1}{2} \mathrm{e}^{\pm \mathrm{i}\omega t}\left(\sin \omega t \mp \mathrm{i} \cos \omega t \pm \mathrm{i} \mathrm{e}^{\mp \mathrm{i}\omega t} \right) \\
&= \frac{1}{2} \mathrm{e}^{\pm \mathrm{i}\omega t}\left(\sin \omega t \mp \mathrm{i} \cos \omega t \pm \mathrm{i} \cos \omega t + \sin \omega t \right) \\
&= \mathrm{e}^{\pm \mathrm{i}\omega t} \sin \omega t.
\end{aligned}
$$

Similarly, it holds that

$$
\begin{aligned}
\frac{\mathrm{d}}{\mathrm{d}t}\left[\mp \frac{\mathrm{i}}{2\omega} \mathrm{e}^{\pm \mathrm{i}\omega t} \cos \omega t + \frac{1}{2} t \right] &= \mp \frac{\mathrm{i}}{2\omega}\left(\pm \mathrm{i}\omega \mathrm{e}^{\pm \mathrm{i}\omega t} \cos \omega t - \omega \mathrm{e}^{\pm \mathrm{i}\omega t} \sin \omega t \right) + \frac{1}{2} \\
&= \frac{1}{2} \mathrm{e}^{\pm \mathrm{i}\omega t} \cos \omega t \pm \frac{\mathrm{i}}{2} \mathrm{e}^{\pm \mathrm{i}\omega t} \sin \omega t + \frac{1}{2} \\
&= \frac{1}{2} \mathrm{e}^{\pm \mathrm{i}\omega t}\left(\cos \omega t \pm \mathrm{i} \sin \omega t + \mathrm{e}^{\mp \mathrm{i}\omega t} \right) \\
&= \frac{1}{2} \mathrm{e}^{\pm \mathrm{i}\omega t}\left(\cos \omega t \pm \mathrm{i} \sin \omega t + \cos \omega t \mp \mathrm{i} \sin \omega t \right) \\
&= \mathrm{e}^{\pm \mathrm{i}\omega t} \cos \omega t.
\end{aligned}
$$

□

1.2 Rotational factor maps and harmonic limits

We let X be a metric space, and consider the dynamical system that is given by the map $T : X \to X$, i.e., we consider trajectories $x, Tx, T^2x, \ldots$ for all starting points $x \in X$. There are several ways to describe the rotational behaviour of such dynamical systems, e.g., different notions of rotation numbers [Fra92; Fra03; KH06; Poi85; Rob99]. In this thesis, we will analyze the concept of rotational factor maps instead, which was introduced in [MB04, Section 3.1] and further pursued, e.g., in [LM08; Row09].

We first define these rotational factor maps, i.e., semi-conjugations to a rotation in the complex plane, in Subsection 1.2.1. Then we define the concept of harmonic limits in Subsection 1.2.2. Finally, we discuss the relation between these two concepts in Subsection 1.2.3. In fact, they are equivalent in a sense, which will be stated more precisely in Theorem 1.2.13.

1.2.1 Rotational factor maps

We want to analyze the rotational behaviour of a discrete-time dynamical system by investigating if the system admits a rotational factor map by some angle. With rotational factor maps, we mean the following:

Definition 1.2.1 (Rotational factor maps). Let $\mathcal{F}$ be a class of functions mapping $X \to \mathbb{C}$. Suppose that there is a map $F \in \mathcal{F}$, such that $F \not\equiv 0$ and $F \circ T = e^{i\omega} \cdot F$ holds for some $\omega \in [0, 2\pi)$. Then we say that T *admits the rotational factor map F in $\mathcal{F}$ by angle ω.* ⌟

By this definition, a system T has a rotational factor map if and only if it is semi-conjugate to a rotation in the complex plane, or in other words, if the system can be mapped into the complex plane such that the dynamics simply become a rotation around the origin.

Remark 1.2.2. The condition $F \not\equiv 0$ in Definition 1.2.1 is necessary, because with $F \equiv 0$, the semi-conjugacy $F \circ T = e^{i\omega} \cdot F$ trivially holds for all systems T and all angles ω. Consider the two simplest cases where $F \not\equiv 0$, namely $F \equiv c \neq 0$ and $F = \mathbf{1}_{\{x_0\}}$ for some $x_0 \in X$. In the first case, the semi-conjugacy equation reduces to $c = e^{i\omega} \cdot c$, which is true if and only if $\omega = 0$. This shows that every system has a rotational factor map by angle 0. In the second case, the equation reduces to $\mathbf{1}_{\{x_0\}}(Tx_0) = e^{i\omega}$, which can only be satisfied for $\omega = 0$ and a fixed point x_0. Sometimes, we will exclude this special case $\omega = 0$. ⌟

Remark 1.2.3. If there is an invariant measure μ on X, we will consider μ-measurable rotational factor maps. In this case, we will additionally assume that $\mu(F \neq 0) > 0$, i.e., that F does not vanish almost everywhere. We can subsume these assumptions by setting $\mathcal{F} := \mathcal{F}_\mu := \{F : X \to \mathbb{C} \mid \mu\text{-measurable}, \mu(F \neq 0) > 0\}$. In this context, it is reasonable to consider maps $F \in \mathcal{F}_\mu$, for which the semi-conjugacy $F \circ T = e^{i\omega} \cdot F$ only holds almost everywhere. We will call them *rotational factor maps μ-almost everywhere.* ⌐

Remark 1.2.4. If a rotational factor map F by angle ω is continuous, it is a *topological semi-conjugacy* with factor $\mathbb{C} \mapsto \mathbb{C}$, $z \mapsto e^{i\omega}z$, in terms of [KH06, Definition 2.3.2]. If there is an invariant measure μ on X, and the rotational factor map F is a metric homomorphism with respect to the Lebesgue measure λ on $\mathbb{C}$, i.e., $\mu(F^{-1}(A)) = \lambda(A)$ for all Lebesgue mesurable sets $A \subset \mathbb{C}$, then the map $\mathbb{C} \mapsto \mathbb{C}$, $z \mapsto e^{i\omega}z$, is a *metric factor* of T in the sense of [KH06, Definition 4.1.21]. ⌐

In general, it is not easy to find a rotational factor map analytically. In the next section, we will introduce harmonic limits, which can be used to construct rotational factor maps. Meanwhile, let us have a look at some examples, where a rotational factor map can explicitly be given.

Example 1.2.5. Consider the system given by $T_\alpha : [0, 2\pi) \to [0, 2\pi)$, $\theta \mapsto \theta + \alpha$ mod 2π, for an arbitrary $\alpha \in (0, 2\pi)$. If we identify $[0, 2\pi)$ with the circle S^1, this simply is a rotation by angle α. In fact, the map $F : [0, 2\pi) \to \mathbb{C}$ given by $F(\theta) := e^{i\theta}$ is a rotational factor map by the angle α, because $F(T_\alpha\theta) = e^{i(\theta+\alpha)} = e^{i\alpha}e^{i\theta} = e^{i\alpha}F(\theta)$. Compare Figure 1.1. ⌐

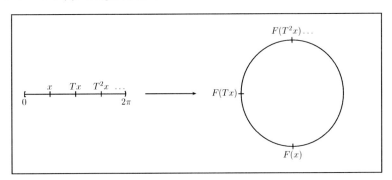

Figure 1.1: Rotational factor map for Example 1.2.5
With $\alpha = x = \pi/2$

Example 1.2.6. Let $X := \mathbb{N}$, and consider the system $T : \mathbb{N} \to \mathbb{N}$, $x \mapsto x + 1$. For every $k \in \mathbb{N}\setminus\{1\}$, the map $F_k : \mathbb{N} \to \mathbb{C}$, $x \mapsto e^{i2\pi x/k}$ is a rotational factor map by the angle $2\pi/k$, because $F_k(Tx) = F_k(x + 1) = e^{i2\pi(x+1)/k} = e^{i2\pi/k} \cdot e^{i2\pi x/k} = e^{i2\pi/k} \cdot F(x)$. See Figure 1.2. ⌐

Example 1.2.7. Consider the quadratic map $T : [0, 1] \to [0, 1]$, $x \mapsto 3.5x(1 - x)$. This quadratic map is studied, e.g., in [LM95, Chapter 1 and Examples 4.1.2 and

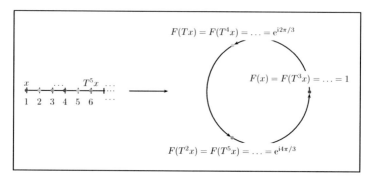

Figure 1.2: Rotational factor map for Example 1.2.6
With $k = 3$ and $x = 1$

6.5.1; Dev03, Sections 1.5 and 3.2]. One can show that there is an attracting 4-periodic orbit x_1, x_2, x_3, x_4 with $x_1 \approx 0.3828, x_2 \approx 0.8269, x_3 \approx 0.5009, x_4 \approx 0.8750$. The measure μ on $[0, 1]$ that is given by $\mu(A) := \#(A \cap \{x_1, x_2, x_3, x_4\})$ for every $A \subset [0, 1]$ is invariant. Hence the map $F : [0, 1] \to \mathbb{C}$ that is given by

$$F(x) = \begin{cases} e^{-i(j \bmod 4)\pi/2} & \text{if } T^j x = x_1 \text{ for some } j \in \mathbb{N}, \\ 0 & \text{otherwise} \end{cases}$$

is a rotational factor map in $\mathcal{F}_\mu$ by angle $\pi/2$.

To see this, first note that, by periodicity, for any $x \in [0, 1]$ there is $j \in \mathbb{N}$ with $T^j x = x_1$ if and only if $T^{j+4} x = x_1$. Hence F is well-defined. Furthermore, if $x \in [0, 1]$ is such that there is $j \in \mathbb{N}$ with $T^j x = x_1$, then $e^{i\pi/2} F(x) = e^{i\pi/2} e^{-i(j \bmod 4)\pi/2} = e^{-i(j+3 \bmod 4)\pi/2} = F(Tx)$, because $T^{j+3} Tx = x_1$. If $x \in [0, 1]$ is such that there is no $j \in \mathbb{N}$ with $T^j x = x_1$, then there also is no $j \in \mathbb{N}$ with $T^j Tx = x_1$. Hence $F(Tx) = 0 = e^{i\pi/2} F(x)$.

Note that T is not invertible. For example, $T^{-1}(x_1) = \{1/2 \pm 1/14\sqrt{49 - 56x_1}\} = \{y, x_4\}$ with $y := 1/2 - 1/14\sqrt{49 - 56x_1} \approx 0.1250$. So $F(x) \neq 0$ also for points different from $x_j, j = 1, \ldots, 4$. Hence the map $\tilde{F} := F \cdot \mathbf{1}_{\{x_1, x_2, x_3, x_4\}}$, i.e.,

$$\tilde{F}(x) = \begin{cases} e^{i(j-1)\pi/2} & \text{if } x = x_j, j = 1, \ldots, 4, \\ 0 & \text{otherwise,} \end{cases}$$

which might have been a more intuitive candidate, is not a rotational factor map, because $F(Ty) = F(x_1) \neq 0$ and $e^{i\pi/2} F(y) = 0$. But we can consider $\tilde{F}$ as a *rotational factor map μ-almost everywhere*, in the sense that $F \circ T = e^{i\pi/2} \cdot F$ holds almost everywhere, see Remark 1.2.3.

Example 1.2.8. Let $X := (S^1)^m \subset \mathbb{C}^m$ and consider the system $T : X \to X$ given by $T(x_1, \ldots, x_m) := (e^{-i\alpha_1}x_1, \ldots, e^{-i\alpha_m}x_m)$ for pairwise different $\alpha_1, \ldots, \alpha_m \in [0, 2\pi)$. See Figure 1.3. This can be interpreted as a rotation on a torus, see [KH06, Section 1.4]. Consider the map $F : (S^1)^m \to (S^1)^m$ given by $F(x) = \overline{x_k}$ for some $k \in \{1, \ldots, m\}$. This is a rotational factor map by angle α_k, because $F(Tx) = \overline{e^{-i\alpha_k}x_k} = e^{i\alpha_k}\overline{x_k} = e^{i\alpha_k}F(x)$. ⌟

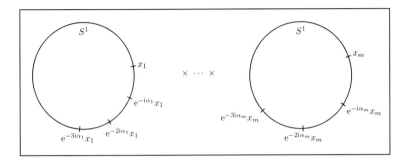

Figure 1.3: System from Example 1.2.8

This concept of rotational factor maps cannot properly describe the kind of rotation that appears, e. g., in $Tx := e^{-(1+I)}x$, because the concept concentrates on the asymptotic behaviour, which, in this example, is described by the fixed point in the origin and hence does not show any rotational behaviour. Compare Subsection 1.3.3 for more details on the influence of asymptotical behaviour of the system on the harmonic limit.

One has to be careful with the choice of $\mathcal{F}$. If we let $\mathcal{F}$ be the class of all functions $X \to \mathbb{C}$ or all Borel measurable functions $X \to \mathbb{C}$, we might encounter unexpected results. See the following example, where we get rotational factor maps to all angles.

Example 1.2.9. Let $X := [0, 1]$ and $Tx := x/2$. Choose $\omega \in [0, 2\pi)$, and let

$$F : X \to \mathbb{C}, x \mapsto \begin{cases} e^{-i\omega \log_2 x} & \text{if } x \neq 0, \\ 0 & \text{if } x = 0. \end{cases}$$

Then it holds for $x \neq 0$ that

$$F(Tx) = F\left(\frac{x}{2}\right) = e^{-i\omega \log_2(x/2)} = e^{-i\omega(\log_2 x - 1)} = e^{i\omega}e^{-i\omega \log_2 x} = e^{i\omega}F(x),$$

and $F(T0) = F(0) = 0 = e^{i\omega}F(0)$, i. e., the system admits rotational factor maps to arbitrary angles.

Note that T preserves the point measure $\mathbf{1}_{\{0\}}$. With this measure, we have $\mathbf{1}_{\{0\}}(F > 0) = \mathbf{1}_{\{0\}}\big((0,1]\big) = 0$, i.e., F vanishes almost everywhere. So if we chose $\mathcal{F} := \mathcal{F}_{\mathbf{1}_{\{0\}}}$ as in Remark 1.2.3, this map F is not admitted.   ⌋

1.2.2 Harmonic limits

In Examples 1.2.5–1.2.8, we presented some rotational factor maps, and verified that they indeed satisfy the semi-conjugacy equation $F \circ T = e^{i\omega} \cdot F$. But how can one find a rotational factor map to a given system, apart from guessing? In this section, we will introduce harmonic limits as a tool for this purpose. There is a strong connection between these harmonic limits and the existence of rotational factor maps, as will be shown in Subsection 1.2.3.

Definition 1.2.10. For $f : X \to \mathbb{C}$, $n \in \mathbb{N}$, $\omega \in [0, 2\pi)$, and $x \in X$, define the *harmonic average* by

$$f_\omega^n(x) := \frac{1}{n} \sum_{j=0}^{n-1} e^{ij\omega} f(T^j x);$$

furthermore, define the *harmonic limit* by

$$f_\omega^*(x) := \lim_{n \to \infty} f_\omega^n(x) = \lim_{n \to \infty} \frac{1}{n} \sum_{j=0}^{n-1} e^{ij\omega} f(T^j x),$$

if the limit exists.   ⌋

Remark 1.2.11. By Definition 1.2.10, the harmonic limit is an ergodic sum of the images of a trajectory under some complex-valued map f. One can interpret this map f as an observation or the output of the system. In particular, the harmonic limit can be computed (or at least approximated by harmonic averages) for measurements of a system, without knowledge of the exact dynamics T. For a given finite trajectory $x, Tx, \ldots, T^{n-1}x$ and $f : X \to \mathbb{C}$, one can easily compute $f_\omega^n(x)$ for several ω using fast Fourier transform methods. Compare, e.g., [Bra86, Chapter 18].   ⌋

Note that for $\omega = 0$, the harmonic limit $f_\omega^*(x)$ is simply the average of f along the trajectory $x, Tx, T^2x, \ldots$, i.e., $f_0^*(x) = \lim_{n \to \infty} \frac{1}{n} \sum_{j=0}^{n-1} f(T^j x)$. Recall Remark 1.2.2, where we already have seen, that $\omega = 0$ plays a special role in this discussion. Let us have a second look at a system, for which we already proposed a rotational factor map in the previous section.

Example 1.2.12. Consider $X := \mathbb{N}$ and $Tx := x + 1$ as in Example 1.2.6.

1. Let $f := \mathrm{id} : \mathbb{N} \to \mathbb{N}$. For $\omega \in (0, 2\pi)$ and any $x \in X$, it holds that

$$\mathrm{id}_\omega^n(x) = \frac{e^{in\omega}}{e^{i\omega} - 1} + \frac{1}{n} \cdot \frac{(e^{i\omega} + x - xe^{i\omega})(1 - e^{in\omega})}{(e^{i\omega} - 1)^2}. \tag{1.2.1}$$

Note that

$$\lim_{n\to\infty} \frac{1}{n} \cdot \frac{(e^{i\omega} + x - xe^{i\omega})(1 - e^{in\omega})}{(e^{i\omega} - 1)^2} = 0,$$

as the fraction is bounded, and that

$$\frac{e^{in\omega}}{e^{i\omega} - 1}$$

does not converge for $n \to \infty$. So $\mathrm{id}^*_\omega(x)$ does not exist.

2. For $k \in \mathbb{N} \setminus \{1\}$, consider $f : X \to \mathbb{C}$ given by $f(x) := x \bmod k$. Then for any $x \in \mathbb{N}$, it holds that

$$f^*_{2\pi/k}(x) = \frac{e^{-ix2\pi/k}}{e^{i2\pi/k} - 1}. \tag{1.2.2}$$

To see that (1.2.1) holds, first note that it holds for $n = 1$ because $\mathrm{id}^1_\omega(x) = e^{i0\omega}x = x$ and

$$\frac{e^{i1\omega}}{e^{i\omega} - 1} + \frac{1}{1} \cdot \frac{(e^{i\omega} + x - xe^{i\omega})(1 - e^{i1\omega})}{(e^{i\omega} - 1)^2} = \frac{e^{i\omega}}{e^{i\omega} - 1} - \frac{e^{i\omega} + x - xe^{i\omega}}{e^{i\omega} - 1} = x.$$

Furthermore, if (1.2.1) holds for n, then it also holds for $n + 1$, because

$$\begin{aligned}
\mathrm{id}^{n+1}_\omega(x) &= \frac{1}{n+1} \sum_{j=0}^{n} e^{ij\omega}(x+j) \\
&= \frac{1}{n+1} \sum_{j=0}^{n-1} e^{ij\omega}(x+j) + \frac{1}{n+1} e^{in\omega}(x+n) \\
&= \frac{n}{n+1} \mathrm{id}^n_\omega(x) + \frac{1}{n+1} e^{in\omega}(x+n) \\
&= \frac{n}{n+1} \left[\frac{e^{in\omega}}{e^{i\omega} - 1} + \frac{1}{n} \cdot \frac{(e^{i\omega} + x - xe^{i\omega})(1 - e^{in\omega})}{(e^{i\omega} - 1)^2} \right] \\
&\quad + \frac{1}{n+1} e^{in\omega}(x+n) \\
&= \left[ne^{in\omega}(e^{i\omega} - 1) + (e^{i\omega} + x - xe^{i\omega})(1 - e^{in\omega}) \right. \\
&\quad \left. + e^{in\omega}(x+n)(e^{i\omega} - 1)^2 \right] / \left[(n+1)(e^{i\omega} - 1)^2 \right] \\
&= \left[ne^{i(n+1)\omega} - ne^{in\omega} + e^{i\omega} + x - xe^{i\omega} - e^{i(n+1)\omega} - xe^{in\omega} \right. \\
&\quad + xe^{i(n+1)\omega} + xe^{i(n+2)\omega} + ne^{i(n+2)\omega} - 2xe^{i(n+1)\omega} - 2ne^{i(n+1)\omega} \\
&\quad \left. + xe^{in\omega} + ne^{in\omega} \right] / \left[(n+1)(e^{i\omega} - 1)^2 \right]
\end{aligned}$$

$$= \left[e^{i\omega} + x - x e^{i\omega} - e^{i(n+1)\omega} + x e^{i(n+2)\omega} + n e^{i(n+2)\omega} - x e^{i(n+1)\omega} \right.$$
$$\left. - n e^{i(n+1)\omega} \right] \Big/ \left[(n+1)(e^{i\omega} - 1)^2 \right]$$
$$= \frac{(n+1)(e^{i\omega} - 1)e^{i(n+1)\omega} + (e^{i\omega} + x - x e^{i\omega})(1 - e^{i(n+1)\omega})}{(n+1)(e^{i\omega} - 1)^2}$$
$$= \frac{e^{i(n+1)\omega}}{e^{i\omega} - 1} + \frac{1}{n+1} \frac{(e^{i\omega} + x - x e^{i\omega})(1 - e^{i(n+1)\omega})}{(e^{i\omega} - 1)^2}.$$

In order to show that (1.2.2) holds, we anticipate a result that will be shown in Subsection 1.3.4. By Proposition 1.3.17, it holds that

$$f^*_{2\pi/k}(x) = \frac{1}{k} \sum_{j=0}^{k-1} e^{ij2\pi/k} f(T^j x) = \frac{1}{k} \sum_{j=0}^{k-1} e^{ij2\pi/k} (x + j \bmod k),$$

because $T^k x = x$. Hence $f^*_{2\pi/k}(x) = 1/k \cdot \sum_{j=0}^{k-1} e^{i(j-x)2\pi/k} j = e^{-ix2\pi/k} \mathrm{id}^k_{2\pi/k}$. So (1.2.1) implies

$$f^*_{2\pi/k}(x) = e^{-ix2\pi/k} \frac{e^{i2\pi}}{e^{i2\pi/k} - 1} + \frac{1}{k} \cdot \frac{(e^{i2\pi/k} + x - x e^{i2\pi/k})(1 - e^{i2\pi})}{(e^{i2\pi/k} - 1)^2}$$
$$= e^{-ix2\pi/k} \frac{1}{e^{i2\pi/k} - 1} + 0,$$

i.e., (1.2.2) holds. ⌐

1.2.3 Nonvanishing harmonic limits and rotational factor maps

As mentioned earlier, there is a strong connection between harmonic limits and rotational factor maps. In fact, there is a map $f : X \to \mathbb{C}$ such that $f^*_\omega \not\equiv 0$ if and only if there is a rotational factor map by angle $\omega \in [0, 2\pi)$. It turns out that the complex conjugate of the harmonic limit itself is a rotational factor map.

Theorem 1.2.13. *Let $f : X \to \mathbb{C}$ and $\omega \in [0, 2\pi)$. If f^*_ω exists on some nonvoid set $M \subset X$ and is not constant zero on M, then T admits the rotational factor map $\mathbf{1}_M \cdot \overline{f^*_\omega}$ by angle ω. If, on the other hand, T admits the rotational factor map $F : X \to \mathbb{C}$ by some angle $\omega \in [0, 2\pi)$, then $\overline{F} = \overline{F^*_\omega}$ almost everywhere.*

Proof. Assume that f^*_ω exists on some nonvoid set $M \subset X$. For all $x \in M$, it holds that

$$e^{-i\omega} f^*_\omega(x) = e^{-i\omega} \lim_{n \to \infty} f^n_\omega(x) \overset{(*)}{=} e^{-i\omega} \lim_{n \to \infty} \frac{1}{n} \sum_{j=1}^{n} e^{ij\omega} f(T^j x)$$

$$= \lim_{n \to \infty} \frac{1}{n} \sum_{j=1}^{n} e^{i(j-1)\omega} f(T^j x) = \lim_{n \to \infty} \frac{1}{n} \sum_{j=0}^{n-1} e^{ij\omega} f(T^{j+1} x)$$

$$= \lim_{n \to \infty} \frac{1}{n} \sum_{j=0}^{n-1} e^{ij\omega} f(T^j T x) = f_\omega^*(Tx).$$

Thus with $F := \mathbf{1}_M \cdot \overline{f_\omega^*}$, it holds that $F \circ T = e^{i\omega} \cdot F$. Furthermore, $F \not\equiv 0$ by assumption. So F is a rotational factor map by angle ω.

To see that the equation marked with $(*)$ holds, note that

$$\frac{1}{n} \sum_{j=1}^{n} e^{ij\omega} f(T^j x) - f_\omega^n(x) = \frac{1}{n} \sum_{j=0}^{n} e^{ij\omega} f(T^j x) - \frac{1}{n} f(x) - f_\omega^n(x)$$

$$= \frac{n+1}{n} f_\omega^{n+1}(x) - \frac{1}{n} f(x) - f_\omega^n(x),$$

which tends to 0 as $n \to \infty$, because $f_\omega^*(x)$ exists.

For the second part, assume that the map $F : X \to \mathbb{C}$ and the point $x \in X$ are such that $F(Tx) = e^{i\omega} F(x)$. Then $F(T^j x) = e^{ij\omega} F(x)$, and hence

$$\overline{F}_\omega^*(x) = \lim_{n \to \infty} \frac{1}{n} \sum_{j=0}^{n-1} e^{ij\omega} \overline{F(T^j x)}$$

$$= \lim_{n \to \infty} \frac{1}{n} \sum_{j=0}^{n-1} e^{ij\omega} e^{-ij\omega} \overline{F}(x)$$

$$= \overline{F}(x). \qquad \square$$

So we can construct a rotational factor map to the angle $\omega \in [0, 2\pi)$ by computing the harmonic limit f_ω^*, if the limit is not constant zero. In fact, by Theorem 1.2.13, we can get every rotational factor map in this manner. Because of this equivalence, we will concentrate on the properties of the harmonic limit in the following.

If we only have access to a measurement of a trajectory, i. e., to $f(x), f(Tx), \dots$ for some $f : X \to \mathbb{C}$, then we still can compute the harmonic limit (see Remark 1.2.11) and thus show the existence of rotational factor maps.

The following three examples illustrate this connection between rotational factor maps and harmonic limits.

Example 1.2.14. Consider the system given by $T_\alpha : [0, 2\pi) \to [0, 2\pi)$, $\theta \mapsto \theta + \alpha$ mod 2π for arbitrary $\alpha \in (0, 2\pi)$. As we have seen in Example 1.2.5, the system has a rotational factor map by the angle α. For $f : [0, 2\pi) \to \mathbb{R}$, $\theta \mapsto \cos \theta$, one can easily show that

$$
f_\omega^*(\theta) = \begin{cases}
\cos\theta & \text{if } \omega = \alpha = \pi, \\
\frac{1}{2}\mathrm{e}^{-\mathrm{i}\theta} & \text{if } \omega = \alpha \neq \pi, \\
\frac{1}{2}\mathrm{e}^{\mathrm{i}\theta} & \text{if } \omega + \alpha = 2\pi \text{ and } \alpha \neq \pi, \\
0 & \text{otherwise},
\end{cases}
$$

by writing

$$
\mathrm{e}^{\mathrm{i}j\omega}\cos(\theta + j\alpha) = \mathrm{e}^{\mathrm{i}j\omega}\frac{1}{2}\left(\mathrm{e}^{\mathrm{i}(\theta+j\alpha)} + \mathrm{e}^{-\mathrm{i}(\theta+j\alpha)}\right)
$$

$$
= \frac{1}{2}\mathrm{e}^{\mathrm{i}\theta}\left(\mathrm{e}^{\mathrm{i}(\omega+\alpha)}\right)^j + \frac{1}{2}\mathrm{e}^{-\mathrm{i}\theta}\left(\mathrm{e}^{\mathrm{i}(\omega-\alpha)}\right)^j
$$

and using the geometric series formula. In particular, the harmonic limit does not vanish for $\omega = \alpha$.

According to Theorem 1.2.13, consider the complex conjugate of the harmonic limit for $\omega = \alpha$:

$$
\overline{f_\alpha^*}(\theta) = \begin{cases}
\cos\theta & \text{if } \alpha = \pi, \\
\frac{1}{2}\mathrm{e}^{\mathrm{i}\theta} & \text{otherwise.}
\end{cases}
$$

For $\alpha = \pi$, it holds that $\overline{f_\alpha^*}(T\theta) = \cos(T\theta) = \cos(\theta + \pi) = -\cos\theta = \mathrm{e}^{\mathrm{i}\pi}\overline{f_\alpha^*}(\theta)$. For $\alpha \neq \pi$, the rotational factor map $\overline{f_\alpha^*}$ equals the rotational factor map from Example 1.2.5 up to the constant factor $1/2$. ⌐

Example 1.2.15. Consider the map $T : [-1, 1] \to [-1, 1]$ given by $x \mapsto \operatorname{frac}(-2x)$. See [MB04, Example 10]. With the map $f : [-1, 1] \to \mathbb{C}$, $f(x) = \operatorname{sgn} x$, we have $f(T^j x) = (-1)^j \operatorname{sgn} x$. Hence $f_\pi^*(x) = \operatorname{sgn} x$, which is zero only for $x = 0$. According to Theorem 1.2.13, this system has the rotational factor map $F = \operatorname{sgn}$ by angle π. ⌐

Example 1.2.16. The rotational factor map in Example 1.2.6 coincides with the complex conjugate of the harmonic limit in part 2. of Example 1.2.12 up to the constant factor $1/(\mathrm{e}^{\mathrm{i}2\pi/k}-1)$. ⌐

Because only the property $f_\omega^*(x) \neq 0$ is of importance when establishing rotational factor maps, the following proposition is handy. It shows that we can also admit functions f with values in $\mathbb{C}^n$, $\mathbb{R}$, or $\mathbb{R}^n$. So, e.g., in systems with $X = \mathbb{C}^n$, we can have a look at harmonic limits for $f = \mathrm{id}_X$ in order to show the existence of rotational factor maps. But usually, we will continue assuming that f has values in $\mathbb{C}$.

Proposition 1.2.17. *Let $x \in X$ and $\omega \in [0, 2\pi)$. For $V \in \{\mathbb{R}, \mathbb{C}, \mathbb{R}^n, \mathbb{C}^n\}$, denote the assertion "There is $f : X \to V$ with $f_\omega^*(x) \neq 0$." by $E(V)$. Then:*

  1. It holds that $E(\mathbb{R}^n) \Leftrightarrow E(\mathbb{R}) \Rightarrow E(\mathbb{C}) \Leftrightarrow E(\mathbb{C}^n)$.

2. *Furthermore, if it additionally holds that, for every $f : X \to \mathbb{C}$, the existence of $f_\omega^*(x)$ implies existence of $(\Re f)_\omega^*(x)$ and $(\Im f)_\omega^*(x)$, then $E(\mathbb{R}^n) \Leftrightarrow E(\mathbb{R}) \Leftrightarrow E(\mathbb{C}) \Leftrightarrow E(\mathbb{C}^n)$.*

3. *Parts 1. and 2. also hold if we only consider continuous functions f. In particular, if $f_\omega^*(x)$ exists for all continuous $f : X \to \mathbb{C}$, then $E(\mathbb{R}^n) \Leftrightarrow E(\mathbb{R}) \Leftrightarrow E(\mathbb{C}) \Leftrightarrow E(\mathbb{C}^n)$.*

Proof.

1. Clearly, if $E(\mathbb{R})$ is true, then also $E(\mathbb{C})$ holds, as $\mathbb{R} \subset \mathbb{C}$.

   Assume that $E(\mathbb{K})$ holds for $\mathbb{K} \in \{\mathbb{R}, \mathbb{C}\}$. Then there is a map $f : X \to \mathbb{K}$ with $f_\omega^*(x) \neq 0$. Define a function $g : X \to \mathbb{K}^n$ by $g := (f, \ldots, f)$. Then, clearly, $g_\omega^*(x) = \big(f_\omega^*(x), \ldots, f_\omega^*(x)\big)$. This implies $E(\mathbb{K}^n)$.

   Assume that $E(\mathbb{K}^n)$ holds for $\mathbb{K} \in \{\mathbb{R}, \mathbb{C}\}$. Then there is $f : X \to \mathbb{K}^n$ with $f_\omega^*(x) \neq 0$. Hence, there is $v \in \mathbb{K}^n$ such that $v^T f_\omega^*(x) \neq 0$, e.g., $v := f_\omega^*(x)$. Define a function $g : X \to \mathbb{K}$ by $g := v^T f$. Then, clearly, $g_\omega^*(x) = v^T f_\omega^*(x) \neq 0$. This implies $E(\mathbb{K})$.

2. It only remains to show that $E(\mathbb{C}) \Rightarrow E(\mathbb{R})$. Assume that $E(\mathbb{C})$ holds. Then there is $f : X \to \mathbb{C}$ with $f_\omega^*(x) \neq 0$. Then $f_\omega^*(x) = (\Re f)_\omega^*(x) + \mathrm{i}(\Im f)_\omega^*(x)$, because these separate limits exist by assumption. So $(\Re f)_\omega^*(x) \neq 0$ or $(\Im f)_\omega^*(x) \neq 0$ must be true. This implies $E(\mathbb{R})$.

3. If, for $\mathbb{K} \in \{\mathbb{R}, \mathbb{C}\}$, a map $f : X \to \mathbb{K}$ is continuous, then also $(f, \ldots, f) : X \to \mathbb{K}^n$ is continuous. Similarly, if, for $\mathbb{K} \in \{\mathbb{R}, \mathbb{C}\}$, a map $f : X \to \mathbb{K}^n$ is continuous, then also $v^T f$ is continuous for any $v \in \mathbb{K}^n$. Furthermore, if $f : X \to \mathbb{C}$ is continuous, then also $\Re f$ and $\Im f$ are continuous. With this, 3. follows from the proof of 1. and 2. $\qquad\square$

The additional condition in part 2. of this proposition in particular holds if f is integrable with respect to a finite invariant measure, as we will see later in Subsection 1.3.2. It is also satisfied at periodic or almost periodic orbits, see Subsection 1.3.4.

1.3 Properties of the harmonic limit

In the previous section, we have seen that harmonic limits can be used as a tool to analyze the rotational behaviour of dynamical systems, because they indicate the existence of rotational factor maps. Now we will have a closer look on these limits and their properties.

First, we note some basic properties in Subsection 1.3.1, including boundedness, and a condition on f for the set $\{\omega \in [0, 2\pi) \mid f_\omega^*(x) \neq 0\}$ to be countable. Then we investigate existence of f_ω^* in Subsection 1.3.2 under the presence of an invariant measure. In this case, the harmonic limit exists almost everywhere (Theorem 1.3.10), which can be shown using the Birkhoff Ergodic Theorem. If the system is ergodic, the harmonic limit exists almost everywhere, independently of $\omega \in [0, 2\pi)$. This result is known as the Wiener-Wintner Ergodic Theorem, see Theorem 1.3.14. Finally, if the system is uniquely ergodic, the harmonic limit exists everywhere (Proposition 1.3.11).

Knowledge of the asymptotic properties of the dynamics can help with the computation of harmonic limits, see Subsection 1.3.3. Likewise, the existence of periodic or almost periodic orbits has some implications on the properties of the harmonic limit, see Subsection 1.3.4. For example, if the system has a periodic orbit with period $\tau \in \mathbb{N}$, then for every $k \in \mathbb{Z}$, there is a continuous map $f : X \to \mathbb{C}$ such that $f_{2k\pi/\tau} \not\equiv 0$.

1.3.1 Basic properties

Boundedness is the first property we will look at. The harmonic average is bounded above by $\sup_{x \in X} |f(x)|$. The same holds for the harmonic limit, if it exists.

Proposition 1.3.1 (Boundedness). *For every $f : X \to \mathbb{C}$, it holds that*

$$\sup_{(n,\omega,x) \in \mathbb{N} \times [0,2\pi) \times X} |f_\omega^n(x)| \leq \sup_{x \in X} |f(x)|. \tag{1.3.1}$$

If $\omega \in [0, 2\pi)$ and $x_0 \in X$ are such that $f_\omega^*(x_0)$ exists, then

$$|f_\omega^*(x_0)| \leq \sup_{x \in X} |f(x)|. \tag{1.3.2}$$

In particular, if f is bounded, then f_ω^n and (in the case of existence) f_ω^* is bounded.

Proof. In order to show (1.3.1), let $n \in \mathbb{N}$, $\omega \in [0, 2\pi)$ and $x_0 \in X$. Then it holds that

$$
\begin{aligned}
|f_\omega^n(x_0)| &= \left| \frac{1}{n} \sum_{j=0}^{n-1} \mathrm{e}^{\mathrm{i}j\omega} f(T^j x_0) \right| \leq \frac{1}{n} \sum_{j=0}^{n-1} \left| \mathrm{e}^{\mathrm{i}j\omega} f(T^j x_0) \right| = \frac{1}{n} \sum_{j=0}^{n-1} \left| f(T^j x_0) \right| \\
&\leq \frac{1}{n} \sum_{j=0}^{n-1} \sup_{x \in X} |f(x)| = \sup_{x \in X} |f(x)|.
\end{aligned}
\tag{1.3.3}
$$

For (1.3.2), let $\omega \in [0, 2\pi)$ and $x_0 \in X$ be such that $f_\omega^*(x_0)$ exists. Then by (1.3.3), it is true that

$$|f_\omega^*(x_0)| = \left| \lim_{n \to \infty} f_\omega^n(x_0) \right| \leq \sup_{(n,\omega,x) \in \mathbb{N} \times [0,2\pi) \times X} |f_\omega^n(x)| \leq \sup_{x \in X} |f(x)|. \qquad \square$$

The following result is a corollary to Theorem 1.2.13, and shows how the harmonic limit behaves along trajectories. In particular, we get that if $f_\omega^*(x)$ exists at some point $x \in X$ for $\omega \in [0, 2\pi)$, then the harmonic limit exists at every point of the trajectory starting in x, and beyond that, the modulus of the harmonic limit is constant along the trajectory.

Corollary 1.3.2 (Harmonic limits along trajectories). *Assume that a map $f : X \to \mathbb{C}$, an angle $\omega \in [0, 2\pi)$, and a point $x \in X$ are such that $f_\omega^*(x)$ exists. Then $f_\omega^*(T^j x) = \mathrm{e}^{-\mathrm{i}j\omega} f_\omega^*(x)$ for all $j \in \mathbb{N}_0$. Particularly, $f_\omega^*(T^j x)$ exists for all $j \in \mathbb{N}_0$, and $|f_\omega^*(T^j x)|$ is independent of j.*

Proof. By Theorem 1.2.13, it holds that $f_\omega^*(Tx) = \mathrm{e}^{-\mathrm{i}\omega} f_\omega^*(x)$ if $f_\omega^*(x)$ exists. So it follows that $f_\omega^*(T^j x) = \mathrm{e}^{-\mathrm{i}j\omega} f_\omega^*(x)$ for all $j \in \mathbb{N}_0$. Hence $f_\omega^*(T^j x)$ exists for all $j \in \mathbb{N}_0$, and $|f_\omega^*(T^j x)| = |f_\omega^*(x)|$ is independent of j. $\qquad\square$

Recall from Theorem 1.2.13 that, in order to prove the existence of a rotational factor map, one has to show that $f_\omega^* \not\equiv 0$. This is equivalent to $|f_\omega^*| \not\equiv 0$. So we could limit our view to the modulus of the harmonic limit, as Levnajić and Mezić do in [LM08]. But because the actual value of f_ω^* gives us a deeper insight into the dynamics (compare, e. g., Section 1.5), we will continue to analyze the harmonic limit itself.

Let the following examples illustrate these two results.

Example 1.3.3. Let $X := S^1 \subset \mathbb{C}$, and consider the system $T : X \to X$ given by $Tx := \mathrm{e}^{-\mathrm{i}\alpha}x$ for some $\alpha \in [0, 2\pi)$. See Figure 1.4. Then $T^j x = \mathrm{e}^{-\mathrm{i}j\alpha}x$. Further consider $f := \mathrm{id} : X \to \mathbb{C}$. It holds that $\mathrm{id}_\omega^n(x) = 1/n \cdot \sum_{j=0}^{n-1} \mathrm{e}^{\mathrm{i}j\omega}\mathrm{e}^{-\mathrm{i}j\alpha}x = 1/n \cdot \sum_{j=0}^{n-1} \mathrm{e}^{\mathrm{i}j(\omega - \alpha)}x$ for every $n \in \mathbb{N}$ and all $x \in X$. So if $\omega = \alpha$, then $\mathrm{id}_\alpha^n(x) = 1/n \cdot \sum_{j=0}^{n-1} x = x$, and hence $\mathrm{id}_\alpha^*(x) = x$. If $\omega \in [0, 2\pi) \setminus \{\alpha\}$, then

$$\mathrm{id}_\omega^n(x) = \frac{1}{n} \sum_{j=0}^{n-1} \left(\mathrm{e}^{\mathrm{i}(\omega - \alpha)}\right)^j x = \frac{1}{n} \cdot \frac{\mathrm{e}^{\mathrm{i}n(\omega - \alpha)} - 1}{\mathrm{e}^{\mathrm{i}(\omega - \alpha)} - 1} x,$$

and hence $\mathrm{id}_\omega^*(x) = 0$.

This is in line with Proposition 1.3.1, as $|f(x)| = |x| = 1$ for all $x \in X$. Furthermore, it holds that $\mathrm{id}_\alpha^*(T^j x) = T^j x = \mathrm{e}^{-\mathrm{i}j\alpha}x$, which matches Corollary 1.3.2. ⌐

Example 1.3.4. Let $X := [0, 1]$, and consider the system $T : X \to X$ that is given by $Tx := \alpha x(1 - x)$ for $\alpha \in [0, 4]$. This quadratic map is studied, e. g., in [LM95, Chapter 1 and Examples 4.1.2 and 6.5.1; Dev03, Sections 1.5 and 3.2]. Consider $f := \mathrm{id}$. By Proposition 1.3.1, $|f_\omega^n(x_0)| \leq \sup_{x \in X} |f(x)| = \sup_{x \in [0,1]} |x| = 1$ for all $n \in \mathbb{N}$, $x_0 \in X$, $\omega \in [0, 2\pi)$, and $\alpha \in [0, 4]$. In fact, e. g., for $\alpha = 3.5$ one can numerically compute that $|f_{\pi/2}^*(0.3)| \approx 0.03 \leq 1$. Compare also Examples 1.2.7 and 1.3.18. ⌐

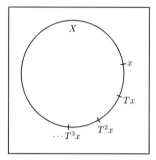

Figure 1.4: System from Example 1.3.3

If one thinks of the harmonic limit as a tool to detect rotational factor maps, a natural question is, which angles ω occur in the system. We will address this question later in Subsection 1.3.4. At this point, we can only give a result on how many angles can occur. The following theorem shows that, under a weak condition, which, e.g., holds for square-integrable f (see Corollary 1.3.7), there can only be countably many $\omega \in [0, 2\pi)$ for which the harmonic limit is different from zero.

Theorem 1.3.5. *Let* $f : X \to \mathbb{C}$ *and* $x \in X$ *be such that*

$$\lim_{N \to \infty} \frac{1}{N} \sum_{n=0}^{N-1} |f(T^n x)|^2$$

exists. Then there are at most countably many $\omega \in [0, 2\pi)$ *for which* $f_\omega^*(x) \neq 0$.

Proof. This proof follows the ideas in [Boh47, Sections 57 and 58], where a similar result is shown in continuous time for almost periodic functions.

First, we show that, for arbitrary distinct angles $\omega_1, \ldots, \omega_M \in [0, 2\pi)$ and arbitrary $c_1, \ldots, c_M \in \mathbb{C}$, it holds that

$$\lim_{N \to \infty} \frac{1}{N} \sum_{n=0}^{N-1} \left| f(T^n x) - \sum_{m=1}^{M} c_m e^{-in\omega_m} \right|^2$$
$$= \lim_{N \to \infty} \frac{1}{N} \sum_{n=0}^{N-1} |f(T^n x)|^2 - \sum_{n=1}^{M} |f_{\omega_n}^*(x)|^2 + \sum_{n=1}^{M} |c_n - f_{\omega_n}^*(x)|^2 \quad (1.3.4)$$

if $f_{\omega_n}^*(x)$ exists for all $n = 1, \ldots, M$, which we will assume. In order to show this, first note that, for all $\alpha, \beta \in \mathbb{C}$, it is true that

$$\begin{aligned}
|\alpha - \beta|^2 &= (\alpha - \beta)(\overline{\alpha} - \overline{\beta}) \\
&= \alpha\overline{\alpha} - \alpha\overline{\beta} - \overline{\alpha}\beta + \beta\overline{\beta} \\
&= |\alpha|^2 - \alpha\overline{\beta} - \overline{\alpha}\beta + |\beta|^2.
\end{aligned} \tag{1.3.5}$$

This implies that

$$\left| f(T^n x) - \sum_{m=1}^{M} c_m e^{-in\omega_m} \right|^2$$

$$= |f(T^n x)|^2 - f(T^n x) \overline{\sum_{m=1}^{M} c_m e^{-in\omega_m}} - \overline{f(T^n x)} \sum_{m=1}^{M} c_m e^{-in\omega_m} + \left| \sum_{m=1}^{M} c_m e^{-in\omega_m} \right|^2$$

$$= |f(T^n x)|^2 - \sum_{m=1}^{M} \overline{c_m} f(T^n x) e^{in\omega_m} - \sum_{m=1}^{M} c_m \overline{f(T^n x)} e^{-in\omega_m}$$

$$+ \left(\sum_{m=1}^{M} c_m e^{-in\omega_m} \right) \left(\sum_{m=1}^{M} \overline{c_m} e^{in\omega_m} \right)$$

$$= |f(T^n x)|^2 - \sum_{m=1}^{M} \overline{c_m} f(T^n x) e^{in\omega_m} - \sum_{m=1}^{M} c_m \overline{f(T^n x)} e^{-in\omega_m} + \sum_{m=1}^{M} \sum_{l=1}^{M} c_m \overline{c_l} e^{-in\omega_m} e^{in\omega_l}.$$

By linearity of the limit and the sum, and because by assumption the limits

$$\lim_{N \to \infty} \frac{1}{N} \sum_{n=0}^{N-1} |f(T^n x)|^2$$

and $f^*_{\omega_n}(x)$, $n = 1, \ldots, M$, exist, it follows that

$$\lim_{N \to \infty} \frac{1}{N} \sum_{n=0}^{N-1} \left| f(T^n x) - \sum_{m=1}^{M} c_m e^{-in\omega_m} \right|^2$$

$$= \lim_{N \to \infty} \frac{1}{N} \sum_{n=0}^{N-1} |f(T^n x)|^2 - \sum_{m=1}^{M} \overline{c_m} \lim_{N \to \infty} \frac{1}{N} \sum_{n=0}^{N-1} f(T^n x) e^{in\omega_m} \tag{1.3.6}$$

$$- \sum_{m=1}^{M} c_m \lim_{N \to \infty} \frac{1}{N} \sum_{n=0}^{N-1} \overline{f(T^n x)} e^{-in\omega_m} + \sum_{m=1}^{M} \sum_{l=1}^{M} c_m \overline{c_l} \lim_{N \to \infty} \frac{1}{N} \sum_{n=0}^{N-1} e^{-in\omega_m} e^{in\omega_l}$$

$$= \lim_{N \to \infty} \frac{1}{N} \sum_{n=0}^{N-1} |f(T^n x)|^2 - \sum_{m=1}^{M} \overline{c_m} f^*_{\omega_m}(x)$$

$$- \sum_{m=1}^{M} c_m \overline{f^*_{\omega_m}(x)} + \sum_{m=1}^{M} \sum_{l=1}^{M} c_m \overline{c_l} \lim_{N \to \infty} \frac{1}{N} \sum_{n=0}^{N-1} e^{-in\omega_m} e^{in\omega_l}.$$

Note that, by assumption, $\omega_m = \omega_l$ only holds if $m = l$. So for $m \neq l$, it holds that $e^{i(\omega_l - \omega_m)} \neq 1$, and thus

$$
\sum_{n=0}^{N-1} e^{-in\omega_m} e^{in\omega_l} = \sum_{n=0}^{N-1} \left[e^{i(\omega_l - \omega_m)} \right]^n
$$
$$
= \frac{\left[e^{i(\omega_l - \omega_m)} \right]^N - 1}{e^{i(\omega_l - \omega_m)} - 1},
$$

which is bounded in N. For $m = l$, it holds that $\sum_{n=0}^{N-1} e^{-in\omega_m} e^{in\omega_l} = N$. So

$$
\lim_{N \to \infty} \frac{1}{N} \sum_{n=0}^{N-1} e^{-in\omega_m} e^{in\omega_l} = \begin{cases} 1 & \text{if } m = l, \\ 0 & \text{otherwise.} \end{cases} \tag{1.3.7}
$$

Equations (1.3.6) and (1.3.7) together imply

$$
\lim_{N \to \infty} \frac{1}{N} \sum_{n=0}^{N-1} \left| f(T^n x) - \sum_{m=1}^{M} c_m e^{-in\omega_m} \right|^2
$$
$$
= \lim_{N \to \infty} \frac{1}{N} \sum_{n=0}^{N-1} |f(T^n x)|^2 - \sum_{m=1}^{M} \overline{c_m} f_{\omega_m}^*(x) - \sum_{m=1}^{M} c_m \overline{f_{\omega_m}^*(x)} + \sum_{m=1}^{M} |c_m|^2
$$
$$
= \lim_{N \to \infty} \frac{1}{N} \sum_{n=0}^{N-1} |f(T^n x)|^2 - \sum_{m=1}^{M} |f_{\omega_n}^*(x)|^2 + \sum_{m=1}^{M} |c_m - f_{\omega_m}^*(x)|^2,
$$

where the last equality holds due to (1.3.5). This proves equation (1.3.4).

Setting $c_n := f_{\omega_n}^*(x)$ in equation (1.3.4) yields

$$
\lim_{N \to \infty} \frac{1}{N} \sum_{n=0}^{N-1} \left| f(T^n x) - \sum_{m=1}^{M} f_{\omega_m}^*(x) e^{-in\omega_m} \right|^2 = \lim_{N \to \infty} \frac{1}{N} \sum_{n=0}^{N-1} |f(T^n x)|^2 - \sum_{n=1}^{M} |f_{\omega_n}^*(x)|^2.
$$

As the left-hand side of this equation is nonnegative, it follows that

$$
\sum_{n=1}^{M} |f_{\omega_n}^*(x)|^2 \leq \lim_{N \to \infty} \frac{1}{N} \sum_{n=0}^{N-1} |f(T^n x)|^2 =: \mu_0. \tag{1.3.8}
$$

Note that μ_0 does not depend on the choice of the ω_n.

Assume that, for some $\varepsilon > 0$, there are M distinct values $\omega_1, \ldots, \omega_M \in [0, 2\pi)$ for $M > \mu_0/\varepsilon^2$, such that $|f_{\omega_n}^*(x)| > \varepsilon$. Then $\sum_{n=1}^{M} |f_{\omega_n}^*(x)|^2 > \sum_{n=1}^{M} \varepsilon^2 = M\varepsilon^2 > \mu_0$, which contradicts (1.3.8).

So there are only finitely many $\omega \in [0, 2\pi)$ with $|f_\omega^*(x)| > 1$. Similarly, for every $n \in \mathbb{N}$ there are only finitely many $\omega \in [0, 2\pi)$ with $1/n \geq |f_\omega^*(x)| > 1/(n+1)$. This implies that there can be only countably many $\omega \in [0, 2\pi)$ with $f_\omega^*(x) \neq 0$. $\qquad\square$

Remark 1.3.6. Note that this theorem also holds, more generally, for sequences $(a_j)_{j \in \mathbb{N}} \subset \mathbb{C}$, i.e., if $\lim_{N \to \infty} 1/N \cdot \sum_{n=0}^{N-1} |a_n|^2$ exists, it holds that there are at most countably many $\omega \in [0, 2\pi)$ for which $\lim_{n \to \infty} 1/n \cdot \sum_{j=0}^{n-1} e^{ij\omega} a_j \neq 0$. The proof is completely analogous. ⌐

Next, we give a corollary to this theorem, which shows that it applies to square-integrable f, if there is an invariant measure.

Corollary 1.3.7. *Let* $f : X \to \mathbb{C}$ *be square-integrable with respect to some finite invariant measure. Then there is a null set* $\Xi \subset X$ *such that for all* $x \in X \setminus \Xi$ *there are at most countably many* $\omega \in [0, 2\pi)$ *such that* $f_\omega^*(x) \neq 0$.

Proof. If $f : X \to \mathbb{C}$ is square-integrable, then $|f|^2$ is integrable. Hence by the Birkhoff Ergodic Theorem [KH06, Theorem 4.1.2], there is a null set $\Xi \subset X$ such that the limit $\lim_{N \to \infty} 1/N \cdot \sum_{n=0}^{N-1} |f(T^n x)|^2$ exists for all $x \in X \setminus \Xi$. So Theorem 1.3.5 implies the assertion. □

Let us have a look at two examples again, where there are in fact only finitely many ω for which the harmonic limit does not vanish.

Example 1.3.8. Recall Example 1.3.3. There, $\mathrm{id}_\omega^*(x) \neq 0$ only holds for $\omega = \alpha$. ⌐

Example 1.3.9. Consider the system from Example 1.2.8 again. Here, it holds that $T^j(x_1, \ldots, x_m) = (e^{-ij\alpha_1} x_1, \ldots, e^{-ij\alpha_m} x_m)$. Let $f : X \to \mathbb{C}$ be given by $f(x_1, \ldots, x_m) := \sum_{k=1}^m x_k$. Then $f_\omega^n(x_1, \ldots, x_m) = 1/n \cdot \sum_{j=0}^{n-1} e^{ij\omega} \sum_{k=1}^m e^{-ij\alpha_k} x_k = \sum_{k=1}^m 1/n \cdot \sum_{j=0}^{n-1} e^{ij(\omega - \alpha_k)} x_k$. Analogously to Example 1.3.3, one can show that $f_\omega^*(x_1, \ldots, x_m) \neq 0$ if and only if $\omega \in \{\alpha_1, \ldots, \alpha_k\}$. ⌐

1.3.2 Existence

In the previous section, we proposed a first result on the existence of harmonic limits (Corollary 1.3.2). It showed that the harmonic limit exists all along a trajectory if it exists at the starting point. The big drawback of this result is that we have to require existence in one point.

Under the presence of an invariant measure, we get stronger results on existence, though: If we have a finite invariant measure μ on X, then for every μ-integrable function f and all $\omega \in [0, 2\pi)$ the harmonic limit exists almost everywhere. If μ is ergodic, the null set, where f_ω^* does not exist, can be chosen independently on ω. Finally, if μ is the unique ergodic measure for our system, then $f_\omega^*(x)$ exists for all $x \in X$, all $\omega \in [0, 2\pi)$, and all integrable f.

Theorem 1.3.10. *Assume that there is a finite invariant measure* μ *on* X *such that the map* $f : X \to \mathbb{C}$ *is of class* $L^p(\mu)$, $1 \leq p < \infty$. *Then for every* $\omega \in [0, 2\pi)$, *there is a null set* $\Xi_\omega \in X$ *such that the harmonic limit* $f_\omega^*(x)$ *exists for all* $x \in X \setminus \Xi_\omega$, *and the map* $x \mapsto f_\omega^*(x)$ *is of class* $L^p(\mu, \mathbb{C})$.

Proof. This follows from the Birkhoff Ergodic Theorem [KH06, Theorem 4.1.2] applied to the extended system U_ω on $S^1 \times X$, given by $(z,x) \mapsto (e^{i\omega}z, Tx)$ for $\omega \in [0, 2\pi)$. More precisely, let $f : X \to \mathbb{C}$ be of class $L^p(\mu)$ and $\omega \in [0, 2\pi)$. Then f particularly is of class $L^1(\mu)$, as μ is finite, compare [Bau92, Korollar 14.7]. Define $g : S^1 \times X \to \mathbb{C}$ by $(z,x) \mapsto zf(x)$. Consider the finite invariant measure $\nu := \lambda \times \mu$ on $S^1 \times X$, where λ is the Lebesgue measure on S^1. Then both $\Re g$ and $\Im g$ are of class $L^1(\nu)$. By the Birkhoff Ergodic Theorem, it follows that both $\lim_{n\to\infty} 1/n \cdot \sum_{j=0}^{n-1} \Re g(U_\omega^j(z,x))$ and $\lim_{n\to\infty} 1/n \cdot \sum_{j=0}^{n-1} \Im g(U_\omega^j(z,x))$ exist for ν-almost all $(z,x) \in S^1 \times X$. Thus also $\lim_{n\to\infty} 1/n \cdot \sum_{j=0}^{n-1} g(U_\omega^j(z,x)) = \lim_{n\to\infty} 1/n \cdot \sum_{j=0}^{n-1} e^{ij\omega} zf(T^j x) = zf_\omega^*(x)$ exists for almost all $(z,x) \in S^1 \times X$. If $z_0 f_\omega^*(x_0)$ exists for some $(z_0, x_0) \in S^1 \times X$, then clearly also $zf_\omega^*(x_0)$ exists for all $z \in S^1$. Thus f_ω^* exists μ-almost everywhere.

By Lebesgue's Dominated Convergence Theorem [Bau92, Satz 15.6], the map $S^1 \times X \to \mathbb{C}$, $(z,x) \mapsto zf_\omega^*(x)$ is of class $L^p(\nu)$. By Fubini's Theorem [Bau92, Korollar 23.7], this implies that $f_\omega^*(x)$ is of class $L^p(\mu)$. □

Note that in Theorem 1.3.10, the set $\Xi_\omega \subset X$ of points, where f_ω^* does not exist, depends on ω. We will use the Wiener-Wintner Ergodic Theorem—see Theorem 1.3.14—later to show that, for ergodic systems, Ξ_ω does not depend on ω.

Existence of a finite invariant measure, which is needed in Theorem 1.3.10, can be shown for a fairly large class of systems by the Theorem of Krylov and Bogolubov [KH06, Theorem 4.1.1], i.e., for continuous T and compact X. Unfortunately, even for those systems, the null set $\Xi_\omega \subset X$ of points where f_ω^* does not exist can depend on ω. But if the system is uniquely ergodic, i.e., if there is only one ergodic measure, then the harmonic limit exists for all angles $\omega \in [0, 2\pi)$ and all points $x \in X$ by the following proposition.

Proposition 1.3.11. *Assume that X is compact, and that $T : X \to X$ and $f : X \to \mathbb{C}$ are continuous. If μ is uniquely ergodic, then $\Xi_\omega = \emptyset$ for all $\omega \in [0, 2\pi)$, i.e., $f_\omega^*(x)$ exists for all $\omega \in [0, 2\pi)$ and all $x \in X$.*

Proof. Compare [Mañ87, Theorem I.9.2]. □

Let us recall two examples from Subsection 1.3.1, which have an invariant measure.

Example 1.3.12. Consider the system from Example 1.3.3 again, i.e., $Tx := e^{-i\alpha}x$ on $S^1 \subset \mathbb{C}$. This system preserves the Lebesgue measure. Hence by Theorem 1.3.10, $f_\omega^*(x)$ exists for all Lebesgue integrable functions $f : S^1 \to \mathbb{C}$ and all $\omega \in [0, 2\pi)$ almost everywhere. Note that $f_\omega^n(x) = f_\omega^n(1) \cdot x$ for all $n \in \mathbb{N}$, $\omega \in [0, 2\pi)$ and $x \in S^1$. Hence $f_\omega^*(x) = f_\omega^*(1) \cdot x$ if one of the limits exist. As the harmonic limit exists almost everywhere, this implies that it actually exists everywhere. ⌟

Example 1.3.13. Consider the quadratic map $Tx := \alpha x(1 - x)$ on $[0, 1]$ from Example 1.3.4 again, and let $\alpha := 4$. Then there is an invariant measure μ with density $^1/\!\left(\pi\sqrt{x(1-x)}\right)$, i.e., $\mu(A) = \int_A {}^1/\!\left(\pi\sqrt{x(1-x)}\right)\mathrm{d}x$ for every Borel measurable set $A \subset [0,1]$, compare [LM95, Example 4.1.2]. Hence by Theorem 1.3.10, the harmonic limit $f_\omega^*(x)$ exists for every μ-integrable function $f : [0,1] \to \mathbb{C}$, all $\omega \in [0,2\pi)$, and almost all $x \in [0,1]$. The measure μ actually is the *unique* ergodic measure of T, i.e., T is uniquely ergodic, compare [LY78; LM95, Example 6.5.1]. So by Proposition 1.3.11, for continuous maps f, the harmonic limit $f_\omega^*(x)$ exists for all $\omega \in [0,2\pi)$ and all $x \in [0,1]$. ⌐

As already noted, for ergodic systems, there is a stronger result on existence of the harmonic limit than Theorem 1.3.10 known as the Wiener-Wintner Ergodic Theorem. This theorem states that f_ω^* exists almost everywhere, independently of $\omega \in [0,2\pi)$. Wiener and Wintner presented this theorem in 1941, see [WW41]. Unfortunately, their proof was wrong [Ass03, p. 24]. The first correct proof was published by Furstenberg in 1960, see [Fur60].

Theorem 1.3.14. *Assume that there is a finite ergodic measure μ on X such that the map $f : X \to \mathbb{C}$ is of class $L^1(\mu)$. Then there is a null set $\mathcal{X} \subset X$, such that the harmonic limit $f_\omega^*(x)$ exists for every $\omega \in [0,2\pi)$ and all $x \in X \setminus \mathcal{X}$.*

Proof. See [Ass92, Theorem 6]. In the case that T is weakly mixing, a simpler proof can be found in [Web09, Theorem 4.7.2]. □

Recall from Proposition 1.3.11, that if we have a *uniquely* ergodic measure, then $f_\omega^*(x)$ exists for all $\omega \in [0,2\pi)$ and all $x \in X$. For further results, related to the Wiener-Wintner Ergodic Theorem, refer to [Ass03].

1.3.3 Asymptotics

Knowledge of the asymptotic behaviour of the system can help in the analysis of harmonic limits. If, e.g., the trajectory starting in $x \in X$ converges to a fixed point $x_0 \in X$, then $f_\omega^*(x) = f_\omega^*(x_0) = f(x_0)$ for continuous f. Similarly, if a trajectory asymptotically approaches a periodic orbit, one can apply results from the following Subsection 1.3.4.

Proposition 1.3.15. *Let $x \in X$ and $f : X \to \mathbb{C}$. Assume that $f(T^j x) - g_j \to 0$ as $j \to \infty$ for some sequence $(g_j)_{j\in\mathbb{N}_0} \subset \mathbb{C}$. Then $f_\omega^*(x) = \lim_{n\to\infty} \frac{1}{n} \sum_{j=0}^{n-1} \mathrm{e}^{\mathrm{i}j\omega} g_j$ for all $\omega \in [0,2\pi)$, provided that the limits exist.*

Proof. Due to the convergence $f(T^j x) - g_j \to 0$, for every $\varepsilon > 0$, there is $n_\varepsilon \in \mathbb{N}$ such that, for all $j \geq n_\varepsilon$, it holds that $|f(T^j x) - g_j| < \varepsilon$. So for any $\varepsilon > 0$ and all

$n \geq n_\varepsilon$, it follows that

$$\left| f_\omega^n(x) - \frac{n_\varepsilon}{n} f_\omega^{n_\varepsilon}(x) - \frac{1}{n} \sum_{j=0}^{n-1} \mathrm{e}^{\mathrm{i}j\omega} g_j + \frac{1}{n} \sum_{j=0}^{n_\varepsilon-1} \mathrm{e}^{\mathrm{i}j\omega} g_j \right|$$

$$= \left| \frac{1}{n} \sum_{j=n_\varepsilon}^{n-1} \mathrm{e}^{\mathrm{i}j\omega} f(T^j x) - \frac{1}{n} \sum_{j=n_\varepsilon}^{n-1} \mathrm{e}^{\mathrm{i}j\omega} g_j \right| \leq \frac{1}{n} \sum_{j=n_\varepsilon}^{n-1} \underbrace{|f(T^j x) - g_j|}_{<\varepsilon} < \frac{n - n_\varepsilon}{n} \varepsilon,$$

which tends to ε for $n \to \infty$. This means,

$$\varepsilon \geq \lim_{n \to \infty} \left| f_\omega^n(x) - \frac{n_\varepsilon}{n} f_\omega^{n_\varepsilon}(x) - \frac{1}{n} \sum_{j=0}^{n-1} \mathrm{e}^{\mathrm{i}j\omega} g_j + \frac{1}{n} \sum_{j=0}^{n_\varepsilon-1} \mathrm{e}^{\mathrm{i}j\omega} g_j \right|$$

$$= \left| f_\omega^*(x) - 0 - \lim_{n \to \infty} \frac{1}{n} \sum_{j=0}^{n-1} \mathrm{e}^{\mathrm{i}j\omega} g_j + 0 \right|,$$

if the limits exist.

As this is true for every $\varepsilon > 0$, it holds that $f_\omega^*(x) = \lim_{n \to \infty} {}^1\!/\!_n \cdot \sum_{j=0}^{n-1} \mathrm{e}^{\mathrm{i}j\omega} g_j$, if the limits exist. $\qquad \square$

Corollary 1.3.16. *Let $f : X \to \mathbb{C}$ be continuous, and let $x \in X$. For all $y \in X$ with $d(T^j x, T^j y) \to 0$ for $j \to \infty$, and all $\omega \in [0, 2\pi)$, it holds that $f_\omega^*(x) = f_\omega^*(y)$, provided that one of the limits exists.*

Proof. Let $x, y \in X$ be such that $d(T^j x, T^j y) \to 0$. Define $g_j := f(T^j y)$. Then by continuity, $f(T^j x) - g_j = f(T^j x) - f(T^j y) \to 0$. So Proposition 1.3.15 implies the assertion. $\qquad \square$

1.3.4 Periodicity

If a system shows some periodic behaviour (if, e. g., it has a periodic orbit), one can show that the harmonic limit exists without using invariant measures, as we did in Theorem 1.3.10 and Theorem 1.3.14. Furthermore, the periods determine the set of angles that can occur.

First, we give two existence results. Proposition 1.3.17, which we have already used in Example 1.2.12, is concerned with periodic points, where the harmonic limit always exists and can be given without a limit. For almost periodic points, Proposition 1.3.19 also shows existence of the harmonic limit.

Proposition 1.3.17. *If $x \in X$ is such that $T^\tau x = x$ for some $\tau \in \mathbb{N}$, then $f_{2k\pi/\tau}^*(x) = {}^1\!/\!_\tau \cdot \sum_{j=0}^{\tau-1} \mathrm{e}^{\mathrm{i}j2k\pi/\tau} f(T^j x)$ for every $f : X \to \mathbb{C}$ and all $k \in \mathbb{Z}$. Particularly, $f_{2k\pi/\tau}^*(x)$ exists.*

Proof. For every $n \in \mathbb{N}$, it holds that

$$
\frac{1}{n} \sum_{j=0}^{n-1} \mathrm{e}^{\mathrm{i}j2k\pi/\tau} f(T^j x)
$$

$$
= \frac{1}{n} \sum_{j=0}^{\lfloor n/\tau \rfloor \tau - 1} \mathrm{e}^{\mathrm{i}j2k\pi/\tau} f(T^j x) + \frac{1}{n} \sum_{j=\lfloor n/\tau \rfloor \tau}^{n-1} \mathrm{e}^{\mathrm{i}j2k\pi/\tau} f(T^j x)
$$

$$
= \frac{1}{n} \sum_{l=1}^{\lfloor n/\tau \rfloor} \sum_{j=(l-1)\tau}^{l\tau-1} \mathrm{e}^{\mathrm{i}j2k\pi/\tau} f(T^j x) + \frac{1}{n} \sum_{j=\lfloor n/\tau \rfloor \tau}^{n-1} \mathrm{e}^{\mathrm{i}j2k\pi/\tau} f(T^j x) \qquad (1.3.9)
$$

$$
= \frac{1}{n} \sum_{l=1}^{\lfloor n/\tau \rfloor} \sum_{j=0}^{\tau-1} \mathrm{e}^{\mathrm{i}j2k\pi/\tau} f(T^j x) + \frac{1}{n} \sum_{j=\lfloor n/\tau \rfloor \tau}^{n-1} \mathrm{e}^{\mathrm{i}j2k\pi/\tau} f(T^j x)
$$

$$
= \frac{\lfloor \frac{n}{\tau} \rfloor}{n} \sum_{j=0}^{\tau-1} \mathrm{e}^{\mathrm{i}j2k\pi/\tau} f(T^j x) + \frac{1}{n} \sum_{j=\lfloor n/\tau \rfloor \tau}^{n-1} \mathrm{e}^{\mathrm{i}j2k\pi/\tau} f(T^j x)
$$

$$
= \frac{\lfloor \frac{n}{\tau} \rfloor}{\frac{n}{\tau}} \frac{1}{\tau} \sum_{j=0}^{\tau-1} \mathrm{e}^{\mathrm{i}j2k\pi/\tau} f(T^j x) + \frac{1}{n} \sum_{j=\lfloor n/\tau \rfloor \tau}^{n-1} \mathrm{e}^{\mathrm{i}j2k\pi/\tau} f(T^j x).
$$

Note that

$$
\lim_{n \to \infty} \frac{\lfloor \frac{n}{\tau} \rfloor}{\frac{n}{\tau}} = 1. \qquad (1.3.10)
$$

Further note that

$$
\left| \frac{1}{n} \sum_{j=\lfloor n/\tau \rfloor \tau}^{n-1} \mathrm{e}^{\mathrm{i}j2k\pi/\tau} f(T^j x) \right| \le \frac{1}{n} \sum_{j=\lfloor n/\tau \rfloor \tau}^{n-1} |f(T^j x)| \le \frac{1}{n} \sum_{j=0}^{\tau-1} |f(T^j x)|
$$

is bounded, and hence

$$
\lim_{n \to \infty} \frac{1}{n} \sum_{j=\lfloor n/\tau \rfloor \tau}^{n-1} \mathrm{e}^{\mathrm{i}j2k\pi/\tau} f(T^j x) = 0. \qquad (1.3.11)
$$

Equations (1.3.9), (1.3.10) and (1.3.11) together imply the assertion. $\qquad \square$

Example 1.3.18. Consider again the quadratic map $x \mapsto 3.5x(1-x)$ from Example 1.3.4. Recall that there is an attracting 4-periodic orbit x_1, x_2, x_3, x_4. Its points are $x_1 \approx 0.3828$, $x_2 \approx 0.8269$, $x_3 \approx 0.5009$, and $x_4 \approx 0.8750$. The orbit starting in 0.3 quickly approaches this periodic orbit, see Figure 1.5. So by Corollary 1.3.16 and Proposition 1.3.17, for any continuous map $f : [0,1] \to \mathbb{C}$ and every $k \in \mathbb{Z}$, it holds that

$$|f^*_{k\pi/2}(0.3)| = \left|\frac{1}{4}[f(x_1) + \mathrm{e}^{\mathrm{i}k\pi/2}f(x_2) + \mathrm{e}^{\mathrm{i}k\pi}f(x_3) + \mathrm{e}^{\mathrm{i}3k\pi/2}f(x_4)]\right|$$

$$\approx \left|\frac{1}{4}[f(0.3828) + \mathrm{i}^k f(0.8269) + (-1)^k f(0.5009) + (-\mathrm{i})^k f(0.8750)]\right|.$$

So for $f = $ id and $k = 1$, we get $|f^*_{\pi/2}(0.3)| \approx |1/4[0.3828 + 0.8269\mathrm{i} + -0.5009 - 0.8750\mathrm{i}]| \approx |-0.02952 - 0.01201\mathrm{i}| \approx 0.03187 < 1$, which is in line with the result in Example 1.3.4. ⌐

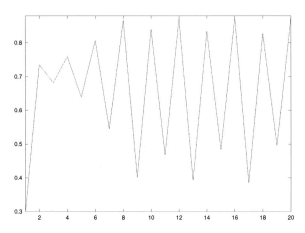

Figure 1.5: Orbit of the quadratic map $x \mapsto 3.5x(1-x)$ starting in 0.3
See Example 1.3.18.

At almost periodic orbits, the harmonic limit also always exists. With *almost periodic orbit* we mean an orbit such that there is an almost periodic function (see Definition 1.1.1) that maps the nonnegative integers onto the orbit, i.e., $p : \mathbb{R} \to X$ with $T^j x = p(j)$ for all $j \in \mathbb{N}_0$. This is also true in the more general case that $f(T^j x)$ is almost periodic, in the sense that there is $p : \mathbb{R} \to \mathbb{C}$ almost periodic with $f(T^j x) = p(j)$ for all $j \in \mathbb{N}_0$.

Proposition 1.3.19.

1. *If $f : X \to \mathbb{C}$ is continuous, and $x \in X$ is such that there is an almost periodic function $p : \mathbb{R} \to X$ with $T^j x = p(j)$ for all $j \in \mathbb{N}_0$, then $f^*_\omega(x)$ exists for all $\omega \in [0, 2\pi)$.*

2. *If $x \in X$ and $f : X \to \mathbb{C}$ are such that there is an almost periodic function $p : \mathbb{R} \to \mathbb{C}$ with $f(T^j x) = p(j)$ for all $j \in \mathbb{N}_0$, then $f_\omega^\star(x)$ exists for all $\omega \in [0, 2\pi)$.*

To show this proposition, we need the following discrete-time variant of the Mean Value Theorem for almost periodic functions.

Lemma 1.3.20. *Let $f : \mathbb{R} \to \mathbb{C}$ be almost periodic. Then $\lim_{n \to \infty} {}^1\!/n \cdot \sum_{j=0}^{n-1} f(j)$ exists.*

Proof. This can be shown analogously to the proof of the Mean Value Theorem in [LZ82, Section 2.3.2] for almost periodic functions. Basically, we only have to replace integrals by sums. But as Levitan and Zhikov show the Mean Value Theorem for bilateral limits $\lim_{T \to \infty} {}^1\!/(2T) \cdot \int_{-T}^{T} f(t) dt$, and for the sake of completeness, we give a full proof.

For every $\omega \in [0, 2\pi)$ and all $n \in \mathbb{N}$, it holds that

$$\frac{1}{n} \sum_{j=0}^{n-1} e^{ij\omega} = \begin{cases} 1 & \text{if } \omega = 0, \\ \frac{e^{in\omega} - 1}{n(e^{i\omega} - 1)} & \text{otherwise.} \end{cases}$$

So

$$\lim_{n \to \infty} \frac{1}{n} \sum_{j=0}^{n-1} e^{ij\omega} = \begin{cases} 1 & \text{if } \omega = 0, \\ 0 & \text{otherwise.} \end{cases}$$

Hence the limit $\lim_{n \to \infty} {}^1\!/n \cdot \sum_{j=0}^{n-1} P(j)$ exists for every trigonometric polynomial $P(j) := \sum_{k=1}^{N} a_k e^{ij\lambda_k}$ with $N \in \mathbb{N}$, $a_k \in \mathbb{C}$, $\lambda_k \in [0, 2\pi)$, $k = 1, \ldots, N$.

As f is almost periodic, for every $\varepsilon > 0$, there is a trigonometric polynomial $P_\varepsilon(j)$ such that $\sup_{j \in \mathbb{N}} \|f(j) - P_\varepsilon(j)\| < \varepsilon$, compare the Approximation Theorem in [Boh47, Section 84]. Note that we can assume that the exponents λ_k of P_ε are in $[0, 2\pi)$ as above, because we only evaluate P_ε at integer values. For every $n, m \in \mathbb{N}$, it holds that

$$\left\| \frac{1}{n} \sum_{j=0}^{n-1} f(j) - \frac{1}{m} \sum_{j=0}^{m-1} f(j) \right\| \leq \left\| \frac{1}{n} \sum_{j=0}^{n-1} f(j) - \frac{1}{n} \sum_{j=0}^{n-1} P_\varepsilon(j) \right\|$$

$$+ \left\| \frac{1}{n} \sum_{j=0}^{n-1} P_\varepsilon(j) - \frac{1}{m} \sum_{j=0}^{m-1} P_\varepsilon(j) \right\|$$

$$+ \left\| \frac{1}{m} \sum_{j=0}^{m-1} P_\varepsilon(j) - \frac{1}{m} \sum_{j=0}^{m-1} f(j) \right\|$$

$$\leq \frac{1}{n} \sum_{j=0}^{n-1} \|f(j) - P_\varepsilon(j)\| + \left\| \frac{1}{n} \sum_{j=0}^{n-1} P_\varepsilon(j) - \frac{1}{m} \sum_{j=0}^{m-1} P_\varepsilon(j) \right\|$$

$$+ \frac{1}{m} \sum_{j=0}^{m-1} \|P_\varepsilon(j) - f(j)\|$$

$$\leq \left\| \frac{1}{n} \sum_{j=0}^{n-1} P_\varepsilon(j) - \frac{1}{m} \sum_{j=0}^{m-1} P_\varepsilon(j) \right\| + 2\varepsilon.$$

As P_ε is a trigonometric polynomial, $1/n \cdot \sum_{j=0}^{n-1} P_\varepsilon(j)$ converges for $n \to \infty$, and hence $1/n \cdot \sum_{j=0}^{n-1} P_\varepsilon(j)$ is a Cauchy sequence. So for every $\varepsilon > 0$, there is $N_\varepsilon \in \mathbb{N}$ such that for all $n, m \geq N_\varepsilon$, it holds that

$$\left\| \frac{1}{n} \sum_{j=0}^{n-1} P_\varepsilon(j) - \frac{1}{m} \sum_{j=0}^{m-1} P_\varepsilon(j) \right\| < \varepsilon.$$

Together, this means that, for every $\varepsilon > 0$, there is $N_\varepsilon \in \mathbb{N}$ such that, for all $n, m \geq N_\varepsilon$, it holds that

$$\left\| \frac{1}{n} \sum_{j=0}^{n-1} f(j) - \frac{1}{m} \sum_{j=0}^{m-1} f(j) \right\| \leq 3\varepsilon.$$

So $1/n \cdot \sum_{j=0}^{n-1} f(j)$ is a Cauchy sequence and hence converges for $n \to \infty$. $\qquad\square$

Proof of Proposition 1.3.19. If $p : \mathbb{R} \to X$ is almost periodic such that $T^j x = p(j)$ for all $j \in \mathbb{N}_0$, and if $f : X \to \mathbb{C}$ is continuous, then $t \mapsto f(p(t))$ is almost periodic by Lemma 1.1.2. So it suffices to show part 2. of this proposition.

Part 2 follows from Theorem IV in [Boh47, p. 38] and Lemma 1.3.20. More precisely, by Theorem IV in [Boh47, p. 38], for every $\omega \in \mathbb{R}$ the map $t \mapsto e^{it\omega} p(t)$ is almost periodic. So by Lemma 1.3.20, the limit exists. $\qquad\square$

If $f(T^j x)$ is almost periodic, then there is the following connection between the harmonic limit and the Fourier coefficients of $f(T^j x)$. Proposition 1.3.21 also reveals a connection between the harmonic limits for discrete-time dynamical systems and a harmonic limit of the type we will analyze in Chapter 2.

Proposition 1.3.21. *Assume that $x \in X$ and $f : X \to \mathbb{C}$ are such that there is an almost periodic function $p : \mathbb{R} \to \mathbb{C}$ with $f(T^j x) = p(j)$ for all $j \in \mathbb{N}_0$. Let $\alpha_k \in \mathbb{R}$ be the Fourier exponents of p, and let $c_k \in \mathbb{C}$ be its Fourier coefficients, $k \in \mathbb{N}_0$, i.e., $p(t) = \sum_{k \in \mathbb{N}_0} c_k e^{i\alpha_k t}$. Then for every $\omega \in [0, 2\pi)$,*

$$f_\omega^*(x) = \lim_{T \to \infty} \frac{1}{T} \int_0^T \sum_{k \in \mathbb{N}_0} c_k e^{i(\alpha_k + \omega \bmod 2\pi)t} dt = \sum_{\substack{k \in \mathbb{N}_0: \\ \alpha_k + \omega \equiv 0 \bmod 2\pi}} c_k. \qquad (1.3.12)$$

Proof. Note that $t \mapsto \mathrm{e}^{\mathrm{i}t\omega}p(t) = \sum_{k\in\mathbb{N}_0} c_k \mathrm{e}^{\mathrm{i}(\alpha_k+\omega)t}$ is almost periodic with Fourier exponents $\alpha_k + \omega =: \beta_k$ and coefficients c_k. Further note that

$$\mathrm{e}^{\mathrm{i}j\omega}p(j) = \sum_{k\in\mathbb{N}_0} c_k \mathrm{e}^{\mathrm{i}(\alpha_k+\omega \bmod 2\pi)t}$$

for all $j \in \mathbb{Z}$, as $\mathrm{e}^{\mathrm{i}(\alpha+\omega)j} = \mathrm{e}^{\mathrm{i}(\alpha+\omega \bmod 2\pi)j}$ for integer j. So we may assume that the Fourier exponents β_k lie in $[0, 2\pi)$. We may additionally assume that $\beta_0 = 0$, and that all β_k, $k \in \mathbb{N}_0$, are pairwise different. Under these assumptions, it suffices to show

$$f_\omega^*(x) = \lim_{T\to\infty} \frac{1}{T} \int_0^T \mathrm{e}^{\mathrm{i}j\omega}p(t)\mathrm{d}t = c_0 \tag{1.3.13}$$

instead of (1.3.12).

The equation $\lim_{T\to\infty} 1/T \cdot \int_0^T \mathrm{e}^{\mathrm{i}j\omega}p(t)\mathrm{d}t = c_0$ in (1.3.13) holds by the definition of the Fourier coefficients, compare [Zha03, p. 6].

Consider $t \mapsto \mathrm{e}^{\mathrm{i}\beta t}$ with $\beta \in [0, 2\pi)$. It holds that

$$\frac{1}{n}\sum_{j=0}^{n-1} \mathrm{e}^{\mathrm{i}\beta j} = \begin{cases} 1 & \text{if } \beta = 0, \\ \frac{\mathrm{e}^{\mathrm{i}\beta n}-1}{n(\mathrm{e}^{\mathrm{i}\beta}-1)} & \text{otherwise.} \end{cases}$$

So

$$\lim_{n\to\infty} \frac{1}{n}\sum_{j=0}^{n-1} \mathrm{e}^{\mathrm{i}\beta j} = \begin{cases} 1 & \text{if } \beta = 0, \\ 0 & \text{otherwise.} \end{cases}$$

Hence for every trigonometric polynomial $P(t) = \sum_{k=0}^m b_k \mathrm{e}^{\mathrm{i}\beta_k t}$ with coefficients $b_k \in \mathbb{C}$, (pairwise different) exponents $\beta_k \in [0, 2\pi)$, $k = 0, \ldots, m$, and $\beta_0 = 0$, we have

$$\lim_{n\to\infty} \frac{1}{n}\sum_{j=0}^{n-1} P(j) = b_0. \tag{1.3.14}$$

As $\mathrm{e}^{\mathrm{i}t\omega}p(t)$ is almost periodic with Fourier exponents β_k in $[0, 2\pi)$, it can be approximated by trigonometric polynomials $P_m(t) = \sum_{k=1}^m b_{k,m} \mathrm{e}^{\mathrm{i}\beta_k t}$, $m \in \mathbb{N}$, $b_{k,m} \in \mathbb{C}$ such that $\sup_{t\in\mathbb{R}} |\mathrm{e}^{\mathrm{i}t\omega}p(t) - P_m(t)| \to 0$ and $b_{k,m} \to c_k$ for $m \to \infty$, see [Zha03, Theorem 4.3]. So for every $\varepsilon > 0$, there is $m_\varepsilon \in \mathbb{N}$ such that $\sup_{t\in\mathbb{R}} |\mathrm{e}^{\mathrm{i}t\omega}p(t) - P_{m_\varepsilon}(t)| \leq \varepsilon/2$ and $|b_{0,m_\varepsilon} - c_0| \leq \varepsilon/2$.

By (1.3.14), we have that $\lim_{n\to\infty} 1/n \cdot \sum_{j=0}^{n-1} P_m(j) = b_{0,m}$. So for every $\varepsilon > 0$, it holds that

$$
\left| \lim_{n \to \infty} \frac{1}{n} \sum_{j=0}^{n-1} e^{ij\omega} p(j) - c_0 \right| \leq \left| \lim_{n \to \infty} \frac{1}{n} \sum_{j=0}^{n-1} e^{ij\omega} p(j) - \lim_{n \to \infty} \frac{1}{n} \sum_{j=0}^{n-1} P_{m_\varepsilon}(j) \right|
$$

$$
+ \left| \lim_{n \to \infty} \frac{1}{n} \sum_{j=0}^{n-1} P_{m_\varepsilon}(j) - c_0 \right|
$$

$$
\leq \lim_{n \to \infty} \frac{1}{n} \sum_{j=0}^{n-1} |e^{ij\omega} p(j) - P_{m_\varepsilon}(j)| + |b_{0,m_\varepsilon} - c_0|
$$

$$
\leq \frac{\varepsilon}{2} + |b_{0,m_\varepsilon} - c_0|
$$

$$
\leq \varepsilon.
$$

Hence $\lim_{n \to \infty} 1/n \cdot \sum_{j=0}^{n-1} e^{ij\omega} p(j) = c_0$. $\qquad\square$

Now, we turn to the question, which angles can occur at periodic and quasi-periodic orbits. With *quasi-periodic orbit* we mean an orbit such that $f(T^j x)$ is quasi-periodic, in the sense that there is a quasi-periodic function (see Definition 1.1.3) that maps the nonnegative integers onto $f(T^j x)$, i.e., $q : \mathbb{R} \to X$ with $T^j x = q(j)$ for all $j \in \mathbb{N}_0$. Using this concept, we can also consider periodic orbits with noninteger periods. In both cases, we get that the harmonic limit $f_\omega^*(x)$ vanishes if ω and the periods of the orbit are independent in some sense.

In the case, where $f(T^j x)$ is quasi-periodic with periods $\tau_1, \ldots, \tau_n$, the following proposition shows that $f_\omega^*(x) = 0$ if the numbers $\omega, 2\pi, 2\pi/\tau_1, \ldots, 2\pi/\tau_n$ are rationally independent, i.e., if there is no nonzero $(c, c_0, c_1, \ldots, c_n) \in \mathbb{Q}^{n+1}$ such that $2\pi c + c_0 \omega + c_1 \cdot 2\pi/\tau_1 + \cdots + c_n \cdot 2\pi/\tau_n = 0$. To put it the other way round, $f_\omega^*(x) \neq 0$ can only hold if the numbers $\omega, 2\pi, 2\pi/\tau_1, \ldots, 2\pi/\tau_n$ are rationally dependent, i.e., if $2\pi, 2\pi/\tau_1, \ldots, 2\pi/\tau_n$ are rationally dependent or $\omega = 2\pi c + c_1 \cdot 2\pi/\tau_1 + \cdots + c_n \cdot 2\pi/\tau_n$ for some $c, c_1, \ldots, c_n \in \mathbb{Q}$.

Proposition 1.3.22. *Assume that $x \in X$ and $f : X \to C$ are such that there is a quasi-periodic function $q : \mathbb{R} \to \mathbb{C}$ with $f(T^j x) = q(j)$ for all $j \in \mathbb{N}_0$. Let $\tau_j, \ j = 1, \ldots, n$, be the periods of q. If $\omega \in [0, 2\pi)$ is such that the numbers $\omega, 2\pi, 2\pi/\tau_1, \ldots, 2\pi/\tau_n$ are rationally independent, then $f_\omega^*(x) = 0$.*

Proof. Define the map $\rho = (\rho^1, \rho^2)$ on $[0, 2\pi) \times [0, 2\pi)^n$ by $\rho^1(\alpha) = \alpha + \omega \mod 2\pi$ and $\rho^2(\beta_1, \ldots, \beta_n) = (\beta_1 + 2\pi/\tau_1 \mod 2\pi, \ldots, \beta_n + 2\pi/\tau_n \mod 2\pi)$. This map preserves the Lebesgue measure. Note that $[0, 2\pi) \times [0, 2\pi)^n$ can be interpreted as an $(n+1)$-dimensional torus. This flow is uniquely ergodic, as $\omega, 2\pi, 2\pi/\tau_1, \ldots, 2\pi/\tau_n$ are rationally independent, compare [KH06, Propositions 4.2.3 and 4.2.2].

Let $Q : \mathbb{R}^n \to \mathbb{C}$ be a continuous generating function for q, i.e., a function which is τ_j-periodic in its j-th component, $j = 1, \ldots, n$, such that $Q(t, \ldots, t) = q(t)$. Let

$g : [0, 2\pi) \times [0, 2\pi)^n \to \mathbb{C}$ be given by

$$(\alpha, \beta_1, \ldots, \beta_n) \mapsto e^{i\alpha} Q\big({\tau_1}/{2\pi} \cdot \beta_1, \ldots, {\tau_n}/{2\pi} \cdot \beta_n\big).$$

In the following, if $\beta \in \mathbb{R}^n$, interpret $Q\big({\tau\beta}/{2\pi}\big)$ as $Q\big({\tau_1}/{2\pi} \cdot \beta_1, \ldots, {\tau_n}/{2\pi} \cdot \beta_n\big)$.

The map g is Lebesgue integrable, as it has compact domain and is continuous. Thus by the Ergodic Theorem [KH06, Theorem 4.1.2] and by Fubini's Theorem [Bau92, Korollar 23.7], it holds for almost all $(\alpha^0, \beta^0) \in [0, 2\pi) \times [0, 2\pi)^n$ that

$$
\begin{aligned}
\lim_{n \to \infty} \frac{1}{n} \sum_{j=0}^{n-1} g\big(\rho^j(\alpha^0, \beta^0)\big) &= \frac{1}{\mu\big([0,2\pi) \times [0,2\pi)^n\big)} \int_{[0,2\pi) \times [0,2\pi)^n} g(\alpha, \beta) \mathrm{d}(\alpha, \beta) \\
&= \frac{1}{(2\pi)^{1+n}} \int_{[0,2\pi) \times [0,2\pi)^n} e^{i\alpha} Q\big({\tau\beta}/{2\pi}\big) \mathrm{d}(\alpha, \beta) \\
&= \frac{1}{(2\pi)^{1+n}} \int_{[0,2\pi)^n} \int_0^{2\pi} e^{i\alpha} Q\big({\tau\beta}/{2\pi}\big) \mathrm{d}\alpha \mathrm{d}\beta \qquad (1.3.15) \\
&= \frac{1}{(2\pi)^{1+n}} \int_{[0,2\pi)^n} \underbrace{\left(\int_0^{2\pi} e^{i\alpha} \mathrm{d}\alpha \right)}_{=0} \cdot Q\big({\tau\beta}/{2\pi}\big) \mathrm{d}\beta \\
&= 0.
\end{aligned}
$$

As ρ is uniquely ergodic with respect to a Borel measure on a compact space, and f and ρ are continuous, equation (1.3.15) actually holds everywhere, compare [Mañ87, Theorem I.9.2]. In particular, $f_\omega^*(x) = \lim_{n \to \infty} {1}/{n} \cdot \sum_{j=0}^{n-1} g\big(\rho^j(0, 0, \ldots, 0)\big) = 0$. □

To illustrate Proposition 1.3.22, consider the following example, which is a rotation on a torus.

Example 1.3.23. Let $T : [0, 2\pi)^2 \to [0, 2\pi)^2$ be given by $(s, t) \mapsto (s + 1 \bmod 2\pi, t + \sqrt{2} \bmod 2\pi)$. Consider $f : [0, 2\pi)^2 \to \mathbb{C}$, $(s, t) \mapsto \cos s \cos t$, and the map $q : \mathbb{R} \to \mathbb{C}$ given by $q(t) := \cos t \cdot \cos \sqrt{2} t$. It holds that $f(T^j 0) = q(j)$ for all $j \in \mathbb{N}_0$. The map q is quasi-periodic with generating function $(s, t) \mapsto \cos s \cdot \cos \sqrt{2} t$, which is locally Lebesgue integrable, and periods 2π and $\sqrt{2}\pi$, see Figure 1.6. For every $\omega \in [0, 2\pi)$, the harmonic limit $f_\omega^*(0)$ exists by Proposition 1.3.19. By Proposition 1.3.22, $f_\omega^*(0) = 0$ if $\omega, 2\pi, 1, \sqrt{2}$ are rationally independent. For example, with $\omega = \sqrt{3}$, we have rational independence, and indeed $f_{\sqrt{3}}^*(0) = 0$. But with $\omega = \sqrt{2} - 1$ the numbers are rationally dependent, and indeed $f_{\sqrt{2}-1}^*(0) = {1}/{4}$. Note that $f_\omega^*(x)$ can also be zero if the numbers are rationally dependent, e. g., $f_{\sqrt{2}}^*(0) = 0$. ⌟

In the periodic case, we get a stronger result, i. e., the harmonic limit can only be different from zero if the period (or the smallest integer period) is a multiple of ${2\pi}/{\omega}$. If, e. g., $x \in X$ and $f : X \to \mathbb{C}$ are such that $f(T^j x) = P(j)$ for a 2π-periodic function P, then $f_\omega^*(x) \neq 0$ implies $2\pi = k \cdot {2\pi}/{\omega}$ for some $k \in \mathbb{Z}$, i. e., $\omega \in \mathbb{Z}$.

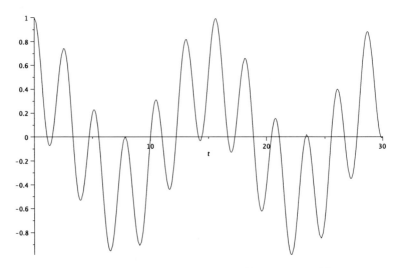

Figure 1.6: The quasi-periodic map $q(t) = \cos t \cdot \cos \sqrt{2} t$
See Example 1.3.23.

Proposition 1.3.24. *Assume that $x \in X$ and $f : X \to \mathbb{C}$ are such that there is a τ-periodic Lebesgue-integrable function $P : \mathbb{R} \to \mathbb{C}$, $\tau > 0$, with $f(T^j x) = P(j)$ for all $j \in \mathbb{N}_0$. Then $f_\omega^*(x) \neq 0$ can only hold if either τ or the smallest integer period of P (if it exists) is a multiple of $2\pi/\omega$.*

We need the following lemma for the proof of this proposition.

Lemma 1.3.25. *Let $s \in \mathbb{Q} \setminus \mathbb{Z}$ and $t \in \mathbb{N}$ such that $st \in \mathbb{Z}$. Then $\sum_{j=0}^{t-1} \mathrm{e}^{\mathrm{i} j 2 \pi s} = 0$.*

Proof. As $s \notin \mathbb{Z}$, it holds that $\mathrm{e}^{\mathrm{i} 2 \pi s} \neq 1$. Hence

$$\sum_{j=0}^{t-1} \mathrm{e}^{\mathrm{i} j 2 \pi s} = \frac{\mathrm{e}^{\mathrm{i} 2 \pi s t} - 1}{\mathrm{e}^{\mathrm{i} 2 \pi s} - 1} = 0. \qquad \square$$

Proof of Proposition 1.3.24. We consider the following cases:

1. $\omega \tau / 2\pi \notin \mathbb{Q}$,

2. $\omega \tau / 2\pi \in \mathbb{Q}$,

    a) $\omega \tau / 2\pi \in \mathbb{Q} \setminus \mathbb{Z}$,

        i. $\tau \in \mathbb{Q}$,

       A. $q_2 \nmid p_1$,

       B. $q_2 \mid p_1$,

     ii. $\tau \notin \mathbb{Q}$,

   b) $\omega\tau/2\pi \in \mathbb{Z}$,

where $q_2 := \mathrm{denom}\left(\omega/2\pi\right)$ and $p_1 := \mathrm{numer}(\tau)$, and show that $f_\omega^*(x) = 0$ for all cases but 2. a) i. B. and 2. b). Case 2. b) is equivalent to the assertion that τ is a multiple of $2\pi/\omega$, and case 2. a) i. B. is equivalent to the assertion that the smallest integer period of P is a multiple of $2\pi/\omega$, which will be shown later in this proof. During this proof, we will use the notation $x \perp y$ to express that x and y are coprime, $x, y \in \mathbb{N}$.

    Note that $\omega = 0$ implies $\omega\tau/2\pi = 0$, i. e., case 2. b). So we can assume that $\omega \neq 0$ in the other cases.

Case 1.: By Proposition 1.3.22, $f_\omega^*(x) = 0$ if $2\pi/\omega$ and τ are rationally independent, i. e., if $\omega\tau/2\pi \notin \mathbb{Q}$.

Case 2. a): Assume that $\omega\tau/2\pi \in \mathbb{Q} \setminus \mathbb{Z}$. Then there are $p \in \mathbb{N}$, $q \in \mathbb{N} \setminus \{1\}$ coprime such that $\omega\tau/2\pi = p/q$, i. e., such that $q\tau = p \cdot 2\pi/\omega =: \sigma > 0$. Clearly, both $\mathrm{e}^{\mathrm{i}\omega t}$ and $P(t)$ are σ-periodic in t.

Case 2. a) i: Further assume that $\tau \in \mathbb{Q}$. Then there are $p_1, q_1 \in \mathbb{N}$ coprime such that $\tau = p_1/q_1$. As $\omega\tau/2\pi \in \mathbb{Q}$, also $\omega/2\pi \in \mathbb{Q}$. So there are $p_2, q_2 \in \mathbb{N}$ coprime such that $\omega/(2\pi) = p_2/q_2$. Let $a := \gcd(p_1, q_2)$, $\tilde{p}_1 := p_1/a \in \mathbb{N}$, and $\tilde{q}_2 := q_2/a \in \mathbb{N}$. Let $b := \gcd(p_2, q_1)$, $\tilde{p}_2 := p_2/b \in \mathbb{N}$, and $\tilde{q}_1 := q_1/b \in \mathbb{N}$. Then $\tilde{p}_1 \perp \tilde{q}_2$ and $\tilde{p}_2 \perp \tilde{q}_1$. Note that also $\tilde{p}_1 \perp \tilde{q}_1$ and $\tilde{p}_2 \perp \tilde{q}_2$, because $p_1 \perp q_1$ and $p_2 \perp q_2$. So

$$\frac{p}{q} = \frac{p_1 p_2}{q_1 q_2} = \frac{\tilde{p}_1 \tilde{p}_2}{\tilde{q}_1 \tilde{q}_2}$$

with $\tilde{p}_1 \tilde{p}_2 \perp \tilde{q}_1 \tilde{q}_2$, and hence $p = \tilde{p}_1 \tilde{p}_2$ and $q = \tilde{q}_1 \tilde{q}_2$, which implies $\sigma = \tilde{q}_1 \tilde{q}_2 \tau = \tilde{q}_1 \tilde{q}_2 \frac{p_1}{q_1} = p_1/b \cdot \tilde{q}_2$. Thus $b\sigma = b\tilde{q}_1 \tilde{q}_2 \tau = \tilde{q}_2 \cdot q_1 \tau$. Note that $b\sigma \in \mathbb{N}$ is the smallest common integer period of $\mathrm{e}^{\mathrm{i}j\omega}$ and $P(j)$, and that $q_1\tau \in \mathbb{N}$ is the smallest integer period of $P(j)$. By Proposition 1.3.17, it holds that

$$f_\omega^*(x) = \lim_{n \to \infty} \frac{1}{n} \sum_{j=0}^{n-1} \mathrm{e}^{\mathrm{i}j\omega} P(j)$$

$$= \frac{1}{b\sigma} \sum_{j=0}^{b\sigma-1} \mathrm{e}^{\mathrm{i}j\omega} P(j) \qquad (1.3.16)$$

$$= \frac{1}{\tilde{q}_2 q_1 \tau} \sum_{j=0}^{\tilde{q}_2 q_1 \tau - 1} \mathrm{e}^{\mathrm{i}j\omega} P(j)$$

$$= \frac{1}{\tilde{q}_2 q_1 \tau} \sum_{j=0}^{\tilde{q}_2-1} \sum_{k=0}^{q_1\tau-1} \mathrm{e}^{\mathrm{i}(jq_1\tau+k)\omega} P(jq_1\tau + k)$$

$$= \frac{1}{\tilde{q}_2 q_1 \tau} \sum_{j=0}^{\tilde{q}_2-1} \mathrm{e}^{\mathrm{i}jq_1\tau\omega} \sum_{k=0}^{q_1\tau-1} \mathrm{e}^{\mathrm{i}k\omega} P(k).$$

Case 2. a) i. A.: Finally, assume that $q_2 \nmid p_1$, which is equivalent to $\tilde{q}_2 \neq 1$. In this case,

$$q_1\tau\omega = 2\pi\frac{q_1 p}{q} = 2\pi\frac{q_1\tilde{p}_1\tilde{p}_2}{\tilde{q}_1\tilde{q}_2} = 2\pi\frac{a\tilde{p}_1\tilde{p}_2}{\tilde{q}_2},$$

and hence $s := q_1 \cdot {}^{\tau\omega}/_{2\pi} \notin \mathbb{Z}$. Furthermore, $t := \tilde{q}_2$ satisfies $st \in \mathbb{Z}$. So we can apply Lemma 1.3.25 and get $\sum_{j=0}^{\tilde{q}_2-1} \mathrm{e}^{\mathrm{i}jq_1\tau\omega} = 0$. Together with (1.3.16), this implies $f_\omega^*(x) = 0$.

Case 2. a) i. B.: In the case that $q_2 \mid p_1$, the harmonic limit can be different from zero, see Example 1.3.26. This case occurs if and only if the smallest integer period of P is a multiple of $2\pi/\omega$. To see this, note that $q_2 \mid p_1$ is equivalent to $\tilde{q}_2 = 1$, and recall that the smallest integer period of P is given by $q_1\tau$. So if we are in case 2. a) i. B., then

$$\begin{aligned} q_1\tau &= p_1 \\ &= p_1\tilde{q}_2 \\ &= \tilde{p}_1 q_2 \\ &= \tilde{p}_1 p_2 \frac{q_2}{p_2} \\ &= \tilde{p}_1 p_2 \frac{2\pi}{\omega}. \end{aligned}$$

On the other hand, if $p_1 = q_1\tau = k \cdot 2\pi/\omega = k \cdot q_2/p_2$ for some $k \in \mathbb{Z}$, then $p_2 = k \cdot q_2/p_1 = k \cdot \tilde{q}_2/\tilde{p}_1$. As $\tilde{q}_2 \perp \tilde{p}_1$, there has to be $z \in \mathbb{Z}$ such that $k = z\tilde{p}_1$, because otherwise $\tilde{p}_2$ would not be an integer. So $p_2 = z\tilde{q}_2$. As $p_2 \perp \tilde{q}_2$, this implies that $\tilde{q}_2 = 1$.

Case 2. a) ii.: If $\tau \notin \mathbb{Q}$, then the map $S : [0, q\tau) \to [0, q\tau)$, $t \mapsto t + 1 \mod q\tau$, is uniquely ergodic with respect to the Lebesgue measure. So with the Lebesgue integrable map $g : [0, q\tau) \to \mathbb{C}$, $t \mapsto \mathrm{e}^{\mathrm{i}t\omega} P(t)$, we have by the Birkhoff Ergodic Theorem that

$$f_\omega^*(x) = \lim_{n\to\infty} \frac{1}{n} \sum_{j=0}^{n-1} \mathrm{e}^{\mathrm{i}j\omega} P(j) \tag{1.3.17}$$

$$= \lim_{n\to\infty} \frac{1}{n} \sum_{j=0}^{n-1} g(j)$$

$$= \lim_{n\to\infty} \frac{1}{n} \sum_{j=0}^{n-1} g\big(S^j(0)\big)$$

$$= \frac{1}{q\tau} \int_0^{q\tau} g(t)\mathrm{d}t$$

$$= \frac{1}{q\tau} \int_0^{q\tau} \mathrm{e}^{\mathrm{i}t\omega} P(t)\mathrm{d}t$$

$$= \frac{1}{q\tau} \sum_{j=0}^{q-1} \int_{j\tau}^{(j+1)\tau} \mathrm{e}^{\mathrm{i}t\omega} P(t)\mathrm{d}t$$

$$= \frac{1}{q\tau} \sum_{j=0}^{q-1} \int_0^{\tau} \mathrm{e}^{\mathrm{i}(t+j\tau)\omega} P(t+j\tau)\mathrm{d}t$$

$$= \frac{1}{q\tau} \sum_{j=0}^{q-1} \mathrm{e}^{\mathrm{i}j\tau\omega} \int_0^{\tau} \mathrm{e}^{\mathrm{i}t\omega} P(t)\mathrm{d}t.$$

As $q > 1$, $p \neq 0$ and p, q coprime, it holds that $s := {}^{\tau\omega}\!/_{2\pi} = {}^p\!/_q \notin \mathbb{Z}$. Furthermore, $t := q$ satisfies $st \in \mathbb{Z}$. Hence we can apply Lemma 1.3.25, and get $\sum_{j=0}^{q-1} \mathrm{e}^{\mathrm{i}j\tau\omega} = 0$.

Case 2. b): If ${}^{\omega\tau}\!/_{2\pi} \in \mathbb{Z}$, then the limit can be different from zero. See Example 1.3.27. $\qquad\square$

This proposition gives two criteria for the limit to be zero. We give three examples, where those criteria fail and the limit is different from zero. Note that the cases, where the criteria fail, are actually the cases we are interested in, because those define the angles that can occur.

Example 1.3.26. Let $T : [0, 2\pi) \to [0, 2\pi)$ be given by $t \mapsto t + 3\pi \bmod 2\pi$. Consider $f : [0, 2\pi) \to \mathbb{C}$, $t \mapsto \cos t$, and the map $P : \mathbb{R} \to \mathbb{C}$ given by $P(t) := \cos(3\pi t)$. Let $\omega := \pi$. It holds that $f(T^j 0) = P(j)$ for all $j \in \mathbb{N}_0$. The map P is τ-periodic with $\tau = {}^2\!/_3 \in \mathbb{Q}$ and ${}^\omega\!/_{2\pi} = {}^1\!/_2$. Hence ${}^{\omega\tau}\!/_{2\pi} = {}^1\!/_3$. So $p_1 = q_2 = 2$, and hence $q_2 \mid p_1$. Thus we are in case 2. a) i. B. of Proposition 1.3.24. It holds that

$$f_\pi^*(0) = \lim_{n\to\infty} \frac{1}{n} \sum_{j=0}^{n-1} \mathrm{e}^{\mathrm{i}j\pi} \cos(3\pi j) = \lim_{n\to\infty} \frac{1}{n} \sum_{j=0}^{n-1} (-1)^j (-1)^j = 1. \qquad \lrcorner$$

Example 1.3.27. Let $T : [0, 2\pi) \to [0, 2\pi)$ be given by $t \mapsto t + 1 \bmod 2\pi$. Consider $f : [0, 2\pi) \to \mathbb{C}$, $t \mapsto \cos t$, the map $P : \mathbb{R} \to \mathbb{C}$ given by $P(t) := \cos t$, and $\omega := 1$. By definition, it holds that $f(T^j 0) = P(j)$ for all $j \in \mathbb{N}_0$. The map P is τ-periodic with $\tau = 2\pi$ and $\omega\tau/2\pi = 1 \in \mathbb{Z}$. So we are in case 2. b) of Proposition 1.3.24. It holds that

$$
\begin{aligned}
f_1^*(0) &= \lim_{n\to\infty} \frac{1}{n} \sum_{j=0}^{n-1} \mathrm{e}^{\mathrm{i}j} \cos j \\
&\overset{(*)}{=} \frac{1}{2\pi} \int_0^{2\pi} \mathrm{e}^{\mathrm{i}t} \cos t \, \mathrm{d}t \\
&\overset{\text{Lemma 1.1.4}}{=} \frac{1}{2\pi} \left[-\frac{\mathrm{i}}{2}\mathrm{e}^{\mathrm{i}t} \cos t + \frac{1}{2}t \right]_{t=0}^{2\pi} \\
&= \frac{1}{2\pi} \left(-\frac{\mathrm{i}}{2}\mathrm{e}^{\mathrm{i}2\pi} \cos 2\pi + \frac{1}{2} \cdot 2\pi + \frac{\mathrm{i}}{2}\mathrm{e}^{\mathrm{i}0} \cos 0 - \frac{1}{2} \cdot 0 \right) \\
&= \frac{1}{2\pi} \left(-\frac{\mathrm{i}}{2} + \pi + \frac{\mathrm{i}}{2} \right) \\
&= \frac{1}{2},
\end{aligned}
$$

where the equation marked with $(*)$ can be shown analogously to the computations for case 2. a) ii. in the proof of Proposition 1.3.24. ⌟

Proposition 1.3.24 only gives a necessary condition for the harmonic limit to be different from zero. This condition is not sufficient. Consider the following example.

Example 1.3.28. Let $T : [0, 2\pi) \to [0, 2\pi)$ be given by $t \mapsto t + 2\pi/3 \mod 2\pi$. Consider $f : [0, 2\pi) \to \mathbb{C}$, $t \mapsto 1/3(1 - \mathrm{e}^{\mathrm{i}2\pi/3})(1 - \mathrm{e}^{\mathrm{i}t})$, and the map $P : \mathbb{R} \to \mathbb{C}$ that is given by $P(t) := 1/3(1 - \mathrm{e}^{\mathrm{i}2\pi/3})(1 - \mathrm{e}^{\mathrm{i}2\pi/3 \cdot t})$. Let $\omega := 2\pi/3$. It holds that $f(T^j 0) = P(j)$ for all $j \in \mathbb{N}_0$. The map P is τ-periodic with prime period $\tau = 3$. Note that $P(0) = 0$, $P(1) = -\mathrm{e}^{\mathrm{i}2\pi/3}$ and $P(2) = 1$. It holds that

$$
\begin{aligned}
f_\omega^*(0) &= \lim_{n\to\infty} \frac{1}{n} \sum_{j=0}^{n-1} \mathrm{e}^{\mathrm{i}j\omega} P(j) \\
&= \lim_{n\to\infty} \frac{1}{n} \sum_{j=0}^{n-1} \mathrm{e}^{\mathrm{i}j2\pi/3} P(j) \\
&= \frac{1}{3} \sum_{j=0}^{2} \mathrm{e}^{\mathrm{i}j2\pi/3} P(j) \\
&= \frac{1}{3} \left[0 - \mathrm{e}^{\mathrm{i}2\pi/3}\mathrm{e}^{\mathrm{i}2\pi/3} + \mathrm{e}^{\mathrm{i}4\pi/3} \right] \\
&= 0,
\end{aligned}
$$

although $2\pi/\omega = 3 = \tau$. ⌟

Proposition 1.3.24 used the notion of a periodic orbit in the sense that $f(T^j x)$ is periodic. If x actually is a periodic point, i.e., if $T^\tau x = x$ for some $\tau \in \mathbb{N}$, then the following corollary holds.

Corollary 1.3.29. *If $x \in X$ is such that $T^\tau x = x$ for some $\tau \in \mathbb{N}$, then $f_\omega^*(x) \neq 0$ can only hold for a map $f : X \to \mathbb{C}$ if τ is a multiple of $2\pi/\omega$.*

Proof. Let $P : \mathbb{R} \to \mathbb{C}$ be τ-periodic such that $P(j) = f(T^j x)$ for each $j \in \mathbb{N}_0$, e.g., $P(t) := p_{\lfloor t \rfloor}$. Then we can apply Proposition 1.3.24. As $\tau \in \mathbb{N}$, we do not have to distinguish between τ and the smallest integer period of P. □

We will now give an interpretation of Corollary 1.3.29 using rotational factor maps. One consequence of the corollary is that, if there is $\tau \in \mathbb{N}$ such that $T^\tau = \mathrm{id}$, i.e., if all points are τ-periodic, then only rotational factor maps to angles ω with $\tau\omega/2\pi \in \mathbb{N}_0$ can exist. In the more common case that we just have a periodic orbit $x, Tx, \dots, T^{\tau-1}x$ with prime period $\tau \in \mathbb{N}$, we can interpret this dynamic as a rotation if we identify these points with the τ-th roots of unity, see Figure 1.7. If we choose $F : X \to \mathbb{C}$ such that $F(T^j x) = e^{i2\pi j/\tau}$, $j = 0, \dots, \tau - 1$, as in Figure 1.7, then this is a rotational factor map to the angle $2\pi/\tau$. If there is $\sigma \in \{2, \dots, \tau - 1\}$ with $\sigma \mid \tau$, then we can also map the periodic orbit to the σ-th roots of unity, see Figure 1.8. So any map $F : X \to \mathbb{C}$ with $F(T^j x) = e^{i2\pi(j \bmod \sigma)/\sigma}$, $j = 0, \dots, \tau - 1$, is a rotational factor map to the angle $2\pi/\sigma$. In this way, we get rotational factor maps to all angles $\omega_k := 2\pi/k$ with $k \in \{2, \dots, \tau\}$ such that $k \mid \tau$, i.e., all angles ω such that τ is a multiple of $2\pi/\omega$.

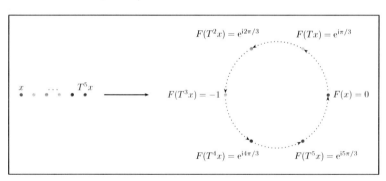

Figure 1.7: Identifying a 6-periodic orbit with the sixth roots of unity

Proposition 1.3.22 and Proposition 1.3.24 only gave necessary criteria for the harmonic limit $f_\omega^*(x)$ to be different from zero. In Example 1.3.28, we have seen, that even in the periodic case the criterion is not sufficient. Using the explicit

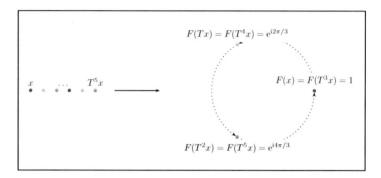

$$F(Tx) = F(T^4x) = \mathrm{e}^{\mathrm{i}2\pi/3}$$

$$x \quad \ldots \quad T^5x$$

$$F(x) = F(T^3x) = 1$$

$$F(T^2x) = F(T^5x) = \mathrm{e}^{\mathrm{i}4\pi/3}$$

Figure 1.8: Identifying a 6-periodic orbit with the third roots of unity

construction of a rotational factor map we have just discussed, the following theorem will fill this gap for periodic points. At a τ-periodic point $x \in X$, for every angle where a nonvanishing harmonic limit can possibly exist by Corollary 1.3.29, i.e., for every ω that is a multiple of $2\pi/\tau$, one can find a continuous function $f : X \to \mathbb{C}$ such that $f_\omega^*(x) \neq 0$.

Theorem 1.3.30. *Let X be a metric space, and $x \in X$ a periodic point of a dynamical system $T : X \to X$ with prime period $\tau \in \mathbb{N}$. Then for every $k \in \mathbb{Z}$, there is $f : X \to \mathbb{R}$ continuous, such that $f_{k2\pi/\tau}^*(x) \neq 0$.*

Proof. Let $k \in \mathbb{Z}$. As τ is the prime period of the periodic orbit $x, Tx, \ldots, T^{\tau-1}x$, this orbit consists of τ different points. So we can define a complex-valued map g on $\mathcal{O} := \{x, Tx, \ldots, T^{\tau-1}x\}$ by $g(T^jx) := \mathrm{e}^{-\mathrm{i}2k\pi j/\tau}$ for $j = 0, \ldots, \tau - 1$. Note that g is continuous and that $\mathcal{O}$ is closed. So by the Tietze extension theorem [Kön00, p. 24], there is a continuous map $f : X \to \mathbb{C}$ with $f|_\mathcal{O} = g$. By Proposition 1.3.17,

$$
\begin{aligned}
f_{2k\pi/\tau}^*(x) &= \frac{1}{\tau} \sum_{j=0}^{\tau-1} \mathrm{e}^{\mathrm{i}2k\pi/\tau} f(T^jx) \\
&= \frac{1}{\tau} \sum_{j=0}^{\tau-1} \mathrm{e}^{\mathrm{i}2k\pi/\tau} g(T^jx) \\
&= \frac{1}{\tau} \sum_{j=0}^{\tau-1} \mathrm{e}^{\mathrm{i}2k\pi/\tau} \mathrm{e}^{-\mathrm{i}2k\pi j/\tau} \\
&= \frac{1}{\tau} \sum_{j=0}^{\tau-1} 1 \\
&= 1.
\end{aligned}
$$

$\square$

Example 1.3.31. Consider $X := \mathbb{C}$ and $T : \mathbb{C} \to \mathbb{C}$ given by $Tx := \mathrm{e}^{\mathrm{i}2\pi/3}(x-1)$. This map has a fixed point in $1/2 - \mathrm{i}\sqrt{3}/6$ and is 3-periodic in all other points. Note that $T^j(0) = P(j)$ for all $j \in \mathbb{N}_0$ and the map $P : \mathbb{R} \to \mathbb{C}$ from Example 1.3.28, which is given by $P(t) := 1/3(1 - \mathrm{e}^{\mathrm{i}2\pi/3})(1 - \mathrm{e}^{\mathrm{i}2\pi/3 \cdot t})$. So from that example, it is known that $\mathrm{id}_{2\pi/3}(0)^* = 0$. Let $f : \mathbb{C} \to \mathbb{C}$ be given by $x \mapsto x^2$. Then

$$
\begin{aligned}
f^*_{2\pi/3}(0) &= \frac{1}{3} \sum_{j=0}^{2} \mathrm{e}^{\mathrm{i}j2\pi/3} f\left(T^j(0)\right) \\
&= \frac{1}{3} \left[0^2 + \mathrm{e}^{\mathrm{i}2\pi/3}(-\mathrm{e}^{\mathrm{i}2\pi/3})^2 + \mathrm{e}^{\mathrm{i}4\pi/3}1^2 \right] \\
&= \frac{1}{3} \left[\mathrm{e}^{\mathrm{i}6\pi/3} + \mathrm{e}^{\mathrm{i}4\pi/3} \right] \\
&= \frac{1}{6} - \mathrm{i}\frac{\sqrt{3}}{6} \\
&\neq 0
\end{aligned}
$$

All results in this section also hold if we consider general sequences $(p_j)_{j\in\mathbb{N}_0} \subset \mathbb{C}$ instead of $f(T^jx)$, i.e., if we look at $\lim_{n\to\infty} 1/n \cdot \sum_{j=0}^{n-1} \mathrm{e}^{\mathrm{i}j\omega} p_j$ instead of $f^*_\omega(x)$. With the exception of Theorem 1.3.30, one only has to replace $f(T^jx)$ by p_j in the proofs.

Remark 1.3.32. The following generalization of Theorem 1.3.30 holds: Let X be a metric space, and $(p_j)_{j\in\mathbb{Z}} \subset X$ a nonconstant sequence with prime period $\tau \in \mathbb{N}$. Then for every $k \in \mathbb{Z}$, there is $f : X \to \mathbb{R}$ continuous, such that $1/\tau \cdot \sum_{j=0}^{\tau-1} \mathrm{e}^{\mathrm{i}jk2\pi/\tau} f(p_j) \neq 0$.

To show this, consider the equivalence relation $\sim$ on $\{0,\dots,\tau-1\}$ given by $j \sim k :\Leftrightarrow p_j = p_k$. Denote the equivalence class of a number $j_0 \in \{0,\dots,\tau-1\}$ with respect to $\sim$ by $[j_0]$. As τ is the prime period of p, and p is not constant, there is $j_0 \in \{0,\dots,\tau-1\}$ such that

$$
\sum_{j\in[j_0]} \mathrm{e}^{\mathrm{i}j2\pi/\tau} \neq 0. \tag{1.3.18}
$$

To see this, note that there can not be a $k \in \{1,\dots,\tau-1\}$ with $k\,|\,\tau$ such that $j_1 + lk \in [j_1]$ for all $l = 0,\dots,\tau/k-1$ and all $j_1 \in \{0,\dots,k-1\}$, because otherwise, p would be k-periodic, which contradicts the assumption that τ is the prime period of p. Hence a result on the sum of roots of unity implies (1.3.18).

Let $U \subset X$ be an open neighbourhood of p_{j_0} such that $p_j \notin U$ for all $j \in \{0,\dots,\tau-1\}\setminus\{j_0\}$. Such a neighbourhood exists, as X is a metric space and hence separable. Define $f : \{p_{j_0}\} \cup X \setminus U \to \mathbb{C}$ by $f(p_{j_0}) := 1$ and $f(x) := 0$ for $x \in X \setminus U$. As $\{p_{j_0}\} \cup X \setminus U$ is closed, we can extend f to a continuous function on X by the Tietze extension theorem [Kön00, p. 24].

For this map $f : X \to \mathbb{R}$, it holds that

$$\frac{1}{\tau} \sum_{j=0}^{\tau-1} e^{ijk2\pi/\tau} f(p_j) = \frac{1}{\tau} \left[\sum_{j \in [j_0]} e^{ijk2\pi/\tau} f(p_j) + \sum_{j \in \{0,\ldots,\tau-1\} \setminus [j_0]} e^{ijk2\pi/\tau} f(p_j) \right]$$

$$= \frac{1}{\tau} \left[\sum_{j \in [j_0]} e^{ijk2\pi/\tau} + 0 \right]$$

$$= \frac{1}{\tau} \sum_{j \in [j_0]} e^{ijk2\pi/\tau},$$

which is different from zero by (1.3.18).

1.4 Koopman and Perron-Frobenius Operators

Rotational factor maps can also be interpreted as eigenfunctions of the Koopman operator. A system has a rotational factor map to the angle $\omega \in [0, 2\pi)$ if and only if $e^{-i\omega}$ is an eigenvalue of the Koopman operator (Proposition 1.4.3). For square-integrable functions f, we will be able to show that the operator $f \mapsto f_\omega^*$ is the orthogonal projection onto the space of eigenfunctions of the Koopman operator associated with eigenvalue $e^{-i\omega}$ (Proposition 1.4.5). We will also show that $f_\omega^*(x) = \int f d\mu$ for some complex measure μ on X (Proposition 1.4.7). These complex measures are connected to eigenfunctions of the Perron-Frobenius operator (Proposition 1.4.12).

Let us first define the Koopman operator.

Definition 1.4.1 (Koopman operator). Let μ be a measure on X. Then the Koopman operator U on $L^\infty(\mu)$ is defined by $Uf := f \circ T$.

Remark 1.4.2. If μ is a finite invariant measure, then U maps $L^p(\mu)$ onto itself for every $1 \le p \le \infty$.

In the following, we will assume that there is a finite invariant measure μ on X. Then we can consider the Koopman operator on $L^p(\mu)$, $1 \le p \le \infty$, by Remark 1.4.2. Furthermore, we know from Theorem 1.3.10, that for $f \in L^p$ and $\omega \in [0, 2\pi)$, the harmonic limit f_ω^* exists almost everywhere. With this, we get the following connection between nonvanishing harmonic limits and eigenfunctions of the Koopman operator.

Proposition 1.4.3. Let $\omega \in [0, 2\pi)$.

    1. Let $f \in L^p$, $1 \le p \le \infty$. If the harmonic limit $f_\omega^* \not\equiv 0$, then f_ω^* is an eigenfunction of U associated with the eigenvalue $e^{-i\omega}$.

2. *Let* $g \in L^p$, $1 \le p \le \infty$, *be an eigenfunction of* U *associated with the eigenvalue* $\mathrm{e}^{-\mathrm{i}\omega}$. *Then it holds that* $g = g_\omega^* \not\equiv 0$.

Proof. Let $f \in L^p$, $1 \le p \le \infty$, such that $f_\omega^* \not\equiv 0$. By Corollary 1.3.2, $Uf_\omega^* = f_\omega^* \circ T = \mathrm{e}^{-\mathrm{i}\omega} f_\omega^*$. This proves part 1.

Let $g \in L^p$, $1 \le p \le \infty$, be an eigenfunction of U associated with the eigenvalue $\mathrm{e}^{-\mathrm{i}\omega}$. Then $\overline{g} \circ T = \overline{g \circ T} = \overline{\mathrm{e}^{-\mathrm{i}\omega} g} = \mathrm{e}^{\mathrm{i}\omega} \overline{g}$, i.e., $\overline{g}$ is a rotational factor map by the angle ω. By Theorem 1.2.13, it holds that $g = g_\omega^*$. As g is an eigenfunction, it is not constant zero. Hence $g_\omega^* \not\equiv 0$. □

In terms of rotational factor maps, this means the following:

Corollary 1.4.4. *Let* $\omega \in [0, 2\pi)$, $1 \le p \le \infty$ *and* $h : M \to \mathbb{C}$ *be* L^p. *Then* h *is an eigenfunction of* U *associated with the eigenvalue* $\mathrm{e}^{-\mathrm{i}\omega}$, *if and only if* T *admits the rotational factor map* $\overline{h}$ *by angle* ω.

Proof. This follows from Proposition 1.4.3 and Theorem 1.2.13. □

In Proposition 1.4.3, we have seen, that f_ω^* is an eigenfunction of the Koopman operator U associated with the eigenvalue $\mathrm{e}^{-\mathrm{i}\omega}$. If f is square-integrable, this connection is even stronger. In this case, f_ω^* is the orthogonal projection onto the space of eigenfunctions of the Koopman operator U associated with the eigenvalue $\mathrm{e}^{-\mathrm{i}\omega}$.

Proposition 1.4.5. *Let* $\omega \in [0, 2\pi)$. *The linear operator* $P_\omega : L^2(\mu) \to L^2(\mu)$, $f \mapsto f_\omega^*$ *is the orthogonal projection onto the space of eigenfunctions of* $U : L^2 \to L^2$ *associated with the eigenvalue* $\mathrm{e}^{-\mathrm{i}\omega}$.

Proof. This follows from the Mean Ergodic Theorem (compare [Hal06, p. 16; RS80, Theorem II.11]). More precisely, we apply this theorem to the Hilbert space L^2 and the linear operator $\mathrm{e}^{\mathrm{i}\omega} U$ on L^2. Then the Mean Ergodic Theorem implies that the operator

$$\lim_{n \to \infty} \frac{1}{n} \sum_{j=0}^{n-1} (\mathrm{e}^{\mathrm{i}\omega} U)^j = \lim_{n \to \infty} \frac{1}{n} \sum_{j=0}^{n-1} \mathrm{e}^{\mathrm{i}j\omega} U^j = P_\omega$$

on $L^2(\mathbb{C})$ is the orthogonal projection onto the space of $\mathrm{e}^{\mathrm{i}\omega} U$-invariant functions, which equals the space of eigenfunctions of U associated with the eigenvalue $\mathrm{e}^{-\mathrm{i}\omega}$. To be able to apply the Mean Ergodic Theorem here, we need to show that $\mathrm{e}^{\mathrm{i}\omega} U$ is an isometry. This is clear by invariance of μ, as, for every $f \in L^2$, it holds that

$$\|\mathrm{e}^{\mathrm{i}\omega} Uf\|^2 = \|Uf\|^2 = \int |Uf|^2 \mathrm{d}\mu$$
$$= \int |f \circ T|^2 \mathrm{d}\mu = \int |f|^2 \mathrm{d}T(\mu) = \int |f|^2 \mathrm{d}\mu = \|f\|^2. \quad \square$$

If X is compact and T is continuous, there exists an invariant measure μ by the theorem of Krylov and Bogolubov [KH06, Theorem 4.1.1]. It can be shown that, for μ-almost every $x \in X$, there is an ergodic invariant measure μ_x for T such that, for any continuous $f : X \to \mathbb{C}$, the time-average can be written as $f^*(x) = \int_X f \mathrm{d}\mu_x$, compare [Mañ87, Theorem 8.4]. A similar result (Proposition 1.4.7) can be shown using complex measures.

Definition 1.4.6 (Complex-valued measure). Let $\mathcal{A}$ be a σ-algebra and $\nu : \mathcal{A} \to \mathbb{C}$. If $\nu(\emptyset) = 0$ and ν is σ-additive then ν is called a *complex-valued measure*. ⌟

Proposition 1.4.7. *Assume that X is compact. For every $x \in X$ and $\omega \in [0, 2\pi)$, there is a complex-valued measure μ_x^ω such that $f_\omega^*(x) = \int_X f \mathrm{d}\mu_x^\omega$ for any continuous $f : X \to \mathbb{C}$.*

Proof. If X is compact, then for every $x \in X$ and $\omega \in [0, 2\pi)$, define $L_x^\omega : C(X) \to \mathbb{C}$ by $L_x^\omega f := f_\omega^*(x)$. Let $C(X)$ be endowed with the supremum norm. Then the functional L_x^ω is bounded and linear. Thus the claim follows by [HS75, Theorem 20.47].

Note that the set $\mathfrak{C}_0(X)$ of continuous functions vanishing at infinity used by Hewitt and Stromberg coincides with the set of continuous functions in our case, as X is compact. ☐

The measure μ_x^ω from Proposition 1.4.7 is an eigenmeasure in the following sense.

Definition 1.4.8 (Eigenmeasure). Let ν be a complex-valued Borel measure on X satisfying $\nu(T^{-1}E) = \lambda \nu(E)$ for some $\lambda \in \mathbb{C}$ and every Borel set $E \subset X$. Then ν is called an *eigenmeasure associated with eigenvalue λ*. ⌟

Remark 1.4.9. Consider the space $\mathcal{M}$ of complex-valued Borel measures on X and let $Q : \mathcal{M} \to \mathcal{M}$, $\nu \to Q\nu := \nu \circ T^{-1}$. Then every nonzero eigenmeasure ν associated with eigenvalue λ is an eigenvector of Q associated with eigenvalue λ and vice versa. ⌟

Proposition 1.4.10. *Assume that X is compact. If T is continuous, then μ_x^ω is an eigenmeasure, $x \in X$, $\omega \in [0, 2\pi)$.*

Proof. For every $f \in C(X)$, also $f \circ T \in C(X)$, and thus it holds that

$$\int_X f \circ T \mathrm{d}\mu_x^\omega = P_T^\omega(f \circ T)(x) = \mathrm{e}^{-\mathrm{i}\omega} P_T^\omega f(x) = \mathrm{e}^{-\mathrm{i}\omega} \int_X f \mathrm{d}\mu_x^\omega.$$

This implies $\mu_x^\omega \circ T^{-1} = \mathrm{e}^{-\mathrm{i}\omega} \mu_x^\omega$. ☐

There is a connection between eigenmeasures and eigenfunctions of the Perron-Frobenius operator, see Proposition 1.4.12. The following definition of this operator is taken from [LM95, Definition 3.2.3]. For a proof of existence and uniqueness of the Perron-Frobenius operator, see [LM95, p. 42].

Definition 1.4.11 (Perron-Frobenius operator). Let $(X, \mathcal{A}, \mu)$ be a measure space. If $S : X \to X$ is a nonsingular transformation, i.e., if $\mu\big(S^{-1}(A)\big) = 0$ for all null sets $A \in \mathcal{A}$, then the unique operator $P : L^1 \to L^1$ defined by

$$\int_A P f \mathrm{d}\mu = \int_{S^{-1}(A)} f \mathrm{d}\mu \quad \text{for } A \in \mathcal{A} \tag{1.4.1}$$

is called the *Perron-Frobenius operator* corresponding to S.

This definition can be applied to the measure space $(X, \mathcal{B}, \mu)$, where $\mathcal{B}$ is the Borel sigma algebra on X and μ is a T-invariant measure, because the transformation T is nonsingular due to invariance. The following proposition specifies the connection between eigenfunctions of the Perron-Frobenius operator and eigenmeasures. In fact, the eigenfunctions are density functions of eigenmeasures.

Proposition 1.4.12. *Let Φ be a complex measure on X given by*

$$\Phi(A) := \int_A g \mathrm{d}\mu \tag{1.4.2}$$

for some $g \in L^1(\mu)$ and every Borel set A. Then Φ is an eigenmeasure associated with the eigenvalue λ if and only if g is an eigenfunction of the Perron-Frobenius operator associated with the eigenvalue λ.

Proof. By definition of Φ and the Perron-Frobenius operator, it holds that

$$\Phi(T^{-1}A) = \int_{T^{-1}A} g \mathrm{d}\mu = \int_A P g \mathrm{d}\mu.$$

So if g is an eigenfunction of the Perron-Frobenius operator associated with the eigenvalue λ, then $\Phi(T^{-1}A) = \int_A \lambda g \mathrm{d}\mu = \lambda \Phi(A)$, i.e., Φ is an eigenmeasure.

Conversely, if Φ is an eigenmeasure associated with the eigenvalue λ, then it holds that $\Phi(T^{-1}A) = \lambda \Phi(A)$, which implies $\int_A P g \mathrm{d}\mu = \int_A \lambda g \mathrm{d}\mu$. As this holds for every Borel set A, it follows that $P g = \lambda g$. □

1.5 *h*-partitioned systems

In Corollary 1.3.2, we have seen that $f_\omega^*(T^j x) = \mathrm{e}^{-\mathrm{i}j\omega} f_\omega^*(x)$. This means, along the trajectory starting in $x \in X$ the harmonic limit only takes values in $\{\mathrm{e}^{-\mathrm{i}j\omega} f_\omega^*(x) \mid j \in \mathbb{N}_0\}$. Consider the case that this set is finite. This occurs if and only if ω is such that $\omega/2\pi \in \mathbb{Q}$. In particular, consider $\omega_h := 2\pi/h$ for some $h \in \mathbb{N}$ and assume that $f_{\omega_h}^*$ takes exactly h different values on X (or on a subset of X of full measure). Then the state space can be partitioned into h subsets such that the system "jumps" through these sets in a given order, see Figure 1.9 and Proposition 1.5.7. We will call such systems h-partitioned. On the other hand, we can reconstruct the partition of an ergodic h-partitioned system by looking at the level sets of the harmonic limit, see Theorem 1.5.4.

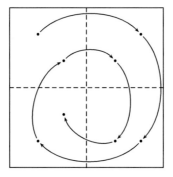

Figure 1.9: A 4-partitioned system

Definition 1.5.1 (h-partitioned systems). A measure-theoretic dynamical system $T : X \to X$ on the metric space X is called h-*partitioned*, $h \in \mathbb{N}$, if there is a partition of X into measurable sets H_j, $j = 0, \ldots, h-1$, which are pairwise disjoint up to null sets and satisfy $T^{-1}H_0 = H_{h-1}$ and $T^{-1}H_j = H_{j-1}$, $j = 1, \ldots, h-1$, up to null sets. ⌟

Remark 1.5.2. Trivially, every system is 1-partitioned. ⌟

Intuitively, one would assume that every h-partitioned system has a rotational factor map to the angle $\omega_h := 2\pi/h$. This indeed is true, and the harmonic average to that angle will be used to reconstruct the partition in Theorem 1.5.4.

Proposition 1.5.3. *Any h-partitioned system admits a rotational factor map with the angle $\omega_h := 2\pi/h$.*

Proof. Let $F : X \to \mathbb{C}$ be defined by $F(x) := \sum_{j=0}^{h-1} \mathbf{1}_{H_j}(x)\mathrm{e}^{\mathrm{i}j\omega_h}$. As the partition elements H_j are measurable, the map F is also measurable and it clearly is not constant 0. This map satisfies

$$F \circ T(x) = \sum_{j=0}^{h-1} \mathbf{1}_{H_j}(Tx)\mathrm{e}^{\mathrm{i}j\omega_h} = \sum_{j=0}^{h-1} \mathbf{1}_{T^{-1}H_j}(x)\mathrm{e}^{\mathrm{i}j\omega_h}$$

$$= \mathbf{1}_{H_{h-1}}(x)\mathrm{e}^{\mathrm{i}0\omega_h} + \sum_{j=1}^{h-1} \mathbf{1}_{H_{j-1}}(x)\mathrm{e}^{\mathrm{i}j\omega_h}$$

$$= \sum_{j=0}^{h-1} \mathbf{1}_{H_{j-1}}(x)\mathrm{e}^{\mathrm{i}j\omega_h} = \sum_{j=0}^{h-1} \mathbf{1}_{H_j}(x)\mathrm{e}^{\mathrm{i}j+1\omega_h}$$

$$= \mathrm{e}^{\mathrm{i}\omega_h} \cdot \sum_{j=0}^{h-1} \mathbf{1}_{H_j}(x)\mathrm{e}^{\mathrm{i}j\omega_h} = \mathrm{e}^{\mathrm{i}\omega_h} \cdot F(x)$$

almost everywhere, and thus the proof is completed. □

Now we can formulate and show the main result of this section, the reconstruction of the partition via level sets of $f_{\omega_h}^*$. For ergodic h-partitioned systems, these level sets coincide with the partition elements H_j.

Theorem 1.5.4. *Consider an ergodic h-partitioned system, and let $f : X \to \mathbb{C}$ be integrable with respect to the ergodic invariant measure μ such that $f_{\omega_h}^* \not\equiv 0$, where $\omega_h := {}^{2\pi}/{}_h$. Such a map always exists by Theorem 1.3.30. Then the level sets of the harmonic limit $f_{\omega_h}^*$ coincide with the partition elements H_j up to sets of zero measure.*

Proof. First, we show that $f_{\omega_h}^*$ is constant on partition elements. Note that the iterate T^h restricted to any partition element is ergodic. Further note that the harmonic average can be written as the sum of h ergodic limits (almost everywhere):

$$f_{\omega_h}^*(x) = \lim_{n \to \infty} \frac{1}{n} \sum_{j=0}^{n-1} \mathrm{e}^{\mathrm{i} j \omega_h} f(T^j x)$$

$$= \lim_{n \to \infty} \frac{1}{hn} \sum_{j=0}^{hn-1} \mathrm{e}^{\mathrm{i} j \omega_h} f(T^j x)$$

$$= \lim_{n \to \infty} \frac{1}{hn} \sum_{j=0}^{hn-1} \mathrm{e}^{\mathrm{i} (j \bmod h) \omega_h} f(T^j x)$$

$$= \frac{1}{h} \sum_{k=0}^{h-1} \lim_{n \to \infty} \frac{1}{n} \sum_{j=0}^{n-1} \mathrm{e}^{\mathrm{i} k \omega_h} f(T^{jh} T^k x)$$

$$= \frac{1}{h} \sum_{k=0}^{h-1} \frac{1}{\mu(H_{\eta(T^k x)})} \int_{H_{\eta(T^k x)}} \mathrm{e}^{\mathrm{i} k \omega_h} f(\xi) \mu(\mathrm{d}\xi),$$

where $\eta(x) \in \{0, \ldots, h-1\}$ denotes the index of the partition element containing x. Note that $\mu(H_j) = {}^1/{}_h \cdot \mu(X) > 0$ for all j. It holds that $\eta(T^k x) = \eta(x) + k \mod h$. Thus

$$f_{\omega_h}^*(x) = \frac{1}{h} \sum_{k=0}^{h-1} \frac{1}{\mu(H_k)} \int_{H_k} \mathrm{e}^{\mathrm{i}\left(k - \eta(x)\right) \omega_h} f(\xi) \mu(\mathrm{d}\xi)$$

$$= \mathrm{e}^{-\mathrm{i}\eta(x)\omega_h} \frac{1}{h} \sum_{k=0}^{h-1} \frac{1}{\mu(H_k)} \int_{H_k} \mathrm{e}^{\mathrm{i} k \omega_h} f(\xi) \mu(\mathrm{d}\xi).$$

As η is constant by definition on the partition elements, this implies that the harmonic limit is constant on partition elements.

Assume that $f^*_{\omega_h} \not\equiv 0$. Thus by the calculation above

$$\frac{1}{h} \sum_{k=0}^{h-1} \frac{1}{\mu(H_k)} \int_{H_k} e^{ik\omega_h} f(\xi)\mu(d\xi) \neq 0.$$

Hence for almost all $x, y \in X$ with $f^*_{\omega_h}(x) = f^*_{\omega_h}(y)$, it follows that $e^{-i\eta(x)\omega_h} = e^{-i\eta(y)\omega_h}$ and thus $\eta(x) = \eta(y)$, as $\eta \cdot \omega_h \in [0, 2\pi)$. This means, x and y lie in the same partition element. $\qquad \square$

Remark 1.5.5. If T is not ergodic, one can combine this result with an ergodic decomposition, i.e., compute the level set for a sequence of functions f, as was done in [MB04]. $\qquad \lrcorner$

Now we have a look at an example, where this reconstruction method is applied.

Example 1.5.6. Consider a map $S : [1, 2] \rightarrow [1, 2]$, and let $T : [-2, -1] \times [1, 2] \rightarrow [-2, -1] \times [1, 2]$ be given by $Tx := -\operatorname{sgn} x \cdot S|x|$, i.e., $T^j x = (-1)^j \operatorname{sgn} x \cdot S^j|x|$. Then this system clearly rotates in the previously defined sense (Definition 1.2.1) by angle π, because it alternates between $[-2, -1]$ and $[1, 2]$, see Figure 1.10, and hence $F := \operatorname{sgn}$ is a rotational factor map. It is possible to reconstruct the partition $[-2, -1], [1, 2]$ by looking at the level sets of a harmonic limit to the angle π.

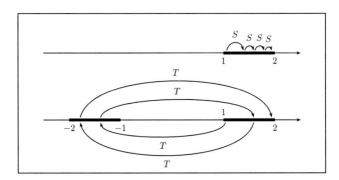

Figure 1.10: Construction of the system from Example 1.5.6

First consider sgn^*_π. Then $\operatorname{sgn}(T^j x) = \operatorname{sgn}\big((-1)^j \operatorname{sgn} x \cdot S^j|x|\big) = (-1)^j \operatorname{sgn} x \cdot \operatorname{sgn}(S^j|x|) = (-1)^j \operatorname{sgn} x$. So $\operatorname{sgn}^*_\pi(x) = \lim_{n\to\infty} 1/n \cdot \sum_{j=0}^{n-1} e^{i\pi j}(-1)^j \operatorname{sgn} x = \operatorname{sgn} x$ for every $x \in [-2, -1] \times [1, 2]$, which clearly has the level sets $[-2, -1], [1, 2]$.

If $S^2 : [1, 2] \rightarrow [1, 2]$ is continuous and uniquely ergodic, then it holds for all

$x_0 \in [-2,-1] \times [1,2]$ and all continuous $f : [-2,-1] \times [1,2] \to \mathbb{C}$, that

$$
\begin{aligned}
f_\pi^*(x_0) &= \lim_{n\to\infty} \frac{1}{n} \sum_{j=0}^{n-1} \mathrm{e}^{\mathrm{i}\pi j} f(T^j x_0) \\
&= \lim_{n\to\infty} \frac{1}{n} \sum_{j=0}^{n-1} (-1)^j f\big((-1)^j \operatorname{sgn} x_0 S^j |x_0|\big) \\
&= \lim_{n\to\infty} \frac{1}{2n} \left[\sum_{j=0}^{n-1} f\big(\operatorname{sgn} x_0 \cdot S^{2j} |x_0|\big) - \sum_{j=0}^{n-1} f\big(-\operatorname{sgn} x_0 \cdot S^{2j+1} |x_0|\big) \right] \\
&= \frac{1}{2} \left[\lim_{n\to\infty} \frac{1}{n} \sum_{j=0}^{n-1} f\big(\operatorname{sgn} x_0 \cdot (S^2)^j |x_0|\big) \right. \\
&\qquad\qquad \left. - \lim_{n\to\infty} \frac{1}{n} \sum_{j=0}^{n-1} f\big(-\operatorname{sgn} x_0 \cdot (S^2)^j (S|x_0|)\big) \right] \\
&= \frac{1}{2} \left[\frac{1}{2} \int_1^2 f(\operatorname{sgn} x_0 \cdot x)\mathrm{d}x - \frac{1}{2} \int_1^2 f(-\operatorname{sgn} x_0 \cdot x)\mathrm{d}x \right] \\
&= \frac{1}{4} \int_1^2 \big(f(\operatorname{sgn} x_0 \cdot x) - f(-\operatorname{sgn} x_0 \cdot x)\big)\mathrm{d}x \\
&= \operatorname{sgn} x_0 \cdot \frac{1}{4} \int_1^2 \big(f(x) - f(-x)\big)\mathrm{d}x
\end{aligned}
$$

by the Birkhoff Ergodic Theorem, compare [Mañ87, Theorem I.9.2]. So, e.g., for $f = \mathrm{id}$, it follows that $\mathrm{id}_\pi^*(x_0) = \operatorname{sgn} x_0 \cdot 1/4 \cdot \int_1^2 2x\mathrm{d}x = \operatorname{sgn} x_0 \cdot 3/4$. So id_π^* has exactly the partition $[-2,-1], [1,2]$ as level sets.   ⌐

The following converse of Theorem 1.5.4 holds. If the harmonic limit to ω_h only takes h values, then the system must be h-partitioned.

Proposition 1.5.7. *Consider a measure-theoretic dynamical system $T : X \to X$ and let $\Xi \subset X$ be of full measure. Assume that $f_{\omega_h}^*$ takes exactly h different values on Ξ for some $h \in \mathbb{N}$. Then the system is h-partitioned.*

Proof. By Remark 1.5.2, we only have to give a proof for $h > 1$. Let $x \in \Xi$ be a point, such that $\phi := f_{\omega_h}^*(x) \neq 0$. Such a point exists because otherwise $f_{\omega_h}^* \equiv 0$ would take only one value on Ξ.

By Corollary 1.3.2, it holds that $f_{\omega_h}^*(T^j x) = \mathrm{e}^{\mathrm{i}j\omega_h}\phi$ for all $j = 0,\ldots,h-1$. These h numbers are pairwise different, so they are exactly the h values that $f_{\omega_h}^*$ takes on Ξ. Let $H_j := \{y \in \Xi \mid f_{\omega_h}^*(y) = \mathrm{e}^{\mathrm{i}j\omega_h}\phi\}$, $j = 0,\ldots,h-1$.

Using Corollary 1.3.2 again, one can see that $T^{-1}H_0 = H_{h-1}$ and $T^{-1}H_j = H_{j-1}$, $j = 1,\ldots,h-1$. So the system is h-partitioned.   □

Now we have a look at an example, where the state space can not be partitioned into finitely many elements by level sets of f^*_ω.

Example 1.5.8. Consider the rotation from Example 1.2.14 again, for $\alpha \in (0, 2\pi)$. We have seen in Example 1.2.14 that $\cos^*_\alpha \theta = \frac{1}{2}e^{-i\theta}$ if $\alpha \neq \pi$. The level sets of that map are $\{\{\theta\} \mid \theta \in [0, 2\pi)\}$, i. e., only points. ⌟

Remark 1.5.9. The concept of h-partitioned systems is related to phase partitions (see [MB04, Definition 13]) and periodic invariant partitions (see [Mez05, Appendix]). In [MB04, Sections 3.4 and 3.5], the authors compute level sets of harmonic limits in two examples. There, they use angles ω with $\omega/2\pi \notin \mathbb{Q}$, i. e., the harmonic limit takes infinitely many values. In order to approximate the harmonic limits, they compute harmonic averages f^N_ω with large N ($N = 10000$ and $N = 50000$ in the first example, and $N = 1000$ in the second example). In the second example, Mezić and Banaszuk also apply a method they discuss earlier in that article for the computation of ergodic partitions, i. e., for $\omega = 0$. Namely, they compute joint level sets of the harmonic limit for different functions f. ⌟

Chapter 2

Harmonic analysis for continuous-time dynamical systems

The concepts of rotational factor maps and harmonic limits, which we have discussed for discrete-time dynamical systems in Chapter 1, can also be defined for continuous-time dynamical systems, e. g., for solutions of ordinary differential equations. So in this chapter, we consider dynamical systems given by a semi-flow $\Phi : \mathbb{R}^+ \times X \to X$ on a metric space X. For most results, we actually do not need the flow property of Φ, so we can, more generally, consider any function $\Phi : \mathbb{R}^+ \times X \to X$, $(t, x) \mapsto \Phi_t x$, that is Borel measurable in t and satisfies $\Phi_0 x = x$. We want to analyze the rotational behaviour of such dynamical systems. More precisely, we investigate, similarly to Chapter 1, if the system is semi-conjugate to a rotation in the complex plane, i. e., if there are $\omega \in \mathbb{R}$ and a map $F : X \to \mathbb{C}$ such that $F \circ \Phi_t = e^{i\omega t} \cdot F$ for all $t \geq 0$. Note that we will not speak of rotations *by the angle* ω as in Chapter 1, but of rotations *with frequency* $\omega/2\pi$. See Subsection 2.2.1 for a discussion of these so-called *rotational factor maps* F.

As in the discrete-time case, one can generalize time-averages

$$\lim_{T \to \infty} \frac{1}{T} \int_0^T f(\Phi_t x) \mathrm{d}t$$

for a map $f : X \to \mathbb{C}$ to harmonic limits

$$f_\omega^*(x) := \lim_{T \to \infty} \frac{1}{T} \int_0^T e^{i\omega t} f(\Phi_t x) \mathrm{d}t$$

for $\omega \in \mathbb{R}$, see Subsection 2.2.2. Harmonic limits turn out to be useful in the analysis of the rotational behaviour of dynamical systems. In fact, in the case of a semi-flow, for $\omega \in \mathbb{R}$, there is a map $f : X \to \mathbb{C}$ such that $f_\omega^* \not\equiv 0$ if and only if the system admits a rotational factor map with frequency $\omega/2\pi$. See Subsection 2.2.3 for a discussion of this important relation.

These concepts were introduced in [MB04] for systems in discrete time, see Chapter 1. In this chapter, we adapt the definitions and results to continuous-time systems. So we first show three simple properties of the harmonic limit in Subsection 2.3.1 and then prove that, for given $x \in X$ and $f : X \to \mathbb{C}$ under a weak condition, there can only be countably many $\omega \in \mathbb{R}$ such that $f_\omega^*(x) \neq 0$, i.e., only countably many frequencies can occur.

In Subsections 2.3.2 and 2.3.3, we show some existence results under the presence of an invariant measure. If there is an invariant measure μ, and if f is μ-integrable, then the harmonic limit f_ω^* exists μ-almost everywhere, see Theorem 2.3.14. Note that the null set of points, where the harmonic limit does not exist, can depend on ω. For ergodic systems, there is a stronger result, known as the Wiener-Wintner Ergodic Theorem, which shows that the set of points, where the harmonic limit does not exist, does not depend on ω, see Theorem 2.3.18. Finally, for uniquely ergodic systems, the harmonic limit exists everywhere, see Proposition 2.3.15.

In the following two sections, we show what impact certain dynamical properties have on the properties of harmonic limits. Particularly, we look at asymptotic behaviour in Subsection 2.3.4, and at periodic behaviour in Subsection 2.3.5. If, e.g., two trajectories $\Phi_t x$ and $\Phi_t y$ asymptotically approach each other, then the harmonic limits $f_\omega^*(x)$ and $f_\omega^*(y)$ coincide for every $\omega \in \mathbb{R}$ and all continuous functions f : $X \to \mathbb{C}$. One important result regarding periodicity is that, along almost periodic trajectories, harmonic limits always exist, see Proposition 2.3.29. Furthermore, one can characterize the set of frequencies that can possibly occur along quasi-periodic and periodic trajectories, see Proposition 2.3.32 and Proposition 2.3.33.

There is a close connection between rotational factor maps and eigenfunctions of the Koopman operator $f \mapsto f \circ T$. This connection and its consequence for the map $f \mapsto f_\omega^*$, which maps a function onto its harmonic limit, will be treated in Section 1.4.

In Section 2.5, we examine linear ordinary differential equations. In Subsection 2.5.2, we look at autonomous equations, where a connection between the occuring frequencies and the imaginary parts of the eigenvalues of the system matrix can be shown. In Subsection 2.5.3, we give similar results in the case of periodic equations. Here, the imaginary parts of the Floquet exponents have to be considered.

In Section 2.6, we introduce and analyze the harmonic growth spectrum which consists of all limit points of the harmonic average. This is a special case of the uniform growth spectrum that was introduced in [Ste09, Definition 2.1.9] as a generalization of the uniform exponential spectrum (see [Grü00]). As shown by Stender, there is a connection to the Morse spectrum, which we introduce in Subsection 2.6.2 and analyze in Subsection 2.6.3. It turns out, that the Morse spectrum is the closed disc around the origin with a radius given by the maximum of $|f_\omega^*(x)|$ for all x in some invariant set, see Theorem 2.6.25. If we consider the Morse spectrum on a limit set $\omega(x)$, then all limit points $\lim_{k \to \infty} f_\omega^{t_k}(x)$ of the harmonic average are contained in the spectrum, see Theorem 2.6.29. In particular, the harmonic limit itself is contained in the Morse spectrum, if it exists. With this result, we can show a connection between the radius of the Morse spectrum and the existence of rotational factor maps, see Theorem 2.6.30 and Corollary 2.6.31.

2.1 Preliminaries

Let us first collect some definitions and results, which we will use later in this chapter.

Recall Definition 1.1.1 and Definition 1.1.3 regarding almost periodicity and quasi-periodicity, and Lemma 1.1.4, which we will use again.

Exposed points In the discussion of the Morse spectrum in Subsection 2.6.3, we use the concept of exposed points of a convex set.

Definition 2.1.1 (Exposed points). Let $K \subset \mathbb{R}^m$ be convex. A point $z \in K$ is called an *exposed point of* K, if there are $c \in \mathbb{R}^m$ and $\beta \in \mathbb{R}$ such that $\{x \in K \mid \langle x, c \rangle = \beta\} = \{z\}$ and $K \subset \{x \in \mathbb{R}^m \mid \langle x, c \rangle \leq \beta\}$. ⌟

Note that we will identify $\mathbb{C}$ with $\mathbb{R}^2$ in Subsection 2.6.3, and hence speak of exposed points of a subset of $\mathbb{C}$.

(ε, T)-chains In Section 2.6, we will need the concept of (ε, T)-chains and chain transitivity, which is defined as follows, compare [CK00, Definitions B.2.17 and B.2.18]. Let $\Phi_t : X \to X$ be a semiflow on the metric space X.

Definition 2.1.2 ((ε, T)-chains). For $x, y \in X$ and $\varepsilon, T > 0$, an (ε, T)-chain from x to y is given by a natural number $n \in \mathbb{N}$, together with points $x_0 = x, x_1, \ldots, x_n = y \in X$, and times $T_0, \ldots, T_{n-1} \geq T$, such that $d(\Phi_{T_j} x_j, x_{j+1}) < \varepsilon$ for $j = 0, \ldots, n-1$. ⌟

Definition 2.1.3 (Chain transitivity). A set $A \subset X$ is chain transitive, if for all $x, y \in A$ and all $\varepsilon, T > 0$, there is an (ε, T)-chain from x to y. ⌟

We will also use products of chains. For this, let $\tilde{\Phi}_t : \tilde{X} \to \tilde{X}$ be a second semi-flow on the metric space $\tilde{X}$. Consider the following definition and lemma.

Definition 2.1.4 (Product of two chains). Let Ψ_t denote the product flow on $X \times \tilde{X}$, i.e., $\Psi_t(x, \tilde{x}) := (\Phi_t x, \tilde{\Phi}_t \tilde{x})$. Let $\varepsilon, T > 0$. Consider an (ε, T)-chain ζ for Φ given by the points $x_0, \ldots, x_n \in X$, and times $T_0, \ldots, T_{n-1} > T$, and an (ε, T)-chain $\tilde{\zeta}$ for $\tilde{\Phi}$ given by the points $\tilde{x}_0, \ldots, \tilde{x}_n \in \tilde{X}$, and times $T_0, \ldots, T_{n-1} > T$. Note that n and the times T_j are the same for both chains. Then the product of ζ and $\tilde{\zeta}$ is the chain for Ψ_t given by the points $(x_0, \tilde{x}_0), \ldots, (x_n, \tilde{x}_n) \in X \times \tilde{X}$, and times $T_0, \ldots, T_{n-1} > T$. We denote the product by $\zeta \times \tilde{\zeta}$. ⌟

Lemma 2.1.5. *If we endow $X \times \tilde{X}$ with the ∞-product metric, i.e., if we set $d\big((x, \tilde{x}), (y, \tilde{y})\big) := \max\{d(x, y), d(\tilde{x}, \tilde{y})\}$, then the product of two (ε, T)-chains again is an (ε, T)-chain.*

Proof. Let $\varepsilon, T > 0$. Let ζ be an (ε, T)-chain for Φ given by the points $x_0, \ldots, x_n \in X$, and times $T_0, \ldots, T_{n-1} > T$, and let $\tilde{\zeta}$ be an (ε, T)-chain for $\tilde{\Phi}$ given by the points $\tilde{x}_0, \ldots, \tilde{x}_n \in \tilde{X}$, and the same times $T_0, \ldots, T_{n-1} > T$. Then for every $j = 0, \ldots, n-1$, it holds that

$$
\begin{aligned}
d\big(\Psi_{T_j}(x_j, \tilde{x}_j), (x_{j+1}, \tilde{x}_{j+1})\big) &= d\big((\Phi_{T_j} x_j, \tilde{\Phi}_{T_j} \tilde{x}_j), (x_{j+1}, \tilde{x}_{j+1})\big) \\
&= \max\{d(\Phi_{T_j} x_j, x_{j+1}), d(\tilde{\Phi}_{T_j} \tilde{x}_j, \tilde{x}_{j+1})\} \\
&< \varepsilon,
\end{aligned}
$$

because ζ and $\tilde{\zeta}$ are (ε, T)-chains. $\qquad\square$

Antiderivatives and integrals   Now we calculate antiderivatives of $e^{i\omega t} \sin \alpha t$ and $e^{i\omega t} \cos \alpha t$, and then use them to compute certain harmonic limits.

Lemma 2.1.6. *Let* $\alpha \in \mathbb{R}$ *and* $\omega \neq \pm\alpha$. *An antiderivative of* $e^{i\omega t} \sin \alpha t$ *with respect to t is given by*

$$
e^{i\omega t} \frac{\alpha \cos \alpha t - i\omega \sin \alpha t}{\omega^2 - \alpha^2},
$$

and an antiderivative of $e^{i\omega t} \cos \alpha t$ *with respect to t is given by*

$$
e^{i\omega t} \frac{\alpha \sin \alpha t + i\omega \cos \alpha t}{\alpha^2 - \omega^2}.
$$

Proof. It holds that

$$
\begin{aligned}
\frac{d}{dt} &\left[e^{i\omega t} \frac{\alpha \cos \alpha t - i\omega \sin \alpha t}{\omega^2 - \alpha^2} \right] \\
&= i\omega e^{i\omega t} \frac{\alpha \cos \alpha t - i\omega \sin \alpha t}{\omega^2 - \alpha^2} + e^{i\omega t} \frac{-\alpha^2 \sin \alpha t - i\alpha\omega \cos \alpha t}{\omega^2 - \alpha^2} \\
&= e^{i\omega t} \frac{i\alpha\omega \cos \alpha t + \omega^2 \sin \alpha t - \alpha^2 \sin \alpha t - i\alpha\omega \cos \alpha t}{\omega^2 - \alpha^2} = e^{i\omega t} \sin \alpha t,
\end{aligned}
$$

and that

$$
\begin{aligned}
\frac{d}{dt} &\left[e^{i\omega t} \frac{\alpha \sin \alpha t + i\omega \cos \alpha t}{\alpha^2 - \omega^2} \right] = i\omega e^{i\omega t} \frac{\alpha \sin \alpha t + i\omega \cos \alpha t}{\alpha^2 - \omega^2} + e^{i\omega t} \frac{\alpha^2 \cos \alpha t - i\alpha\omega \sin \alpha t}{\alpha^2 - \omega^2} \\
&= e^{i\omega t} \frac{i\alpha\omega \sin \alpha t - \omega^2 \cos \alpha t + \alpha^2 \cos \alpha t - i\alpha\omega \sin \alpha t}{\alpha^2 - \omega^2} \\
&= e^{i\omega t} \cos \alpha t.
\end{aligned}
$$

$\qquad\square$

Lemma 2.1.7. *For every $\alpha \in \mathbb{R}$ and all $\omega \in \mathbb{R} \setminus \{\pm\alpha\}$, it holds that*

$$\lim_{T \to \infty} \frac{1}{T} \int_0^T e^{i\omega t} \sin \alpha t \, dt = 0,$$

and

$$\lim_{T \to \infty} \frac{1}{T} \int_0^T e^{i\omega t} \cos \alpha t \, dt = 0.$$

Proof. By Lemma 2.1.6, it holds that

$$\frac{1}{T} \int_0^T e^{i\omega t} \sin \alpha t \, dt = \frac{1}{T} \left[e^{i\omega t} \frac{\alpha \cos \alpha t - i\omega \sin \alpha t}{\omega^2 - \alpha^2} \right]_{t=0}^T$$
$$= \frac{1}{T} \left[e^{i\omega T} \frac{\alpha \cos \alpha T - i\omega \sin \alpha T}{\omega^2 - \alpha^2} - \frac{\alpha}{\omega^2 - \alpha^2} \right]$$
$$\to 0$$

for $T \to \infty$, and that

$$\frac{1}{T} \int_0^T e^{i\omega t} \cos \alpha t \, dt = \frac{1}{T} \left[e^{i\omega t} \frac{\alpha \sin \alpha t + i\omega \cos \alpha t}{\alpha^2 - \omega^2} \right]_{t=0}^T$$
$$= \frac{1}{T} \left[e^{i\omega T} \frac{\alpha \sin \alpha T + i\omega \cos \alpha T}{\alpha^2 - \omega^2} - \frac{i\omega}{\alpha^2 - \omega^2} \right]$$
$$\to 0$$

for $T \to \infty$. $\qquad\square$

To simplify the proof of the following lemma, we anticipate Corollary 2.3.5 and Proposition 2.3.28.

Lemma 2.1.8. *For every $\omega \in \mathbb{R} \setminus \{0\}$ and all $x \in \mathbb{R}$, it is true that*

$$\lim_{T \to \infty} \frac{1}{T} \int_0^T e^{\pm i\omega t} \cos(\omega t + x) dt = \frac{\omega}{2\pi} \int_0^{2\pi/\omega} e^{\pm i\omega t} \cos(\omega t + x) dt = \frac{1}{2} e^{\mp ix}$$

and

$$\lim_{T \to \infty} \frac{1}{T} \int_0^T e^{\pm i\omega t} \sin(\omega t + x) dt = \frac{\omega}{2\pi} \int_0^{2\pi/\omega} e^{\pm i\omega t} \sin(\omega t + x) dt = \pm \frac{i}{2} e^{\mp ix}.$$

Proof. In both cases, the first equation follows from Proposition 2.3.28. Furthermore, it holds that

$$\lim_{T \to \infty} \frac{1}{T} \int_0^T e^{\pm i\omega t} f(\omega t + x) dt = e^{\mp i\omega x} \lim_{T \to \infty} \frac{1}{T} \int_0^T e^{\pm i\omega t} f(\omega t) dt$$

for $f \in \{\cos, \sin\}$ by Corollary 2.3.5 applied to the flow Φ on $\mathbb{R}$ given by $\Phi_t x := x + t$. By periodicity of the integrand, it follows from Proposition 2.3.28 that

$$\mathrm{e}^{\mp \mathrm{i}\omega x} \lim_{T \to \infty} \frac{1}{T} \int_0^T \mathrm{e}^{\pm \mathrm{i}\omega t} f(\omega t) \mathrm{d}t = \mathrm{e}^{\mp \mathrm{i}\omega x} \frac{\omega}{2\pi} \int_0^{2\pi/\omega} \mathrm{e}^{\pm \mathrm{i}\omega t} f(\omega t) \mathrm{d}t$$

for $f \in \{\cos, \sin\}$.

Lemma 1.1.4 implies

$$
\begin{aligned}
\mathrm{e}^{\mp \mathrm{i}\omega x} \frac{\omega}{2\pi} \int_0^{2\pi/\omega} \mathrm{e}^{\pm \mathrm{i}\omega t} \cos \omega t \, \mathrm{d}t &= \mathrm{e}^{\mp \mathrm{i}\omega x} \frac{\omega}{2\pi} \left[\mp \frac{\mathrm{i}}{2\omega} \mathrm{e}^{\pm \mathrm{i}\omega t} \cos \omega t + \frac{1}{2}t \right]_{t=0}^{2\pi/\omega} \\
&= \mathrm{e}^{\mp \mathrm{i}\omega x} \frac{\omega}{2\pi} \left[\mp \frac{\mathrm{i}}{2\omega} + \frac{\pi}{\omega} \pm \frac{\mathrm{i}}{2\omega} - \frac{1}{2} \cdot 0 \right] \\
&= \mathrm{e}^{\mp \mathrm{i}\omega x} \frac{\omega}{2\pi} \cdot \frac{\pi}{\omega} \\
&= \frac{1}{2} \mathrm{e}^{\mp \mathrm{i}\omega x}
\end{aligned}
$$

and

$$
\begin{aligned}
\mathrm{e}^{\mp \mathrm{i}\omega x} \frac{\omega}{2\pi} \int_0^{2\pi/\omega} \mathrm{e}^{\pm \mathrm{i}\omega t} \sin \omega t \, \mathrm{d}t &= \mathrm{e}^{\mp \mathrm{i}\omega x} \frac{\omega}{2\pi} \left[\mp \frac{\mathrm{i}}{2\omega} \mathrm{e}^{\pm \mathrm{i}\omega t} \sin \omega t \pm \frac{\mathrm{i}}{2}t \right]_{t=0}^{2\pi/\omega} \\
&= \mathrm{e}^{\mp \mathrm{i}\omega x} \frac{\omega}{2\pi} \left[\mp \frac{\mathrm{i}}{2\omega} \cdot 0 \pm \frac{\mathrm{i}\pi}{\omega} \pm \frac{\mathrm{i}}{2\omega} \cdot 0 \mp \frac{\mathrm{i}}{2} \cdot 0 \right] \\
&= \pm \mathrm{e}^{\mp \mathrm{i}\omega x} \frac{\omega}{2\pi} \cdot \frac{\mathrm{i}\pi}{\omega} \\
&= \pm \frac{\mathrm{i}}{2} \mathrm{e}^{\mp \mathrm{i}\omega x}. \qquad \square
\end{aligned}
$$

Ergodic theory The well-known Birkhoff Ergodic Theorem (see [CFS82, Theorem 1.2.1]) proves existence of certain ergodic limits almost everywhere. For uniquely ergodic systems, existence actually holds everywhere, as the following theorem shows.

Theorem 2.1.9. *Consider a metric space X, a semi-flow $\Phi : \mathbb{R}^+ \times X \to X$ given by $(t, x) \mapsto \Phi_t x$ that is continuous in x, and a continuous function $f : X \to \mathbb{C}$. Assume that $f(\Phi_t x)$ is Borel measurable in t. If Φ_t is uniquely ergodic with ergodic invariant measure μ, then the limit*

$$\lim_{T \to \infty} \frac{1}{T} \int_0^T f\big(\Phi_t(x)\big) \mathrm{d}t$$

converges for all $x \in X$ to $1/\mu(X) \cdot \int_X f \mathrm{d}\mu$.

Proof. See [CFS82, Theorem 1.2.1 and Lemma 1.2.2; Mañ87, Theorem I.9.2]. □

The following is a continuous-time version of von Neumann's Mean Ergodic Theorem.

Theorem 2.1.10 (Mean Ergodic Theorem). *Let U_t, $t \geq 0$, be a semi-group of linear operators on a Hilbert space $\mathcal{H}$ with $\|U_t\| \leq 1$ for all $t \geq 0$. Then for any $f \in \mathcal{H}$,*

$$\lim_{T \to \infty} \frac{1}{T} \int_0^T U_t f \, \mathrm{d}t$$

equals the orthogonal projection of f onto $\{g \in \mathcal{H} \mid \forall t \geq 0 : U_t g = g\}$.

Proof. See [RS80, Theorem II.11 and Problem II.18], and also compare [Wer05, Aufgabe V.6.38]. □

Continuity   Now we give a definition and a lemma on continuity. Hölder continuity is defined as follows.

Definition 2.1.11 (Hölder continuity). Let X, Y be metric spaces. A map $f : X \to Y$ is called *Hölder continuous*, if there are constants $L, \alpha > 0$ such that for all $x_1, x_2 \in X$, it holds that $d\big(f(x_1), f(x_2)\big) \leq L d(x_1, x_2)^\alpha$. ⌟

The following lemma shows that continuity is equivalent to continuity on compact sets.

Lemma 2.1.12. *Let metric spaces X, Y, and a map $f : X \to Y$ be given. The map f is continuous on X, if and only if f is continuous on every compact subset of X.*

Proof. If f is continuous on X, it clearly is continuous on every subset of X.

Assume that f is continuous on every compact subset of X. Let $(x_n)_{n \in \mathbb{N}} \subset X$ be a series that converges to some point $x_0 \in X$. In order to prove continuity of f on X, we have to show that $f(x_n) \to f(x_0)$ for $n \to \infty$.

Note that the set $M := \{x_j \mid j \in \mathbb{N}_0\}$ is compact. To see this, consider an open cover $\mathcal{A}$ of M. For every $j \in \mathbb{N}_0$, let $A_j \in \mathcal{A}$ denote an open set containing x_j. By convergence, and because $x_0 \in A_0$, the set A_0 contains all but finitely many points in M. So there is $r \in \mathbb{N}$ such that $\{A_0, A_1, \ldots, A_r\}$ is a finite open cover of M.

As M is compact, f is continuous on M by assumption. So $f(x_n) \to f(x_0)$ for $n \to \infty$. □

Remark 2.1.13. Lemma 2.1.12 also holds on topological spaces X, Y, if X is a sequential space, i. e., if sequential continuity is equivalent to continuity in X. ⌟

Error function   We will need the error function and the following lemma in an example later.

Definition 2.1.14. For $z \in \mathbb{C}$, define the *error function*

$$\operatorname{erf}(z) := \frac{2}{\sqrt{\pi}} \int_0^z e^{-t^2} dt.$$

Lemma 2.1.15. *It holds that* $\lim_{T \to \infty} \operatorname{erf}\big(T \cdot (1 - \mathrm{i})\big) = 1$.

Proof. By Equation 1 in [GR07, Section 8.252], it holds that

$$\operatorname{erf}\big(T \cdot (1 - \mathrm{i})\big) = \frac{2(1 - \mathrm{i})}{\sqrt{\pi}} \int_0^T e^{-t^2(1-\mathrm{i})^2} dt = \frac{2(1 - \mathrm{i})}{\sqrt{\pi}} \int_0^T e^{2\mathrm{i}t^2} dt$$

$$= \frac{\sqrt{2}(1 - \mathrm{i})}{\sqrt{\pi}} \int_0^{\sqrt{2}T} e^{\mathrm{i}t^2} dt.$$

By Equation 1 in [GR07, Section 8.256], it follows that $\operatorname{erf}\big(T \cdot (1 - \mathrm{i})\big) = (1 - \mathrm{i})\big[C(\sqrt{2}T) + \mathrm{i}S(\sqrt{2}T)\big]$, where S and C are given by $S(x) := \sqrt{2}/\sqrt{\pi} \cdot \int_0^x \sin t^2 dt$ and $C(x) := \sqrt{2}/\sqrt{\pi} \cdot \int_0^x \cos t^2 dt$. As by Equations 3 and 4 in [GR07, Section 8.257], $\lim_{x \to \infty} S(x) = \lim_{x \to \infty} C(x) = 1/2$, it follows that

$$\lim_{T \to \infty} \operatorname{erf}\big(T \cdot (1 - \mathrm{i})\big) = (1 - \mathrm{i}) \cdot \frac{1}{2}(1 + \mathrm{i}) = 1. \qquad \square$$

2.2 Rotational factor maps and harmonic limits

Let X be a metric space, and consider a map $\Phi : \mathbb{R}^+ \times X \to X$, $(t, x) \mapsto \Phi_t x$ that is Borel measurable in t and satisfies $\Phi_0 x = x$. For some results, we will additionally assume that Φ is a semi-flow, i.e., that $\Phi_s \Phi_t = \Phi_{s+t}$ for all $s, t \in \mathbb{R}^+$. Note that we do not generally assume continuity in t or x, which, e.g., is the case for solutions of ODEs. One example, which will be discussed in detail in Subsection 2.5.2, is the flow given by the solution of a linear autonomous ODE projected onto the unit sphere.

There are several ways to describe the rotational behaviour of such dynamical systems, e.g., different notions of rotation numbers [JM82; Ruf97; San88; Ste09]. We will use the concept of factor maps to rotation instead, which was introduced for discrete-time dynamical systems in [MB04, Section 3.1], and further pursued, e.g., in [LM08; Row09].

2.2.1 Rotational factor maps

We want to analyze the rotational behaviour of a dynamical system by investigating if the system admits a rotational factor map to some frequency. With rotational factor map we mean the following.

Definition 2.2.1. Let $\mathcal{F}$ be a class of functions mapping $X \to \mathbb{C}$. Suppose that there is a map $F \in \mathcal{F}$, such that $F \not\equiv 0$, and

$$F \circ \Phi_t = e^{it\omega} \cdot F \qquad (2.2.1)$$

holds for some $\omega \in \mathbb{R}$ and all $t \geq 0$. Then we say that Φ_t *admits the rotational factor map F in $\mathcal{F}$ with frequency* $\omega/2\pi$. ⌟

By this definition, a system Φ_t has a rotational factor map, if it is semi-conjugate to a rotation in the complex plane, or in other words, if the system can be mapped into the complex plane such that the dynamics simply become a rotation.

Remark 2.2.2. The condition $F \not\equiv 0$ in Definition 2.2.1 is necessary, because, with $F \equiv 0$, the semi-conjugacy (2.2.1) trivially holds for all systems Φ_t and all $\omega \in \mathbb{R}$. Consider the two simplest cases where $F \not\equiv 0$, namely $F \equiv c \neq 0$ and $F = \mathbf{1}_{\{x_0\}}$ for some $x_0 \in X$. In the first case, the semi-conjugacy equation reduces to $c = e^{it\omega} \cdot c$, which is true for all $t \geq 0$ if and only if $\omega = 0$. This shows that every system has a rotational factor map with frequency 0. In the second case, the equation reduces to $\mathbf{1}_{\{x_0\}}(\Phi_t x_0) = e^{it\omega}$, which can only be true for $\omega = 0$ and a fixed point x_0. Sometimes, we will exclude this special case $\omega = 0$. ⌟

Remark 2.2.3. If there is an invariant measure μ on X, we will consider μ-measurable rotational factor maps. In this case, we will additionally assume that $\mu(F \neq 0) > 0$, i.e., we assume that F does not vanish almost everywhere. We can subsume these assumptions by setting $\mathcal{F} := \mathcal{F}_\mu := \{F : X \to \mathbb{C} \mid \mu\text{-measurable}, \mu(F \neq 0) > 0\}$. In this context, it is reasonable to consider maps $F \in \mathcal{F}_\mu$, for which the semi-conjugacy (2.2.1) only holds almost everywhere. We will call them *rotational factor maps μ-almost everywhere.* ⌟

Note that, if $\mathcal{F}$ is invariant under complex conjugation, and if a semiflow Φ_t has a rotational factor map with frequency $\omega/2\pi$, then it also has a rotational factor map with frequency $-\omega/2\pi$. In order to see this, consider the complex conjugate $\overline{F}$ of the original rotational factor map F. So we could restrict our view to nonnegative ω, and thereby to nonnegative frequencies. But we will always consider $\omega \in \mathbb{R}$.

In general, it is not easy to find a rotational factor map analytically. In the next section, we will introduce harmonic limits, which can be used to construct rotational factor maps. Meanwhile, let us look at an example, where a rotational factor map can explicitly be given.

Example 2.2.4. Let $X := \mathbb{R}^2$ and $\alpha \in \mathbb{R}$, and consider the system given by the linear differential equation $\dot{x} = \left(\begin{smallmatrix} 0 & \alpha \\ alpha & 0 \end{smallmatrix}\right)x$, i.e., consider $\Phi_t x := \left(\begin{smallmatrix} \cos\alpha t & \sin\alpha t \\ -\sin\alpha t & \cos\alpha t \end{smallmatrix}\right)x$. This is a rotation around the origin with frequency $\omega/2\pi$. The map $F : \mathbb{R}^2 \to \mathbb{C}$ given by

$F(x_1, x_2) := x_1 - \mathrm{i}x_2$ is a rotational factor map by frequency $\alpha/2\pi$, because

$$
\begin{aligned}
F(\Phi_t x) &= F(x_1 \cos \alpha t + x_2 \sin \alpha t, -x_1 \sin \alpha t + x_2 \cos \alpha t) \\
&= x_1 \cos \alpha t + x_2 \sin \alpha t + \mathrm{i}x_1 \sin \alpha t - \mathrm{i}x_2 \cos \alpha t \\
&= x_1 \mathrm{e}^{\mathrm{i}\alpha t} - \mathrm{i}x_2 \mathrm{e}^{\mathrm{i}\alpha t} \\
&= \mathrm{e}^{\mathrm{i}\alpha t}(x_1 - \mathrm{i}x_2) \\
&= \mathrm{e}^{\mathrm{i}\alpha t} F(x).
\end{aligned}
$$

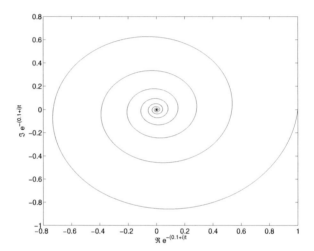

Figure 2.1: Plot of $t \mapsto \mathrm{e}^{-(0.1+I)t}x$
For $x = 1$ and $t \in [0, 100]$

This concept of rotational factor maps cannot properly describe the kind of rotation that appears, e. g., in $t \mapsto \mathrm{e}^{-(0.1+I)t}x$, see Figure 2.1, because the concept concentrates on the asymptotic behaviour, which, in this case, is described by the fixed point in the origin and hence does not show any rotational behaviour. Compare Subsection 2.3.4 for more details on the influence of asymptotics on the harmonic limit.

One has to be careful with the choice of $\mathcal{F}$. If we let $\mathcal{F}$ be the class of all functions $X \to \mathbb{C}$ or all Borel measurable functions $X \to \mathbb{C}$, we might encounter unexpected results. See the following example, where we get rotational factor maps to all frequencies.

Example 2.2.5. Let $X := [0, 1]$ and $\dot{x} = -x$, i. e., $\Phi_t x = \mathrm{e}^{-t}x$. Choose $\omega \in \mathbb{R}$, and let

$$F : X \to \mathbb{C}, x \mapsto \begin{cases} e^{-i\omega \log x} & \text{if } x \neq 0, \\ 0 & \text{if } x = 0. \end{cases}$$

Then Equation (2.2.1) holds, i.e., the system admits rotational factor maps to arbitrary frequencies.

Note that Φ_t preserves the point measure $\mathbf{1}_{\{0\}}$. With this measure, we have $\mathbf{1}_{\{0\}}(F > 0) = \mathbf{1}_{\{0\}}\big((0,1]\big) = 0$, i.e., F vanishes almost everywhere. So if we choose $\mathcal{F} := \mathcal{F}_{\mathbf{1}_{\{0\}}}$ as in Remark 2.2.3, this map F is not a rotational factor map $\mathbf{1}_{\{0\}}$-almost everywhere. ⌟

2.2.2 Harmonic limits

In Example 2.2.4, we presented a rotational factor map and verified that it indeed satisfies the semi-conjugacy (2.2.1). But how can one find a rotational factor map to a given system, apart from guessing? In this section, we will introduce harmonic limits as a tool for this purpose. There is a strong connection between these harmonic limits and the existence of rotational factor maps, as will be shown in Subsection 2.2.3.

Definition 2.2.6 (Harmonic average and limit). For any function $f : X \to \mathbb{C}$, such that $f(\Phi_t x)$ is locally integrable with respect to t, all $\omega \in \mathbb{R}$, and $x \in X$, define the *harmonic average* by

$$f_\omega^T(x) := \frac{1}{T} \int_0^T e^{i\omega t} f(\Phi_t x) \mathrm{d}t;$$

furthermore, define the *harmonic limit* by

$$f_\omega^*(x) := \lim_{T \to \infty} f_\omega^T(x),$$

if the limit exists. ⌟

Remark 2.2.7. By Definition 2.2.6, the harmonic limit is an ergodic integral over the image of a trajectory under some complex-valued map f. One can interpret this map f as an observation of the system, or its output. In particular, the harmonic limit can be computed (or at least approximated by harmonic averages) for measurements of a system, without knowledge of the exact dynamics Φ_t. ⌟

Note that, for $\omega = 0$, the harmonic limit $f_\omega^*(x)$ is simply the average of f along the trajectory $\Phi_t x$, i.e., it holds that $f_0^*(x) = \lim_{T \to \infty} {}^1\!/T \cdot \int_0^T f(\Phi_t x) \mathrm{d}t$. Recall Remark 2.2.2, where we already have seen, that $\omega = 0$ plays a special role in this discussion.

Before we turn to an example, we give some remarks on local integrability, which is assumed in Definition 2.2.6. For any map $g : \mathbb{R} \to \mathbb{C}$, and $\omega \in \mathbb{R}$, it holds that

$e^{i\omega t}g(t)$ is locally integrable if and only if $g(t)$ is locally integrable, compare [Bau92, Satz 12.2]. Thus local integrability of $f(\Phi_t x)$ implies that the integral $\int_0^T e^{i\omega t} f(\Phi_t x)\mathrm{d}t$ is finite for all $T \geq 0$. The map $t \to f(\Phi_t x)$ is locally integrable if, e.g., f is measurable and bounded. This particularly is the case for continuous f, if X is compact.

Example 2.2.8. Consider the system from Example 2.2.4 again, i.e., let $X := \mathbb{R}^2$ and $\Phi_t x := \left(\begin{smallmatrix} \cos \alpha t & \sin \alpha t \\ -\sin \alpha t & \cos \alpha t \end{smallmatrix} \right) x$. Let $f : X \to \mathbb{C}$ be given by $f(x_1, x_2) := x_1$. Then for all $\alpha \in \mathbb{R}$, any $x \in X$, and $\omega = \pm \alpha$, it holds by Lemma 2.1.8 that

$$f_{\pm\alpha}^*(x) = \lim_{T\to\infty} \frac{1}{T} \int_0^T e^{\pm i\alpha t}(x_1 \cos \alpha t + x_2 \sin \alpha t)\mathrm{d}t$$

$$= x_1 \lim_{T\to\infty} \frac{1}{T} \int_0^T e^{\pm i\alpha t} \cos \alpha t \mathrm{d}t + x_2 \lim_{T\to\infty} \frac{1}{T} \int_0^T e^{\pm i\alpha t} \sin \alpha t \mathrm{d}t$$

$$= \frac{1}{2}(x_1 \pm i x_2).$$

Furthermore, for $\omega \neq \pm\alpha$, it holds by Lemma 2.1.7 that $f_\omega^*(x) = 0$. ⌐

2.2.3 Nonvanishing harmonic limits and rotational factor maps

As mentioned earlier, there is a strong connection between harmonic limits and rotational factor maps. In fact, there is a map $f : X \to \mathbb{C}$ such that $f_\omega^* \not\equiv 0$ if and only if there is a rotational factor map by frequency $\omega/2\pi \in \mathbb{R}$. It turns out that the complex conjugate of the harmonic limit itself is a rotational factor map.

Theorem 2.2.9. *Let $f : X \to \mathbb{C}$ and $\omega \in \mathbb{R}$.*

1. *Assume that Φ is a semi-flow. If f_ω^* exists on some nonvoid set $M \subset X$, and if it is not constant zero on M, then Φ_t admits the rotational factor map $\mathbf{1}_M \cdot \overline{f_\omega^*}$ by frequency $\omega/2\pi$.*

2. *If Φ_t admits the rotational factor map $F : X \to \mathbb{C}$ by some frequency $\omega/2\pi$, $\omega \in \mathbb{R}$, then $\overline{F} = \overline{F_\omega^*}$ almost everywhere.*

Remark 2.2.10. Note that Φ does not need to be a semi-flow for Part 2 of this theorem. ⌐

Proof of Theorem 2.2.9.

1. Let $f : X \to \mathbb{C}$, $\omega \in \mathbb{R}$ and $x \in X$ be such that $f_\omega^*(x)$ exists. Then for any $t \geq 0$, it holds that

$$e^{-it\omega} f_\omega^*(x) = e^{-it\omega} \lim_{T\to\infty} f_\omega^T(x) = \lim_{T\to\infty} e^{-it\omega} f_\omega^T(x) = \lim_{T\to\infty} \frac{1}{T} \int_0^T e^{i\omega(\tau-t)} f(\Phi_\tau x)\mathrm{d}\tau$$

$$\overset{(*)}{=} \lim_{T\to\infty} \frac{1}{T} \int_t^{T+t} e^{i\omega(\tau-t)} f(\Phi_\tau x)\mathrm{d}\tau = \lim_{T\to\infty} \frac{1}{T} \int_0^T e^{i\omega\tau} f(\Phi_{\tau+t} x)\mathrm{d}\tau$$

$$= \lim_{T \to \infty} \frac{1}{T} \int_0^T e^{i\omega\tau} f\big(\Phi_\tau(\Phi_t x)\big) d\tau$$
$$= f_\omega^*(\Phi_t x).$$

Hence $\overline{f_\omega^*}(\Phi_t x) = e^{it\omega} \overline{f_\omega^*}(x)$. If $M \subset X$ is the nonvoid set of points x, where the harmonic limit $f_\omega^*(x)$ exists and does not vanish, then this implies that $1_M \cdot \overline{f_\omega^*}$ is a rotational factor map by frequency $\omega/2\pi$.

The equation marked with $(*)$ holds, as $\int_0^t e^{i\omega(\tau-t)} f(\Phi_\tau x) d\tau$ is finite due to local integrability, which implies $\lim_{T\to\infty} 1/T \cdot \int_0^t e^{i\omega(\tau-t)} f(\Phi_\tau x) d\tau = 0$, and

$$\frac{1}{T} \int_T^{T+t} e^{i\omega(\tau-t)} f(\Phi_\tau x) d\tau = \frac{1}{T} e^{-it\omega} \big((T+t) f_\omega^{T+t}(x) - T f_\omega^T(x)\big)$$
$$= e^{-it\omega} \left(f_\omega^{T+t}(x) + \frac{t}{T} f_\omega^{T+t}(x) - f_\omega^T(x) \right),$$

which tends to 0 as $T \to \infty$, because $\lim_{T\to\infty} f_\omega^T(x) = \lim_{T\to\infty} f_\omega^{T+t}(x) = f_\omega^*(x)$.

2. Assume that Φ_t admits the rotational factor map $F : X \to \mathbb{C}$ to rotation with frequency $\omega/2\pi$, i.e., that $F \circ \Phi_t = e^{i\omega t} \cdot F$. Let $f := \overline{F}$. Then for every $x \in X$, it holds that

$$f_\omega^T(x) = \frac{1}{T} \int_0^T e^{i\omega t} \overline{F} \circ \Phi_t x \, dt$$
$$= \frac{1}{T} \int_0^T e^{i\omega t} \overline{F \circ \Phi_t x} \, dt$$
$$= \frac{1}{T} \int_0^T e^{i\omega t} \overline{e^{i\omega t} \cdot F(x)} \, dt$$
$$= \frac{1}{T} \int_0^T \overline{F}(x) \, dt$$
$$= \overline{F}(x),$$

and so $\overline{F}_\omega^* = f_\omega^* = \overline{F}$. $\qquad\qquad\square$

So we can construct a rotational factor map by frequency $\omega/2\pi$, $\omega \in \mathbb{R}$, by computing the harmonic limit f_ω^*, if the limit is not constant zero. In fact, by Theorem 2.2.9, we can get every rotational factor map in this manner. So in the following, we will concentrate on the properties of the harmonic limit. If we only have access to a measurement of a trajectory, i.e., to $f(\Phi_t x)$, $t \geq 0$, for some $f : X \to \mathbb{C}$, then we still can compute the harmonic limit (see Remark 2.2.7) and thus show the existence of rotational factor maps.

Example 2.2.11. Consider the system from Example 2.2.4 and Example 2.2.8 again, i.e., $X := \mathbb{R}^2$ and $\Phi_t x := \left(\begin{smallmatrix} \cos \alpha t & \sin \alpha t \\ -\sin \alpha t & \cos \alpha t \end{smallmatrix} \right) x$. Let $f : X \to \mathbb{C}$ be given by $f(x_1, x_2) := x_1$ as in Example 2.2.8. We have seen that $f_\alpha^*(x) = 1/2 \cdot (x_1 + ix_2)$. So the complex conjugate of this harmonic limit coincides with the rotational factor map presented in Example 2.2.4 up to the constant factor $1/2$. ⌟

Because only the property $f_\omega^*(x) \neq 0$ is of importance when establishing rotational factor maps, the following proposition is handy. It shows that we can also admit functions f with values in $\mathbb{C}^n$, $\mathbb{R}$, or $\mathbb{R}^n$. So, e.g., in systems with $X = \mathbb{C}^n$, we can look at harmonic limits for $f = \mathrm{id}_X$ in order to show the existence of rotational factor maps. But usually, we will continue to assume that f has values in $\mathbb{C}$.

Proposition 2.2.12. *Let $x \in X$ and $\omega \in \mathbb{R}$. For $V \in \{\mathbb{R}, \mathbb{C}, \mathbb{R}^n, \mathbb{C}^n\}$, denote the assertion "There is $f : X \to V$ with $f_\omega^*(x) \neq 0$." by $E(V)$.*

1. *It holds that $E(\mathbb{R}^n) \Leftrightarrow E(\mathbb{R}) \Rightarrow E(\mathbb{C}) \Leftrightarrow E(\mathbb{C}^n)$.*

2. *Furthermore, if it additionally holds that, for every $f : X \to \mathbb{C}$, the existence of $f_\omega^*(x)$ implies existence of $(\Re f)_\omega^*(x)$ and $(\Im f)_\omega^*(x)$, then $E(\mathbb{R}^n) \Leftrightarrow E(\mathbb{R}) \Leftrightarrow E(\mathbb{C}) \Leftrightarrow E(\mathbb{C}^n)$.*

3. *Parts 1. and 2. also hold if we only consider continuous functions f. In particular, if $f_\omega^*(x)$ exists for all continuous $f : X \to \mathbb{C}$, then $E(\mathbb{R}^n) \Leftrightarrow E(\mathbb{R}) \Leftrightarrow E(\mathbb{C}) \Leftrightarrow E(\mathbb{C}^n)$.*

Proof. The proof is completely analogous to its discrete-time counterpart. See Proposition 1.2.17. □

The additional condition in part 2. of this proposition in particular holds if f is integrable with respect to a finite invariant measure, as we will see later in Subsection 2.3.2. It is also satisfied at periodic or almost periodic orbits, see Proposition 2.3.28 and Proposition 2.3.29.

2.3 Properties of the harmonic limit

In the previous section, we have seen that harmonic limits can be used as a tool to analyze the rotational behaviour of dynamical systems, because they indicate the existence of rotational factor maps. Now we will have a closer look at these limits and their properties.

First, we note some basic properties in Subsection 2.3.1. This includes continuity, boundedness, and a condition on f for the set $\{\omega \in \mathbb{R} \mid f_\omega^*(x) \neq 0\}$ to be countable. Then we investigate existence of f_ω^* in Subsection 2.3.2 under the presence of an invariant measure. In this case, the harmonic limit exists almost everywhere (Theorem 2.3.14), which can be shown using the Birkhoff Ergodic Theorem. If the system is ergodic, the harmonic limit exists almost everywhere, independently of $\omega \in [0, 2\pi)$. This result is known as the Wiener-Wintner Ergodic Theorem, which we prove in Subsection 2.3.3, and use in Theorem 2.3.17. If the system is uniquely ergodic, the harmonic limit exists everywhere, see Proposition 2.3.15.

Knowledge of the asymptotic properties of the dynamics can help with the computation of harmonic limits, see Subsection 2.3.4. Likewise, the existence of periodic or almost periodic orbits has some implications on the properties of the harmonic limit, see Subsection 2.3.5. If, e.g., $\Phi_t x$ is τ-periodic in t, $\tau > 0$, or, more generally, if $f(\Phi_t x)$ is τ-periodic in t, then $f_\omega^*(x) \neq 0$ can only hold, if the period τ is a multiple of $2\pi/\omega$.

2.3.1 Basic properties

The first property we will look at is continuity of the harmonic average.

Proposition 2.3.1. *If* $f : X \to \mathbb{C}$ *and* $\Phi : \mathbb{R}^+ \times X \to X$ *are continuous, and if we set* $f_\omega^0(x) := f(x)$, *then* $f_\omega^T(x)$ *is continuous in* $(\omega, T, x) \in \mathbb{R} \times [0, \infty) \times X$.

Proof. First, we show that $f_\omega^T(x)$ is continuous in $(T, x) \in [0, T_0] \times M$ for every $T_0 > 0$ and any compact $M \subset X$. So let $\omega \in \mathbb{R}$ and $T_0 > 0$, and let $M \subset X$ be compact. By continuity of f and Φ, the map $(t, x) \mapsto f(\Phi_t x)$ is continuous. By compactness, this implies uniform continuity of $f(\Phi_t x)$ on $[0, T_0] \times M$. In particular, for every $\varepsilon > 0$, there is $\delta_\varepsilon > 0$ such that, for all $x_1, x_2 \in M$ with $d(x_1, x_2) < \delta_\varepsilon$ and all $t \in [0, T_0]$, it holds that $|f(\Phi_t x_1) - f(\Phi_t x_2)| < \varepsilon$. Let $T \in [0, T_0]$. Then for $x_1, x_2 \in M$ with $d(x_1, x_2) < \delta_\varepsilon$, it follows that

$$
\begin{aligned}
|f_\omega^T(x_1) - f_\omega^T(x_2)| &= \left| \frac{1}{T} \int_0^T e^{i\omega t} f(\Phi_t x_1) \mathrm{d}t - \frac{1}{T} \int_0^T e^{i\omega t} f(\Phi_t x_2) \mathrm{d}t \right| \\
&\leq \frac{1}{T} \int_0^T |f(\Phi_t x_1) - f(\Phi_t x_2)| \mathrm{d}t \\
&< \varepsilon,
\end{aligned}
\tag{2.3.1}
$$

i.e., $f_\omega^T(x)$ is continuous in x. Note that δ_ε does not depend on $T \in [0, T_0]$. So by [Ama95, Lemma 8.1], continuity of $f_\omega^T(x)$ in $(T, x) \in [0, T_0] \times M$ follows, if $f_\omega^T(x)$ is continuous in $T \in [0, T_0]$ for all $x \in M$. To show continuity of $f_\omega^T(x)$ in T, first note that, by the Fundamental Theorem of Calculus (compare [Kön04, p. 200]), $\int_0^T e^{i\omega t} f(\Phi_t x) \mathrm{d}t$ is differentiable in T. So clearly, $f_\omega^T(x)$ is continuous in $T \in (0, \infty)$ and it suffices to show that $\lim_{T \to 0} f_\omega^T(x) = f(x)$. This follows from l'Hôpital's Rule (compare [Kön04, p. 150]):

$$
\lim_{T \to 0} \frac{\int_0^T e^{i\omega t} f(\Phi_t x) \mathrm{d}t}{T} = \lim_{T \to 0} \frac{e^{i\omega T} f(\Phi_T x)}{1} = f(x).
$$

This proves that $f_\omega^T(x)$ is continuous in $(T, x) \in [0, T_0] \times M$ for every $T_0 > 0$ and any compact $M \subset X$.

Next, we show that $f_\omega^T(x)$ is continuous in $(\omega, T, x) \in \mathbb{R} \times [0, T_0] \times M$ for every $T_0 > 0$ and any compact $M \subset X$. Note that, by continuity of Φ, the set $A := \{\Phi_t x \in X \mid (t, x) \in [0, T_0] \times M\}$ is compact. Define $F := \max_{x \in A} f(x)$, which exists due to continuity of f. The harmonic average $f_\omega^T(x)$ is differentiable in ω. It holds that

$$\frac{\mathrm{d}}{\mathrm{d}\omega} f_\omega^T(x) = \frac{1}{T} \int_0^T \mathrm{i} t \mathrm{e}^{\mathrm{i}\omega t} f(\Phi_t x) \mathrm{d}t,$$

compare [Kön00, Section 2.6]. So for any $T \in [0, T_0]$, it holds that

$$\left| \frac{\mathrm{d}}{\mathrm{d}\omega} f_\omega^T(x) \right| \leq \frac{1}{T} \int_0^T t F \mathrm{d}t = \frac{1}{2T} \cdot T^2 F = \frac{1}{2} FT \leq \frac{1}{2} FT_0.$$

So $f_\omega^T(x)$ is Lipschitz continuous in ω with Lipschitz constant $\tfrac{1}{2}FT_0$, which does not depend on T or on x. Hence $f_\omega^T(x)$ is continuous in $\omega \in \mathbb{R}$ uniformly in $(T, x) \in \times [0, T_0] \times M$. So by [Ama95, Lemma 8.1], continuity of $f_\omega^T(x)$ in $(\omega, T, x) \in \mathbb{R} \times [0, T_0] \times M$ follows, because $f_\omega^T(x)$ is continuous in $(T, x) \in [0, T_0] \times M$, as we have already shown.

As $(\omega, T, x) \mapsto f_\omega^T(x)$ is continuous on $\mathbb{R} \times [0, T_0] \times M$ for every $T_0 > 0$, it is continuous on $\mathbb{R} \times L \times M$ for every compact set $L \subset [0, \infty)$. This is true, because every compact set $L \subset [0, \infty)$ is contained in $[0, T_0]$ for some T_0. So by Lemma 2.1.12, it follows that $(\omega, T, x) \mapsto f_\omega^T(x)$ is continuous on $\mathbb{R} \times [0, \infty) \times X$. $\square$

Consider the following two examples regarding continuity. The first one is a trivial example, which shows that $f_\omega^T(x)$ can be discontinuous in x, if f is not continuous.

Example 2.3.2. Let X be an arbitrary metric space, and consider the semi-flow $\Phi_t x := x$ for all $t \in \mathbb{R}$, $x \in X$. Let $f : X \to \mathbb{R}$ be bounded and measurable, but not continuous. Then $f_0^T(x) = f(x)$ for all $x \in X$ and $T > 0$. So $f_0^T(x)$ is not continuous in x. ⌐

The second example shows that $f_\omega^T(x)$ might be continuous even if f is discontinuous.

Example 2.3.3. Let $X := D_1 \subset \mathbb{C}$, and consider the semi-flow Φ given by $\Phi_t(x) := \mathrm{e}^{-\mathrm{i}\omega t} x$ for $\omega \in \mathbb{R} \setminus \{0\}$. We define f by $f(x) := 0$ if $x \in \mathbb{R} \setminus \{0\}$ and $f(x) := 1$ otherwise. See Figure 2.2.

Let $x \in X$. If $x = 0$, clearly $\Phi_t x = 0$ for all t, and so

$$f_\omega^T(0) = \frac{1}{T} \int_0^T \mathrm{e}^{\mathrm{i}\omega t} f(0) \mathrm{d}t$$

$$= \frac{1}{T} \int_0^T \mathrm{e}^{\mathrm{i}\omega t} \mathrm{d}t$$

$$= \frac{1}{\mathrm{i}\omega T} \left[\mathrm{e}^{\mathrm{i}\omega T} - 1 \right].$$

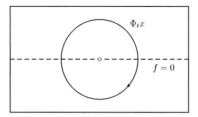

Figure 2.2: Φ and f from Example 2.3.3
Outside the dashed line and in the origin, $f = 1$.

If $x \neq 0$, let $\alpha := \arg x$. Then $\Phi_t x = |x| e^{i(\alpha - t\omega)}$. Thus

$$
\begin{aligned}
f_\omega^T(x) &= \frac{1}{T} \int_0^T e^{i\omega t} f\left(|x| e^{i(\alpha - t\omega)}\right) dt \\
&= \frac{1}{T} \int_0^T e^{i\omega t} \mathbf{1}_{(0,\pi) + \pi\mathbb{Z}}(\alpha - t\omega) dt \\
&= \frac{1}{T} \int_0^T e^{i\omega t} \mathbf{1}_{[0,\pi] + \pi\mathbb{Z}}(\alpha - t\omega) dt \\
&= \frac{1}{T} \int_0^T e^{i\omega t} dt \\
&= \frac{1}{i\omega T} \left[e^{i\omega T} - 1\right].
\end{aligned}
$$

So $f_\omega^T(x)$ is continuous in x, although f is not continuous. Note that it is also continuous in T and ω. ⌟

The next property we will discuss is boundedness. The harmonic average is bounded above by $\sup_{x \in X} |f(x)|$. The same holds for the harmonic limit, if it exists.

Proposition 2.3.4. *For all $x \in X$ and $\omega \in \mathbb{R}$, it holds that*

$$
|f_\omega^T(x)| \leq \sup_{y \in X} |f(y)|. \tag{2.3.2}
$$

Furthermore,

$$
|f_\omega^*(x)| \leq \sup_{y \in X} |f(y)|, \tag{2.3.3}
$$

if $f_\omega^*(x)$ exists.

Proof. In order to show (2.3.2), let $T \geq 0$, $\omega \in \mathbb{R}$, and $x_0 \in X$. Then it holds that

$$|f_\omega^T(x_0)| = \left| \frac{1}{T} \int_0^T e^{i\omega t} f(\Phi_t x_0) dt \right| \leq \frac{1}{T} \int_0^T \left| e^{i\omega t} f(\Phi_t x_0) \right| dt = \frac{1}{T} \int_0^T |f(\Phi_t x_0)| dt$$

$$\leq \frac{1}{T} \int_0^T \sup_{x \in X} |f(x)| dt = \sup_{x \in X} |f(x)|. \tag{2.3.4}$$

For (2.3.3), let $\omega \in \mathbb{R}$ and $x_0 \in X$ be such that $f_\omega^*(x_0)$ exists. Then by (2.3.4), it is true that

$$|f_\omega^*(x_0)| = \left| \lim_{T \to \infty} f_\omega^T(x_0) \right| \leq \sup_{(n,\omega,x) \in \mathbb{N} \times [0,2\pi) \times X} |f_\omega^n(x)| \leq \sup_{x \in X} |f(x)|. \qquad \square$$

The following result is a corollary to Theorem 2.2.9, and shows how the harmonic limit behaves along trajectories. In particular, for semi-flows Φ, we get that, if $f_\omega^*(x)$ exists at some point $x \in X$ for $\omega \in \mathbb{R}$, then the harmonic limit exists at every point of the trajectory starting in x, and, beyond that, the modulus of the harmonic limit is constant along the trajectory. We have already used this corollary in Lemma 2.1.8.

Corollary 2.3.5. *Assume that Φ is a semi-flow. If $f : X \to \mathbb{C}$, $\omega \in \mathbb{R}$, and $x \in X$ are such that $f_\omega^*(x)$ exists, then $f_\omega^*(\Phi_t x) = e^{-it\omega} f_\omega^*(x)$ for all $t \geq 0$. In particular, the harmonic limit $f_\omega^*(\Phi_t x)$ exists for all $t \geq 0$, and its modulus $|f_\omega^*(\Phi_t x)|$ is independent of t.*

Proof. By part 1. of Theorem 2.2.9, it holds that $f_\omega^*(\Phi_t x) = e^{-i\omega t} f_\omega^*(x)$ if $f_\omega^*(x)$ exists. So $f_\omega^*(\Phi_t)$ exists for all $t \geq 0$, and $|f_\omega^*(\Phi_t x)| = |f_\omega^*(x)|$ is independent of j. $\qquad \square$

Let us look at the following example in order to illustrate these results.

Example 2.3.6. Consider the system from Example 2.2.4, Example 2.2.8 and Example 2.2.11 again, i.e., $X := \mathbb{R}^2$ and $\Phi_t x := \left(\begin{smallmatrix} \cos \alpha t & \sin \alpha t \\ -\sin \alpha t & \cos \alpha t \end{smallmatrix} \right) x$. Let $f : X \to \mathbb{C}$ be given by $f(x_1, x_2) := x_1$ as in Example 2.2.8 and Example 2.2.11. We have seen that $f_\alpha^*(x) = 1/2 \cdot (x_1 + ix_2)$.

It holds that $|f_\alpha^*(x)| < \infty = \sup_{y_1 \in \mathbb{R}} |y_1| = \sup_{y \in X} |f(y)|$, which is in line with Proposition 2.3.4. It further holds that

$$f_\alpha^*(\Phi_t x) = 1/2 \cdot \left((x_1 \cos \alpha t + x_2 \sin \alpha t) + i(-x_1 \sin \alpha t + x_2 \cos \alpha) \right) = 1/2 \cdot (x_1 + ix_2) e^{-i\alpha t}.$$

Hence $|f_\alpha^*(\Phi_t x)| = 1/2 |x_1 + ix_2|$ is independent of t, which is in line with Corollary 2.3.5.

If one thinks of the harmonic limit as a tool to detect rotational factor maps, a natural question is, which frequencies $\omega/2\pi$ occur in the system. We will address this question later in Subsection 1.3.4. At this point, we can only give a result on how many frequencies can occur. The following theorem shows that under a weak condition, which, e.g., holds for square-integrable f (see Corollary 2.3.10), there can only be countably many $\omega \in \mathbb{R}$ for which the harmonic limit is different from zero.

Theorem 2.3.7. *Let $f : X \to \mathbb{C}$ and $x \in X$ be such that $\lim_{T\to\infty} 1/T \cdot \int_0^T |f(\Phi_t x)|^2 \mathrm{dt}$ exists. Then there are at most countably many $\omega \in \mathbb{R}$ for which $f_\omega^*(x) \neq 0$.*

Proof. This proof follows the ideas in Sections 57 and 58 of [Boh47], where a similar result is shown for almost periodic functions. Compare also Theorem 1.3.5.

First, we show that for arbitrary distinct $\omega_1, \ldots, \omega_N \in \mathbb{R}$ and arbitrary numbers $c_1, \ldots, c_N \in \mathbb{C}$, it holds that

$$
\lim_{T\to\infty} \frac{1}{T} \int_0^T \left| f(\Phi_t x) - \sum_{n=1}^{N} c_n \mathrm{e}^{-\mathrm{i}\omega_n t} \right|^2 \mathrm{dt}
$$

$$
= \lim_{T\to\infty} \frac{1}{T} \int_0^T |f(\Phi_t x)|^2 \mathrm{dt} - \sum_{n=1}^{N} |f_{\omega_n}^*(x)|^2 + \sum_{n=1}^{N} |c_n - f_{\omega_n}^*(x)|^2, \quad (2.3.5)
$$

if $f_{\omega_n}^*(x)$ exists for all ω_n. In order to show this, first note that for all $\alpha, \beta \in \mathbb{C}$, it is true that

$$
\begin{aligned}
|\alpha - \beta|^2 &= (\alpha - \beta)(\overline{\alpha} - \overline{\beta}) \\
&= \alpha\overline{\alpha} - \alpha\overline{\beta} - \overline{\alpha}\beta + \beta\overline{\beta} \\
&= |\alpha|^2 - \alpha\overline{\beta} - \overline{\alpha}\beta + |\beta|^2.
\end{aligned} \quad (2.3.6)
$$

This implies that

$$
\left| f(\Phi_t x) - \sum_{n=1}^{N} c_n \mathrm{e}^{-\mathrm{i}\omega_n t} \right|^2
$$

$$
= |f(\Phi_t x)|^2 - f(\Phi_t x) \overline{\sum_{n=1}^{N} c_n \mathrm{e}^{-\mathrm{i}\omega_n t}} - \overline{f(\Phi_t x)} \sum_{n=1}^{N} c_n \mathrm{e}^{-\mathrm{i}\omega_n t} + \left| \sum_{n=1}^{N} c_n \mathrm{e}^{-\mathrm{i}\omega_n t} \right|^2
$$

$$
= |f(\Phi_t x)|^2 - \sum_{n=1}^{N} \overline{c_n} f(\Phi_t x) \mathrm{e}^{\mathrm{i}\omega_n t} - \sum_{n=1}^{N} c_n \overline{f(\Phi_t x)} \mathrm{e}^{-\mathrm{i}\omega_n t}
$$

$$
+ \left(\sum_{n=1}^{N} c_n \mathrm{e}^{-\mathrm{i}\omega_n t} \right) \left(\sum_{n=1}^{N} \overline{c_n} \mathrm{e}^{\mathrm{i}\omega_n t} \right)
$$

$$
= |f(\Phi_t x)|^2 - \sum_{n=1}^{N} \overline{c_n} f(\Phi_t x) \mathrm{e}^{\mathrm{i}\omega_n t} - \sum_{n=1}^{N} c_n \overline{f(\Phi_t x)} \mathrm{e}^{-\mathrm{i}\omega_n t} + \sum_{m=1}^{N} \sum_{n=1}^{N} c_m \overline{c_n} \mathrm{e}^{-\mathrm{i}\omega_m t} \mathrm{e}^{\mathrm{i}\omega_n t}.
$$

By assumption,

$$
\lim_{T\to\infty} \frac{1}{T} \int_0^T |f(\Phi_t x)|^2 \mathrm{dt}
$$

and $f^*_{\omega_n}(x)$ exist. So by linearity of the limit and the integral, it follows that

$$\lim_{T \to \infty} \frac{1}{T} \int_0^T \left| f(\Phi_t x) - \sum_{n=1}^N c_n e^{-i\omega_n t} \right|^2 dt$$

$$= \lim_{T \to \infty} \frac{1}{T} \int_0^T |f(\Phi_t x)|^2 dt - \sum_{n=1}^N \overline{c_n} \lim_{T \to \infty} \frac{1}{T} \int_0^T f(\Phi_t x) e^{i\omega_n t} dt$$

$$- \sum_{n=1}^N c_n \lim_{T \to \infty} \frac{1}{T} \int_0^T \overline{f(\Phi_t x)} e^{-i\omega_n t} dt + \sum_{m=1}^N \sum_{n=1}^N c_m \overline{c_n} \lim_{T \to \infty} \frac{1}{T} \int_0^T e^{-i\omega_m t} e^{i\omega_n t} dt$$

$$= \lim_{T \to \infty} \frac{1}{T} \int_0^T |f(\Phi_t x)|^2 dt - \sum_{n=1}^N \overline{c_n} f^*_{\omega_n}(x) - \sum_{n=1}^N c_n \overline{f^*_{\omega_n}(x)}$$

$$+ \sum_{m=1}^N \sum_{n=1}^N c_m \overline{c_n} \lim_{T \to \infty} \frac{1}{T} \int_0^T e^{-i\omega_m t} e^{i\omega_n t} dt. \quad (2.3.7)$$

Note that, by assumption, $\omega_m = \omega_n$ only if $m = n$. So it holds that

$$\lim_{T \to \infty} \frac{1}{T} \int_0^T e^{-i\omega_m t} e^{i\omega_n t} dt = \begin{cases} 1 & \text{if } m = n, \\ 0 & \text{otherwise.} \end{cases} \quad (2.3.8)$$

Equations (2.3.7) and (2.3.8) together imply

$$\lim_{T \to \infty} \frac{1}{T} \int_0^T \left| f(\Phi_t x) - \sum_{n=1}^N c_n e^{-i\omega_n t} \right|^2 dt$$

$$= \lim_{T \to \infty} \frac{1}{T} \int_0^T |f(\Phi_t x)|^2 dt - \sum_{n=1}^N \overline{c_n} f^*_{\omega_n}(x) - \sum_{n=1}^N c_n \overline{f^*_{\omega_n}(x)} + \sum_{n=1}^N |c_n|^2$$

$$= \lim_{T \to \infty} \frac{1}{T} \int_0^T |f(\Phi_t x)|^2 dt - \sum_{n=1}^N |f^*_{\omega_n}(x)|^2 + \sum_{n=1}^N |c_n - f^*_{\omega_n}(x)|^2,$$

where the last equality holds due to (2.3.6). This proves equation (2.3.5).

Setting $c_n := f^*_{\omega_n}(x)$ in equation (2.3.5) yields

$$\lim_{T \to \infty} \frac{1}{T} \int_0^T \left| f(\Phi_t x) - \sum_{n=1}^N f^*_{\omega_n}(x) e^{-i\omega_n t} \right|^2 dt = \lim_{T \to \infty} \frac{1}{T} \int_0^T |f(\Phi_t x)|^2 dt - \sum_{n=1}^N |f^*_{\omega_n}(x)|^2.$$

As the left-hand side of this equation is nonnegative, it follows that

$$\sum_{n=1}^N |f^*_{\omega_n}(x)|^2 \leq \lim_{T \to \infty} \frac{1}{T} \int_0^T |f(\Phi_t x)|^2 dt =: C. \quad (2.3.9)$$

Note that C does not depend on the choice of the ω_n. So for every $\varepsilon > 0$, there can be only finitely many $\omega \in \mathbb{R}$ with $|f_\omega^*(x)| > \varepsilon$. To see this, assume that for some $\varepsilon > 0$ there are N distinct values $\omega_1, \ldots, \omega_N \in \mathbb{R}$ for $N > C/\varepsilon^2$, such that $|f_{\omega_n}^*(x)| > \varepsilon$. Then

$$\sum_{n=1}^{N} |f_{\omega_n}^*(x)|^2 > \sum_{n=1}^{N} \varepsilon^2 = N\varepsilon^2 > C,$$

which contradicts (2.3.9).

So there are only finitely many $\omega \in \mathbb{R}$ with $|f_\omega^*(x)| > 1$. Similarly, for every $n \in \mathbb{N}$ there are only finitely many $\omega \in \mathbb{R}$ with $1/n \geq |f_\omega^*(x)| > 1/(n+1)$. This implies that there can be only countably many $\omega \in \mathbb{R}$ with $f_\omega^*(x) \neq 0$. $\qquad \square$

Remark 2.3.8. Note that this result also holds, more generally, for measurable maps $g : \mathbb{R}^+ \to \mathbb{C}$, i.e., if

$$\lim_{T \to \infty} \frac{1}{T} \int_0^T |g(t)|^2 \mathrm{d}t$$

exists, it holds that there are at most countably many $\omega \in \mathbb{R}$ for which $\lim_{T \to \infty} 1/T \cdot \int_0^T \mathrm{e}^{\mathrm{i}\omega t} g(t) \mathrm{d}t \neq 0$. $\qquad \lrcorner$

Remark 2.3.9. Inequality (2.3.9) is Bessel's Inequality [BN66, Theorem 9.3 (1)] applied to the linear space spanned by f and the functions $t \mapsto \mathrm{e}^{-\mathrm{i}\omega_j t}$, $j = 1, \ldots, n$, endowed with the inner product $\langle g, h \rangle := \lim_{T \to \infty} 1/T \cdot \int_0^T g(t)\overline{h(t)}\mathrm{d}t$. So Theorem 2.3.7 also holds in a much more general setting, compare [BN66, Theorem 9.3 (2)]. Compare also Parseval's Equation in Lemma 2.3.36. $\qquad \lrcorner$

Now we give a corollary to this theorem, which shows that it applies to square-integrable f, if there is an invariant measure.

Corollary 2.3.10. *Assume that Φ is a semi-flow, and let $f : X \to \mathbb{C}$ be square-integrable with respect to some finite invariant measure. Then there is a null set $\Xi \subset X$ such that, for all $x \in X \setminus \Xi$, there are at most countably many $\omega \in \mathbb{R}$ such that $f_\omega^*(x) \neq 0$.*

Proof. If $f : X \to \mathbb{C}$ is square-integrable, then $|f|^2$ is integrable. Hence by the Birkhoff Ergodic Theorem [CFS82, Theorem 1.2.1], there is a null set $\Xi \subset X$ such that the limit $\lim_{T \to \infty} 1/T \cdot \int_0^T |f(\Phi_t x)|^2 \mathrm{d}t$ exists for all $x \in X \setminus \Xi$. So Theorem 2.3.7 implies the assertion. $\qquad \square$

Let us look at an example again, where there are in fact only finitely many ω for which the harmonic limit does not vanish.

Example 2.3.11. Consider the flow Φ from Example 2.3.3 again, i.e., the flow on the closed unit disc $X := D_1 \subset \mathbb{C}$ given by $\Phi_t(x) := \mathrm{e}^{-\mathrm{i}\omega t}x$ for $\omega \in \mathbb{R} \setminus \{0\}$, i.e., Φ is a rotation around the origin with frequency $\omega/2\pi$. We choose $f(x) := \Re x = 1/2 \cdot (x + \overline{x})$. It holds that

$$f_\omega^T(x) = \frac{1}{T}\int_0^T e^{i\omega t} \cdot \frac{1}{2}(e^{-i\omega t}x + e^{i\omega t}\overline{x})dt = \frac{1}{2T}\int_0^T (x + e^{i2\omega t}\overline{x})dt$$

$$= \frac{1}{2T}\left(Tx + \frac{1}{i2\omega}e^{i2\omega T}\overline{x} - \frac{1}{i2\omega}\overline{x}\right) = \frac{1}{2}x + \frac{i\overline{x}}{4\omega T}(1 - e^{i2\omega T})$$

Thus $f_\omega^*(x) = 1/2x$. On the other hand, for $\alpha \neq \omega$, it holds that

$$f_\alpha^T(x) = \frac{1}{T}\int_0^T e^{i\alpha t} \cdot \frac{1}{2}(e^{-i\omega t}x + e^{i\omega t}\overline{x})dt = \frac{1}{2T}\int_0^T \left(e^{i(\alpha-\omega)t}x + e^{i(\alpha+\omega)t}\overline{x}\right)dt$$

$$= \frac{1}{2T}\left(\frac{1}{i(\alpha-\omega)}e^{i(\alpha-\omega)T}x + \frac{1}{i(\alpha+\omega)}e^{i(\alpha+\omega)T}\overline{x} - \frac{1}{i(\alpha-\omega)}x - \frac{1}{i(\alpha+\omega)}\overline{x}\right)$$

$$= \frac{i}{2T}\left(\frac{1}{(\alpha-\omega)}\left(1 - e^{i(\alpha-\omega)T}\right)x + \frac{1}{(\alpha+\omega)}\left(1 - e^{i(\alpha+\omega)T}\right)\overline{x}\right),$$

which implies $f_\alpha^*(x) = 0$.

In general, the countable set $\{\omega \in \mathbb{R} \mid f_\omega^*(x) \neq 0\}$ from Theorem 2.3.7 depends on x. The union of all those sets can be uncountable, as the following example shows.

Example 2.3.12. Consider the differential equation

$$\dot{x}_1 = -\sqrt{x_1^2 + x_2^2} \cdot x_2$$
$$\dot{x}_2 = \sqrt{x_1^2 + x_2^2} \cdot x_1$$

on $X := \{x \in \mathbb{R}^2 \mid \|x\| \leq 1\}$. Its solution is given by

$$\Phi_t x = \exp\begin{pmatrix} 0 & -|x|t \\ |x|t & 0 \end{pmatrix}x$$

for $x \in X$. For every $x \in X \setminus \{0\}$, it holds that $\mathrm{id}_{|x|}^*(x) \neq 0$, because

$$\mathrm{id}_{|x|}^*(x) = \lim_{T\to\infty}\frac{1}{T}\int_0^T e^{i|x|t}\exp\begin{pmatrix} 0 & -|x|t \\ |x|t & 0 \end{pmatrix}x dt$$

$$= \lim_{T\to\infty}\frac{1}{T}\int_0^T e^{i|x|t}\begin{pmatrix} \cos|x|t & -\sin|x|t \\ \sin|x|t & \cos|x|t \end{pmatrix}x dt$$

$$= \begin{pmatrix} 1/2 & -1/2 \cdot i \\ 1/2 \cdot i & 1/2 \end{pmatrix}x,$$

where the last equality holds due to Lemma 2.1.8. As x has only real entries and is not zero, this means that $\mathrm{id}_{|x|}^*(x) \neq 0$. So for every $\omega \in (0,1]$, there is $x \in X$ such that $\mathrm{id}_\omega^*(x) \neq 0$. This means that every set $\Omega \subset \mathbb{R}$ that satisfies $\mathrm{id}_\omega^*(x) = 0$ for all $\omega \in \mathbb{R} \setminus \Omega$ and all $x \in X$ must contain $(0,1]$, and hence is uncountable.

Note that there are systems Φ and functions $f \not\equiv 0$ such that $f_\omega^* = 0$ for all ω, see the following example, which is based on an example in [Boh47, Section 70].

Example 2.3.13. Consider the following ordinary differential equation in $D_1 \times \mathbb{R} \subset \mathbb{R}^3$:

$$\dot{x} = -2yz$$
$$\dot{y} = 2xz$$
$$\dot{z} = 1.$$

The solution to the initial value (x_0, y_0, z_0) can be given by

$$x(t) = x_0 \cdot \cos(t^2 + 2z_0 t) - y_0 \cdot \sin(t^2 + 2z_0 t),$$
$$y(t) = x_0 \cdot \sin(t^2 + 2z_0 t) + y_0 \cdot \cos(t^2 + 2z_0 t),$$
$$z(t) = t + z_0.$$

Let $f : D_1 \times \mathbb{R} \to \mathbb{C}$ be given by $f(x,y,z) := x + \mathrm{i}y$. Then $f\big(x(t), y(t), z(t)\big) = (x_0 + \mathrm{i}y_0) \cdot \mathrm{e}^{\mathrm{i}(t^2 + 2z_0 t)}$.

It holds that $f_\omega^*(x,y,z) = 0$ for all $\omega \in \mathbb{R}$ and all $x,y,z \in D_1 \times \mathbb{R}$. To see this, we show that $\lim_{T \to \infty} 1/T \cdot \int_0^T \mathrm{e}^{\mathrm{i}\omega t} g(t)\mathrm{d}t = 0$ for all $\omega \in \mathbb{R}$, where $g(t) = \mathrm{e}^{\mathrm{i}(t^2 + bt)}$ for $n \in \mathbb{R}$.

For every $\omega \in \mathbb{R}$ and all $b \in \mathbb{R}$, it holds that

$$
\begin{aligned}
\int_0^T \mathrm{e}^{\mathrm{i}\omega t} g(t)\mathrm{d}t &= \int_0^T \mathrm{e}^{\mathrm{i}[t^2 + (b+\omega)t]}\mathrm{d}t \\
&= \int_0^T \mathrm{e}^{1/4 \cdot \mathrm{i}[-(b+\omega)^2 + 4t^2 + 4(b+\omega)t + (b+\omega)^2]}\mathrm{d}t \\
&= \int_0^T \mathrm{e}^{1/4 \cdot \mathrm{i}[-(b+\omega)^2]}\mathrm{e}^{1/4 \cdot \mathrm{i}(2t + b + \omega)^2}\mathrm{d}t \\
&= \mathrm{e}^{1/4 \cdot \mathrm{i}[-(b+\omega)^2]} \cdot \int_0^T \mathrm{e}^{1/4 \cdot \mathrm{i}(2t + b + \omega)^2}\mathrm{d}t.
\end{aligned}
$$

Substituting $s := 1/4\sqrt{2}(1 - \mathrm{i})(2t + b + \omega)$, i.e., $t = 1/2\sqrt{2}(1 + \mathrm{i})s - 1/2 \cdot (b + \omega)$ and thus $\mathrm{d}t = 1/2\sqrt{2}(1 + \mathrm{i})\mathrm{d}s$, one gets

$$
\begin{aligned}
\int_0^T \mathrm{e}^{\mathrm{i}\omega t} f(t)\mathrm{d}t &= \mathrm{e}^{1/4 \cdot \mathrm{i}[-(b+\omega)^2]} \cdot \frac{1}{2}\sqrt{2}(1 + \mathrm{i}) \int_{1/4\sqrt{2}(1-\mathrm{i})(b+\omega)}^{1/4\sqrt{2}(1-\mathrm{i})(2T+b+\omega)} \mathrm{e}^{-s^2}\mathrm{d}s \\
&= \mathrm{e}^{1/4 \cdot \mathrm{i}[-(b+\omega)^2]} \cdot \frac{1}{2}\sqrt{2}(1 + \mathrm{i})\frac{\sqrt{\pi}}{2}\Big[\mathrm{erf}\big(1/4\sqrt{2}(1 - \mathrm{i})(2T + b + \omega)\big) \\
&\quad - \mathrm{erf}\big(1/4\sqrt{2}(1 - \mathrm{i})(b + \omega)\big)\Big],
\end{aligned}
$$

where $\operatorname{erf}(z) := {}^2\!/\!\sqrt{\pi} \cdot \int_0^z \mathrm{e}^{-t^2}\mathrm{d}t$ is the error function, see Definition 2.1.14. By Lemma 2.1.15, it follows that

$$\lim_{T\to\infty} \int_0^T \mathrm{e}^{\mathrm{i}\omega t} f(t)\mathrm{d}t = \mathrm{e}^{1/4 \cdot \mathrm{i}[-(b+\omega)^2]} \cdot \frac{1}{2}\sqrt{2}(1+\mathrm{i})\frac{\sqrt{\pi}}{2}\big[1 - \operatorname{erf}\big({}^1\!/\!4\sqrt{2}(1-\mathrm{i})(b+\omega)\big)\big],$$

which is bounded. Thus $\lim_{T\to\infty} \frac{1}{T}\int_0^T \mathrm{e}^{\mathrm{i}\omega t} g(t)\mathrm{d}t = 0$. ⌟

2.3.2 Existence

In the previous section, we proposed a first result on the existence of harmonic limits (Corollary 2.3.5). It revealed that the harmonic limit exists all along a trajectory if it exists at the starting point. The big drawback of this result is that we have to assume existence in one point.

Under the presence of an invariant measure, we get stronger results on existence, though: If we have a finite invariant measure μ on X, then for every μ-integrable function f and all $\omega \in \mathbb{R}$, the harmonic limit exists almost everywhere. If μ is ergodic, the null set, where f_ω^* does not exist, can be chosen independently on ω. Finally, if μ is the unique ergodic measure for our system, then $f_\omega^*(x)$ exists for all $x \in X$, all $\omega \in \mathbb{R}$, and all integrable f.

Theorem 2.3.14. *Assume that Φ is a semi-flow, and that there is a finite invariant measure μ on X such that the map $f : X \to \mathbb{C}$ is of class $L^p(\mu)$, $1 \leq p < \infty$. Then for every $\omega \in \mathbb{R}$, there is a null set $\Xi_\omega \subset X$ such that the harmonic limit $f_\omega^*(x)$ exists for all $x \in X \setminus \Xi_\omega$, and the map $x \mapsto f_\omega^*(x)$ is of class $L^p(\mu, \mathbb{C})$.*

Proof. This follows from the Birkhoff Ergodic Theorem [CFS82, Theorem 1.2.1] applied to the extended semiflow Ψ_t^ω on $S^1 \times X$ given by $(z,x) \mapsto (\mathrm{e}^{\mathrm{i}t\omega}z, \Phi_t x)$ for $\omega \in \mathbb{R}$. More precisely, let $f : X \to \mathbb{C}$ be of class $L^p(\mu)$, $1 \leq p < \infty$, and $\omega \in \mathbb{R}$. Define $g : S^1 \times X \to \mathbb{C}$ by $(z,x) \mapsto zf(x)$. Consider the finite invariant measure $\nu := \lambda \times \mu$ on $S^1 \times X$, where λ is the Lebesgue measure on S^1. Then g is of class $L^p(\nu)$. By the Birkhoff Ergodic Theorem, it follows that

$$\lim_{T\to\infty} \frac{1}{T} \int_0^T g\big(\Psi_t^\omega(z,x)\big)\mathrm{d}t = \lim_{T\to\infty} \frac{1}{T} \int_0^T \mathrm{e}^{\mathrm{i}t\omega} zf(\Phi_t x)\mathrm{d}t = zf_\omega^*(x)$$

exists for ν-almost all $(z,x) \in S^1 \times X$, and that the map $(z,x) \mapsto zf_\omega^*(x)$ is of class $L^p(\nu)$. By Fubini's Theorem [Bau92, Korollar 23.7], this implies that $f_\omega^*(x)$ is of class $L^p(\mu)$.

If $z_0 f_\omega^*(x_0)$ exists for some $(z_0, x_0) \in S^1 \times X$, then clearly also $z f_\omega^*(x_0)$ exists for all $z \in S^1$. Thus f_ω^* exists μ-almost everywhere. $\square$

Note that, in Theorem 2.3.14, the set $\Xi_\omega \subset X$ of points where f_ω^* does not exist depends on ω. We will use the Wiener-Wintner Ergodic Theorem 2.3.18 later to show that, for ergodic systems, Ξ_ω does not depend on ω.

Existence of a finite invariant measure, which is needed in Theorem 2.3.14, can be shown for a fairly large class of systems by the Theorem of Krylov and Bogolubov, see [NS89, Theorem VI.9.05], i. e., for continuous Φ and compact X. Unfortunately, even for those systems, the null set $\Xi_\omega \subset X$ of points, where f_ω^* does not exist, can depend on ω. But if the system is uniquely ergodic, i. e., if there is only one ergodic measure, then the harmonic limit exists for all frequencies $\omega/2\pi$, $\omega \in \mathbb{R}$, and all points $x \in X$ by the following proposition.

Proposition 2.3.15. *Assume that X is compact, and that $\Phi : X \times \mathbb{R}^+ \to X$ and $f : X \to \mathbb{C}$ are continuous. If μ is uniquely ergodic, then $\Xi_\omega = \emptyset$ for all $\omega \in [0, 2\pi)$, i. e., $f_\omega^*(x)$ exists for all $\omega \in [0, 2\pi)$ and all $x \in X$.*

Proof. Compare Theorem 2.1.9. $\qquad\qquad\qquad\qquad\qquad\qquad\qquad\qquad\qquad\square$

Look at the following example for existence.

Example 2.3.16. Let $X := [0, 1]$ and $\Phi_t s := \mathrm{e}^{-t} s$. The point measure $\mathbf{1}_{\{0\}}$ is invariant. Let
$$f(s) := \begin{cases} 0 & \text{if } s = 0, \\ \frac{1}{s} & \text{otherwise.} \end{cases}$$
The map f clearly is $\mathbf{1}_{\{0\}}$-integrable, as $\int_X f \mathrm{d}\mathbf{1}_{\{0\}} = f(0) = 0$. It holds that
$$\mathrm{e}^{\mathrm{i}\omega t} f(\Phi_t s) = \begin{cases} 0 & \text{if } s = 0, \\ \frac{1}{s} \mathrm{e}^{(1+\mathrm{i}\omega)t} & \text{otherwise.} \end{cases}$$
So
$$f_\omega^T(s) = \begin{cases} 0 & \text{if } s = 0, \\ \frac{\mathrm{e}^{(1+\mathrm{i}\omega)T} - 1}{T(1+\mathrm{i}\omega)s} & \text{otherwise,} \end{cases}$$
which tends to 0 for $T \to \infty$ if $s = 0$ and to ∞ otherwise. Thus $f_\omega^*(s)$ only exists if $s = 0$, i. e., it does not exist on the set $(0, 1]$ which has zero measure. $\qquad\qquad\lrcorner$

As already noted, there is a stronger result on existence of the harmonic limit than Theorem 2.3.14 for ergodic systems, known as the Wiener-Wintner Ergodic Theorem. The theorem states that f_ω^* exists almost everywhere, independently of $\omega \in \mathbb{R}$. We give the existence result here, and prove the Wiener-Wintner Ergodic Theorem in the following subsection.

Theorem 2.3.17. *Assume that there is an ergodic invariant measure μ on X. Let $f : X \to \mathbb{C}$ be integrable. Then there is a null set $\Xi \subset X$ such that for all $x \in X \setminus \Xi$, the harmonic limit $f_\omega^*(x)$ exists for all $\omega \in \mathbb{R}$.*

Proof. This follows from the Wiener-Wintner Ergodic Theorem 2.3.18. $\qquad\qquad\square$

2.3.3 Proof of the Wiener-Wintner Ergodic Theorem

As already noted, there is a strong result on existence of the harmonic limit for ergodic systems, known as the Wiener-Wintner Ergodic Theorem. The theorem states that f_ω^* exists almost everywhere, independently of $\omega \in \mathbb{R}$. Wiener and Wintner presented this theorem in 1941 for discrete-time systems, noting that the proof can easily be modified for continuous-time systems, see [WW41]. Unfortunately, their proof was wrong, see [Ass03, p. 24]. The first correct proof was published by Furstenberg in 1960, see [Fur60].

These references deal with the discrete-time problem. In many places, it is stated that the continuous-time version of the Wiener-Wintner Ergodic Theorem can be proved analogously, but an actual proof is hard to find. So we present a complete proof here. To this end, we had to modify the proof in [Ass03, Chapter 2]. In particular, we had to prove a continuous-time version of van der Corput's inequality, see Lemma 2.3.20, and of Wiener's criterion for continuity of a measure, see Lemma 2.3.24.

Theorem 2.3.18 (Continuous-time Wiener-Wintner ergodic theorem). *Let* Φ : $\mathbb{R}^+ \times X \to X$, $(t, x) \mapsto \Phi_t(x)$ *be a semi-flow that is continuous in t. Assume that there is a finite ergodic invariant measure μ. Let $f \in L^1(\mu, \mathbb{C})$. Then there is a null set Ξ_f such that, for all $\omega \in \mathbb{R}$, and all $x \in X \setminus \Xi_f$, the harmonic limit $f_\omega^*(x)$ exists.*

Remark 2.3.19. For the proof of this theorem, it suffices to assume that X is a measurable space. But as we work on metric spaces throughout this thesis we state the theorem for metric spaces here. ⌟

As a first step towards the proof of this theorem, we adapt Van der Corput's inequality [KN74, Lemma 3.1] to the continuous-time case.

Lemma 2.3.20 (Van der Corput's inequality). *Let $T > 0$, and let $u : [0, T] \to \mathbb{C}$ be Lebesgue integrable. Then*

$$S^2 \left| \int_0^T u(t)\mathrm{d}t \right|^2 \leq 2(T + S) \int_0^S (S - p)\Re \int_0^{T-p} u(q)\overline{u(q + p)}\mathrm{d}q\mathrm{d}p$$

holds for all $S \in [0, T]$.

Proof. Let $S \in [0, T]$. Extend u to $\mathbb{R}$ by setting $u(t) = 0$ for $t < 0$ or $t > T$. Then it holds that

$$S \int_0^T u(t)\mathrm{d}t = \int_0^S \int_0^T u(t)\mathrm{d}t\mathrm{d}s = \int_0^S \int_{-s}^{T+S-s} u(t)\mathrm{d}t\mathrm{d}s$$

$$= \int_0^S \int_0^{T+S} u(t - s)\mathrm{d}t\mathrm{d}s = \int_0^{T+S} \int_0^S u(t - s)\mathrm{d}s\mathrm{d}t. \quad (2.3.10)$$

The Cauchy-Schwarz inequality implies

$$\left| \int_0^{T+S} \int_0^S u(t-s) \mathrm{d}s \mathrm{d}t \right|^2 = \left| \int_0^{T+S} \int_0^S u(t-s) \mathrm{d}s \cdot 1 \mathrm{d}t \right|^2$$

$$\leq \int_0^{T+S} \left| \int_0^S u(t-s) \mathrm{d}s \right|^2 \mathrm{d}t \cdot \int_0^{T+S} 1^2 \mathrm{d}t \qquad (2.3.11)$$

$$= \int_0^{T+S} \left| \int_0^S u(t-s) \mathrm{d}s \right|^2 \mathrm{d}t \cdot (T+S).$$

Equations (2.3.10) and (2.3.11) together imply

$$S^2 \left| \int_0^T u(t) \mathrm{d}t \right|^2$$

$$\leq (T+S) \int_0^{T+S} \left| \int_0^S u(t-s) \mathrm{d}s \right|^2 \mathrm{d}t$$

$$= (T+S) \int_0^{T+S} \int_0^S u(t-s) \mathrm{d}s \int_0^S \overline{u(t-r)} \mathrm{d}r \mathrm{d}t \qquad (2.3.12)$$

$$= (T+S) \int_0^{T+S} \int_0^S \int_0^S u(t-s) \overline{u(t-r)} \mathrm{d}r \mathrm{d}s \mathrm{d}t$$

$$= (T+S) \int_0^{T+S} \left(\int_0^S \int_0^s u(t-s) \overline{u(t-r)} \mathrm{d}r \mathrm{d}s + \int_0^S \int_s^S u(t-s) \overline{u(t-r)} \mathrm{d}r \mathrm{d}s \right) \mathrm{d}t$$

$$\overset{(*)}{=} (T+S) \int_0^{T+S} \left(\int_0^S \int_0^s u(t-s) \overline{u(t-r)} \mathrm{d}r \mathrm{d}s + \int_0^S \int_0^r u(t-s) \overline{u(t-r)} \mathrm{d}s \mathrm{d}r \right) \mathrm{d}t$$

$$= (T+S) \int_0^{T+S} \left(\int_0^S \int_0^s u(t-s) \overline{u(t-r)} \mathrm{d}r \mathrm{d}s + \int_0^S \int_0^s u(t-r) \overline{u(t-s)} \mathrm{d}r \mathrm{d}s \right) \mathrm{d}t$$

$$= (T+S) \int_0^{T+S} \left(\int_0^S \int_0^s u(t-s) \overline{u(t-r)} \mathrm{d}r \mathrm{d}s + \int_0^S \int_0^s \overline{u(t-r)u(t-s)} \mathrm{d}r \mathrm{d}s \right) \mathrm{d}t$$

$$= (T+S) \int_0^{T+S} 2\Re \int_0^S \int_0^s u(t-s) \overline{u(t-r)} \mathrm{d}r \mathrm{d}s \mathrm{d}t$$

$$= 2(T+S) \Re \int_0^{T+S} \int_0^S \int_0^s u(t-s) \overline{u(t-r)} \mathrm{d}r \mathrm{d}s \mathrm{d}t,$$

where the equation marked with $(*)$ holds, because

$$\int_0^S \int_s^S u(t-s) \overline{u(t-r)} \mathrm{d}r \mathrm{d}s = \int_0^S \int_0^S \mathbf{1}_{\{r \geq s\}}(r,s) u(t-s) \overline{u(t-r)} \mathrm{d}r \mathrm{d}s$$

$$= \int_0^S \int_0^S \mathbf{1}_{\{r \geq s\}}(r,s) u(t-s) \overline{u(t-r)} \mathrm{d}s \mathrm{d}r$$

$$= \int_0^S \int_0^r u(t-s)\overline{u(t-r)}\mathrm{d}s\mathrm{d}r.$$

By substituting $q := t - s$ and $p := s - r$, one gets

$$\int_0^{T+S} \int_0^S \int_0^s u(t-s)\overline{u(t-r)}\mathrm{d}r\mathrm{d}s\mathrm{d}t$$

$$= \int_0^S \int_0^s \int_0^{T+S} u(t-s)\overline{u(t-r)}\mathrm{d}t\mathrm{d}r\mathrm{d}s$$

$$= \int_0^S \int_0^s \int_s^{T+s} u(t-s)\overline{u(t-r)}\mathrm{d}t\mathrm{d}r\mathrm{d}s$$

$$= \int_0^S \int_0^s \int_0^T u(q)\overline{u(q+s-r)}\mathrm{d}q\mathrm{d}r\mathrm{d}s$$

$$= \int_0^S \int_0^s \int_0^T u(q)\overline{u(q+p)}\mathrm{d}q\mathrm{d}p\mathrm{d}s \qquad (2.3.13)$$

$$= \int_0^S \int_0^S \mathbf{1}_{p\leq s}(p,s) \int_0^T u(q)\overline{u(q+p)}\mathrm{d}q\mathrm{d}p\mathrm{d}s$$

$$= \int_0^S \int_0^S \mathbf{1}_{p\leq s}(p,s) \int_0^T u(q)\overline{u(q+p)}\mathrm{d}q\mathrm{d}s\mathrm{d}p$$

$$= \int_0^S \int_p^S \int_0^T u(q)\overline{u(q+p)}\mathrm{d}q\mathrm{d}s\mathrm{d}p$$

$$= \int_0^S (S-p) \int_0^T u(q)\overline{u(q+p)}\mathrm{d}q\mathrm{d}p$$

$$= \int_0^S (S-p) \int_0^{T-p} u(q)\overline{u(q+p)}\mathrm{d}q\mathrm{d}p.$$

Equations (2.3.12) and (2.3.13) together imply

$$S^2 \left| \int_0^T u(t)\mathrm{d}t \right|^2 \leq 2(T+S)\Re \int_0^S (S-p) \int_0^{T-p} u(q)\overline{u(q+p)}\mathrm{d}q\mathrm{d}p$$

$$= 2(T+S) \int_0^S (S-p)\Re \int_0^{T-p} u(q)\overline{u(q+p)}\mathrm{d}q\mathrm{d}p. \qquad \square$$

For the proof of Theorem 2.3.18, we need the following corollary of van der Corput's inequality.

Corollary 2.3.21. *Let $T > 0$, and let $u : [0,T] \to \mathbb{C}$ be Lebesgue integrable. Then*

$$\sup_{\omega \in \mathbb{R}} \left| \frac{1}{T} \int_0^T e^{i\omega t} u(t)\mathrm{d}t \right|^2 \leq \frac{4}{S} \int_0^S \left| \frac{1}{T} \int_0^{T-p} u(q)\overline{u(q+p)}\mathrm{d}q \right| \mathrm{d}p$$

holds for all $S \in (0,T]$.

Proof. Let $S \in (0, T]$ and $\omega \in \mathbb{R}$. By van der Corput's inequality (Lemma 2.3.20), it holds that

$$S^2 \left| \int_0^T e^{i\omega t} u(t) dt \right|^2$$

$$\leq 2(T+S) \int_0^S (S-p) \Re \int_0^{T-p} e^{i\omega q} u(q) \overline{e^{i\omega(q+p)} u(q+p)} dq dp$$

$$= 2(T+S) \int_0^S (S-p) \Re \int_0^{T-p} e^{-i\omega p} u(q) \overline{u(q+p)} dq dp$$

$$= 2(T+S) \int_0^S (S-p) \Re e^{-i\omega p} \int_0^{T-p} u(q) \overline{u(q+p)} dq dp$$

$$\leq 2 \cdot 2T \int_0^S S \left| \int_0^{T-p} u(q) \overline{u(q+p)} dq \right| dp$$

$$= 4TS \int_0^S \left| \int_0^{T-p} u(q) \overline{u(q+p)} dq \right| dp.$$

Dividing by $S^2 T^2$ yields

$$\left| \frac{1}{T} \int_0^T e^{i\omega t} u(t) dt \right|^2 \leq \frac{4}{S} \int_0^S \left| \frac{1}{T} \int_0^{T-p} u(q) \overline{u(q+p)} dq \right| dp.$$

As the right-hand side does not depend on ω, this is also true for the supremum. □

We will also need Lemma 2.3.23, which is the continuous-time version of [Ass03, Lemma 2.1]. To simplify our formulations, we follow Assani and introduce the Wiener-Wintner property.

Definition 2.3.22 (Wiener-Wintner property). A function $f \in L^1(\mu)$ has the *Wiener-Wintner property* if there is a null set $\Xi \subset X$ such that, for all $\omega \in \mathbb{R}$ and all $x \in X \setminus \Xi$, the harmonic limit $f_\omega^*(x)$ exists. ⌐

This property is preserved under closure in L^1 norm.

Lemma 2.3.23. *Let $f_k \in L^1(\mu)$, $k \in \mathbb{N}$, be converging in L^1 norm to some $f \in L^1(\mu)$. If all maps f_k have the Wiener-Wintner property, then also f has the Wiener-Wintner property.*

Proof. Let $f_k \in L^1(\mu)$, $k \in \mathbb{N}$, be converging in L^1 norm to some $f \in L^1(\mu)$, and assume that all f_k have the Wiener-Wintner property. For every $k \in \mathbb{N}$, let $\Xi_k \subset X$ be a null set such that, for all $\omega \in [0, 2\pi)$ and all $x \in X \setminus \Xi_k$, the harmonic limit $(f_k)_\omega^*(x)$ exists.

In the following, denote by $R_\omega^T : L^1(\mu) \to L^1(\mu)$ the linear operator $g \mapsto \Re(g_\omega^T)$, $\omega \in \mathbb{R}$, $T > 0$. Note that, by linearity of R_ω^T,

$$\limsup_{T \to \infty} R_\omega^T f(x) - \liminf_{T \to \infty} R_\omega^T f(x)$$

$$= \limsup_{T \to \infty} R_\omega^T (f - f_k)(x) - \liminf_{T \to \infty} R_\omega^T (f - f_k)(x)$$

$$+ \limsup_{T \to \infty} R_\omega^T f_k(x) - \liminf_{T \to \infty} R_\omega^T f_k(x)$$

for all $x \in X$, $\omega \in \mathbb{R}$, and $k \in \mathbb{N}$. For all $k \in \mathbb{N}$ and all $x \in X \setminus \Xi_k$, $\omega \in \mathbb{R}$, it follows from the Wiener-Wintner property that

$$\limsup_{T \to \infty} R_\omega^T f(x) - \liminf_{T \to \infty} R_\omega^T f(x) = \limsup_{T \to \infty} R_\omega^T (f - f_k)(x) - \liminf_{T \to \infty} R_\omega^T (f - f_k)(x)$$

$$\leq \sup_{T \to \infty} R_\omega^T (f - f_k)(x) - \inf_{T \to \infty} R_\omega^T (f - f_k)(x)$$

$$\leq 2 \sup_{T \to \infty} \left| R_\omega^T (f - f_k)(x) \right|$$

$$\leq 2 \sup_{T \to \infty} \frac{1}{T} \int_0^T \left| (f - f_k)(\Phi_t x) \right| \mathrm{d}t.$$

Similarly, one can show that, with the linear operator $I_\omega^T : L^1(\mu) \to L^1(\mu)$ given by $g \mapsto \Re(g_\omega^T)$, $\omega \in \mathbb{R}$, $T > 0$, it is true that

$$\limsup_{T \to \infty} I_\omega^T f(x) - \liminf_{T \to \infty} I_\omega^T f(x) \leq 2 \sup_{T \to \infty} \frac{1}{T} \int_0^T \left| (f - f_k)(\Phi_t x) \right| \mathrm{d}t$$

for every $k \in \mathbb{N}$, and all $x \in X \setminus \Xi_k$, $\omega \in \mathbb{R}$. Hence

$$\mu\left(\left\{ x \in X \mid \forall \omega \in \mathbb{R} : \limsup_{T \to \infty} R_\omega^T f(x) - \liminf_{T \to \infty} R_\omega^T f(x) > \lambda \right\} \right)$$

$$\leq \mu\left(\left\{ x \in X \mid 2 \sup_{T \to \infty} \frac{1}{T} \int_0^T \left| (f - f_k)(\Phi_t x) \right| \mathrm{d}t > \lambda \right\} \right)$$

and

$$\mu\left(\left\{ x \in X \mid \forall \omega \in \mathbb{R} : \limsup_{T \to \infty} I_\omega^T f(x) - \liminf_{T \to \infty} I_\omega^T f(x) > \lambda \right\} \right)$$

$$\leq \mu\left(\left\{ x \in X \mid 2 \sup_{T \to \infty} \frac{1}{T} \int_0^T \left| (f - f_k)(\Phi_t x) \right| \mathrm{d}t > \lambda \right\} \right)$$

for all $k \in \mathbb{N}$, $\omega \in \mathbb{R}$, and $\lambda > 0$. By Markov's inequality [Bau92, Lemma 20.1], the right-hand sides of these two inequalities are bounded above by

$$
\frac{1}{\lambda} \int_X 2 \sup_{T \to \infty} \frac{1}{T} \int_0^T \left| (f - f_k)(\Phi_t x) \right| \mathrm{d}t \mathrm{d}\mu(x) = \frac{1}{\lambda} \left\| 2 \sup_{T \to \infty} \frac{1}{T} \int_0^T \left| (f - f_k) \circ \Phi_t \right| \mathrm{d}t \right\|_{L^1}
$$

$$
\leq \frac{2}{\lambda} \sup_{T \to \infty} \frac{1}{T} \int_0^T \left\| (f - f_k) \circ \Phi_t \right\|_{L^1} \mathrm{d}t
$$

$$
= \frac{2}{\lambda} \sup_{T \to \infty} \frac{1}{T} \int_0^T \left\| f - f_k \right\|_{L^1} \mathrm{d}t
$$

$$
= \frac{2}{\lambda} \| f - f_k \|_{L^1},
$$

i. e.,

$$
\mu\big(\{x \in X \mid \forall \omega \in \mathbb{R} : \limsup_{T \to \infty} R_\omega^T f(x) - \liminf_{T \to \infty} R_\omega^T f(x) > \lambda\}\big) \leq \frac{2}{\lambda} \| f - f_k \|_{L^1}
$$

and

$$
\mu\big(\{x \in X \mid \forall \omega \in \mathbb{R} : \limsup_{T \to \infty} I_\omega^T f(x) - \liminf_{T \to \infty} I_\omega^T f(x) > \lambda\}\big) \leq \frac{2}{\lambda} \| f - f_k \|_{L^1}.
$$

As this holds for all $k \in \mathbb{N}$ and $\lim_{k \to \infty} \| f - f_k \|_{L^1} = 0$, it follows that the limits $\lim_{T \to \infty} R_\omega^T f(x)$ and $\lim_{T \to \infty} I_\omega^T f(x)$ exist for all $\omega \in \mathbb{R}$ and all $x \in X \setminus \Xi_f$ for some null set $\Xi_f \subset X$. Hence $f_\omega^*(x) = \lim_{T \to \infty} R_\omega^T f(x) + \lim_{T \to \infty} I_\omega^T f(x)$ exists for all $\omega \in \mathbb{R}$ and all $x \in X \setminus \Xi_f$. $\qquad \square$

As a last step towards the proof of Theorem 2.3.18, we have to adapt the Theorem in [Kat04, Section I.7.13] to the continuous-time case. This is Wiener's criterion for continuity of a measure.

Lemma 2.3.24. *Let μ be a Borel measure on $[0, 2\pi)$, and $\tau \in [0, 2\pi)$. Then it holds that*

$$
\mu(\{\tau\}) = \lim_{T \to \infty} \frac{1}{T} \int_0^T \hat{\mu}(\xi) \mathrm{e}^{\mathrm{i}\tau\xi} \mathrm{d}\xi,
$$

where $\hat{\mu}(\xi) := \int \mathrm{e}^{-\mathrm{i}t\xi} \mathrm{d}\mu(t)$.

Proof. We adapt the proof of the Theorem in [Kat04, Section I.7.13] to the continuous-time case. Let

$$
\phi_T(t) := \frac{1}{T} \int_0^T \mathrm{e}^{\mathrm{i}(\tau - t)\xi} \mathrm{d}\xi.
$$

It holds that $|\phi_T(t)| \leq {}^{1}/_{T} \cdot \int_0^T |\mathrm{e}^{\mathrm{i}(\tau - t)\xi}| \mathrm{d}\xi = {}^{1}/_{T} \cdot \int_0^T 1 \mathrm{d}\xi = 1$. Furthermore, $\phi_T(\tau) = {}^{1}/_{T} \cdot \int_0^T 1 \mathrm{d}\xi = 1$. Additionally, $\lim_{T \to \infty} \phi_T(t) = 0$ uniformly for t outside any neighbourhood of τ. So $\int \mathbf{1}_{\{\tau\}} \mathrm{d}\mu = \lim_{T \to \infty} \int \phi_T \mathrm{d}\mu$. Hence the lemma follows from

$$\mu(\{\tau\}) = \int \mathbf{1}_{\{\tau\}} \mathrm{d}\mu$$

$$= \lim_{T\to\infty} \int \phi_T \mathrm{d}\mu$$

$$= \lim_{T\to\infty} \int \frac{1}{T} \int_0^T \mathrm{e}^{\mathrm{i}(\tau-t)\xi} \mathrm{d}\xi \mathrm{d}\mu(t)$$

$$= \lim_{T\to\infty} \frac{1}{T} \int_0^T \mathrm{e}^{\mathrm{i}\tau\xi} \int \mathrm{e}^{\mathrm{i}-t\xi} \mathrm{d}\mu(t) \mathrm{d}\xi$$

$$= \lim_{T\to\infty} \frac{1}{T} \int_0^T \mathrm{e}^{\mathrm{i}\tau\xi} \hat{\mu}(\xi) \mathrm{d}\xi. \qquad \square$$

Now we finally can prove Theorem 2.3.18. Note that we already use the concept of the Koopman operator and Lemma 2.4.3 here, which we will discuss later in Section 2.4.

Proof of Theorem 2.3.18. We first show, that we can assume f to be in $L^2(\mu)$ without loss of generality. The space $L^2(\mu)$ is dense in $L^1(\mu)$ with respect to the L^1 norm, because μ is finite. So for every $f \in L^1(\mu)$, there is a sequence $(f_k)_{k\in\mathbb{N}} \subset L^2(\mu)$ such that $\|f - f_k\|_{L^1} \to 0$ for $k \to \infty$. Assume that, for every $k \in \mathbb{N}$, there is a null set Ξ_k such that, for all $\omega \in [0, 2\pi)$ and all $x \in X \setminus \Xi_k$, the harmonic limit $(f_k)_\omega^*(x)$ exists. Then by Lemma 2.3.23, the harmonic limit $f_\omega^*(x)$ exists for all $\omega \in \mathbb{R}$ and all $x \in X \setminus \Xi_f$. So it suffices to show the theorem for $f \in L^2(\mu)$.

Consider the Koopman operator U_t on L^2 defined for every $t \geq 0$ by $U_t f := f \circ \Phi_t$. Let

$$\mathcal{K} := \mathrm{cl}\left\{ \sum_{j=1}^n a_j g_j \;\middle|\; n \in \mathbb{N}, a_j \in \mathbb{C}, g_j \text{ eigenfunction of } U \right\}$$

be the closed linear span in L^2 of eigenfunctions of U, i.e., of functions g satisfying $U_t g = \lambda^t g$ for some λ and all $t \geq 0$. Note that all eigenvalues of U have modulus 1, see Lemma 2.4.3.

One can decompose every function $f \in L^2(\mu)$ into $f_1 \in \mathcal{K}$ and $f_2 \in \mathcal{K}^\perp$ such that $f = f_1 + f_2$. It holds that $f_\omega^*(x) = (f_1)_\omega^*(x) + (f_2)_\omega^*(x)$, whereever the limits exist. So we can look at f_1 and f_2 seperately.

By the definition of $\mathcal{K}$, the map f_1 can be approximated in L^2 by functions of the form $\sum_{j=1}^n a_j g_j$ for some $n \in \mathbb{N}$, $a_j \in \mathbb{C}$ and eigenfunctions g_j. We show that every function of this form $\sum_{j=1}^n a_j g_j$ has the Wiener-Wintner property. Then Lemma 2.3.23 implies that f_1 has the Wiener-Wintner property.

Let $n \in \mathbb{N}$, and $a_j \in \mathbb{C}$, $j = 1, \ldots, n$. Let $g_1, \ldots, g_n$ be eigenfunctions of U_t associated with the eigenvalues $\mathrm{e}^{\mathrm{i}\omega_j t}$, $j = 1, \ldots, n$, i.e., $U_t g_j = \mathrm{e}^{\mathrm{i}\omega_j t} g_j$ for every

$j = 1, \ldots, n$, and all $t \geq 0$. Then

$$
\begin{aligned}
\left(\sum_{j=1}^{n} a_j g_j \right)_{\omega}^{*}(x) &= \lim_{T \to \infty} \frac{1}{T} \int_0^T \mathrm{e}^{\mathrm{i}\omega t} \sum_{j=1}^{n} a_j g_j(\Phi_t x) \mathrm{d}t \\
&= \sum_{j=1}^{n} a_j \lim_{T \to \infty} \frac{1}{T} \int_0^T \mathrm{e}^{\mathrm{i}\omega t} g_j(\Phi_t x) \mathrm{d}t \\
&= \sum_{j=1}^{n} a_j \lim_{T \to \infty} \frac{1}{T} \int_0^T \mathrm{e}^{\mathrm{i}\omega t} \mathrm{e}^{\mathrm{i}\omega_j t} g_j(x) \mathrm{d}t \\
&= \sum_{j=1}^{n} a_j g_j(x) \lim_{T \to \infty} \frac{1}{T} \int_0^T \mathrm{e}^{\mathrm{i}(\omega + \omega_j)t} \mathrm{d}t \\
&= \begin{cases} 0 & \text{if } \omega \neq \omega_j, \\ \sum_{j=1}^{n} a_j g_j(x) & \text{if } \omega = \omega_j, \end{cases}
\end{aligned}
$$

i.e., $\sum_{j=1}^{n} a_j g_j(x)$ has the Wiener-Wintner property. So f_1 has the Wiener-Wintner property, i.e., $(f_1)_{\omega}^{*}(x)$ exists independently of ω for almost all x.

Now we show that $(f_2)_{\omega}^{*}(x) = 0$ almost everywhere independently of ω, or more precisely that

$$
\lim_{T \to \infty} \sup_{\omega \in \mathbb{R}} \left| \frac{1}{T} \int_0^T \mathrm{e}^{\mathrm{i}\omega t} f_2(\Phi_t x) \mathrm{d}t \right| = 0 \tag{2.3.14}
$$

for almost all $x \in X$. Corollary 2.3.21 implies, with $u(t) := f_2(\Phi_t x)$, that

$$
\sup_{\omega \in \mathbb{R}} \left| \frac{1}{T} \int_0^T \mathrm{e}^{\mathrm{i}\omega t} f_2(\Phi_t x) \mathrm{d}t \right|^2 \leq \frac{4}{S} \int_0^S \left| \frac{1}{T} \int_0^{T-p} f_2(\Phi_q x) \overline{f_2(\Phi_{q+p} x)} \mathrm{d}q \right| \mathrm{d}p
$$

holds for all $S \in (0, T]$. So

$$
\begin{aligned}
&\limsup_{T \to \infty} \sup_{\omega \in \mathbb{R}} \left| \frac{1}{T} \int_0^T \mathrm{e}^{\mathrm{i}\omega t} f_2(\Phi_t x) \mathrm{d}t \right|^2 \\
&\leq \lim_{T \to \infty} \frac{4}{S} \int_0^S \left| \frac{1}{T} \int_0^{T-p} f_2(\Phi_q x) \overline{f_2(\Phi_{q+p} x)} \mathrm{d}q \right| \mathrm{d}p \\
&= \frac{4}{S} \int_0^S \left| \lim_{T \to \infty} \frac{1}{T} \int_0^{T-p} f_2(\Phi_q x) \overline{f_2(\Phi_{q+p} x)} \mathrm{d}q \right| \mathrm{d}p \\
&= \frac{4}{S} \int_0^S \left| \lim_{T \to \infty} \frac{1}{T} \int_0^{T-p} f_2(\Phi_q x) \overline{f_2(\Phi_q \Phi_p x)} \mathrm{d}q \right| \mathrm{d}p \\
&= \frac{4}{S} \int_0^S \left| \lim_{T \to \infty} \frac{1}{T} \int_0^{T-p} g(\Phi_q x) \mathrm{d}q \right| \mathrm{d}p
\end{aligned}
$$

holds with $g(x) := f_2(x)\overline{f_2(\Phi_p x)}$, if the limit exists. The Birkhoff Ergodic Theorem ([CFS82, Theorem 1.2.1 and Lemma 1.2.2]) implies

$$\frac{1}{T}\int_0^{T-p} g(\Phi_q x)\mathrm{d}q = \frac{T-p}{T} \cdot \frac{1}{T-p}\int_0^{T-p} g(\Phi_q x)\mathrm{d}q \to \frac{1}{\mu(X)}\int g(x)\mathrm{d}\mu(x)$$

for $T \to \infty$ almost everywhere. So

$$\limsup_{T\to\infty}\sup_{\omega\in\mathbb{R}}\left|\frac{1}{T}\int_0^T e^{i\omega t} f_2(\Phi_t x)\mathrm{d}t\right|^2 \leq \frac{4}{S}\int_0^S \left|\frac{1}{\mu(X)}\int g(x)\mathrm{d}\mu(x)\right|\mathrm{d}p$$

$$= \frac{4}{S\mu(X)}\int_0^S \left|\int f_2(x)\overline{f_2(\Phi_p x)}\mathrm{d}\mu(x)\right|\mathrm{d}p \quad (2.3.15)$$

for μ-almost every $x \in X$.

For $p \in \mathbb{R}$, let

$$a(p) := \begin{cases} \int f_2(x)\overline{f_2(\Phi_p x)}\mathrm{d}\mu(x) = \langle f_2, U_p f_2\rangle & \text{if } p \geq 0, \\ \int f_2(\Phi_{-p} x)\overline{f_2(x)}\mathrm{d}\mu(x) = \langle U_{-p} f_2, f_2\rangle & \text{if } p < 0, \end{cases}$$

where $\langle\cdot,\cdot\rangle$ denotes the inner product in $L^2(\mu)$, and U_p is the Koopman operator given by $U_p f_2(x) := f_2(\Phi_p x)$. Note that $a(p-q) = \langle U_q f_2, U_p f_2\rangle$ for any $p, q \geq 0$, due to invariance of μ. Then for any $N \in \mathbb{N}$, all $\xi_1, \ldots, \xi_N \geq 0$, and all $z_1, \ldots, z_N \in \mathbb{C}$, it holds that

$$\sum_{j=1}^N \sum_{k=1}^N a(\xi_j - \xi_k) z_j \overline{z_k} = \sum_{j=1}^N \sum_{k=1}^N \langle U_{\xi_k} f_2, U_{\xi_j} f_2\rangle z_j \overline{z_k}$$

$$= \left\langle \sum_{k=1}^N U_{\xi_k} f_2 \overline{z_k}, \sum_{j=1}^N U_{\xi_j} f_2 \overline{z_j}\right\rangle$$

$$= \left\|\sum_{j=1}^N U_j f_2 \overline{z_j}\right\|^2$$

$$\geq 0,$$

i.e., the sequence $a(p)$, $p \in \mathbb{R}$, is positive definite in the sense of the definition in [Kat04, Section VI.2.8]. Note that we may assume $\xi_j \geq 0$, $j = 1, \ldots, N$, because otherwise let $\tilde{\xi}_j := \xi_j - \min_{k=1,\ldots,N} \xi_k$, $j = 1, \ldots, N$, and note that $a(\tilde{\xi}_j - \tilde{\xi}_k) = a(\xi_j - \xi_k)$, $j, k = 1, \ldots, N$.

Hence, by the Theorem of Bochner in [Kat04, Section VI.2.8], there is a Borel measure μ_2 on $\mathbb{R}$ such that $a(p)$, $p \in \mathbb{R}$, is the Fourier-Stieltjes transform $\hat{\mu}_2(p)$ of μ_2.

So Equation (2.3.15) implies

$$
\limsup_{T\to\infty} \sup_{\omega\in\mathbb{R}} \left| \frac{1}{T} \int_0^T \mathrm{e}^{\mathrm{i}\omega t} f_2(\Phi_t x)\mathrm{d}t \right|^2 \leq \frac{4}{S\mu(X)} \int_0^S \left| \int f_2(x)\overline{f_2(\Phi_p x)}\mathrm{d}\mu(x) \right| \mathrm{d}p
$$

$$
= \frac{4}{S\mu(X)} \int_0^S |\langle f_2(x), U_p f_2\rangle|\, \mathrm{d}p
$$

$$
= \frac{4}{S\mu(X)} \int_0^S |a(p)|\, \mathrm{d}p
$$

$$
= \frac{4}{S\mu(X)} \int_0^S |\hat{\mu}_2(p)|\, \mathrm{d}p.
$$

By the Theorem in [Kat04, Section VI.2.11] (compare also the Remarks in [Kat04, Section I.7.13]), $\lim_{S\to\infty} 1/s \cdot \int_0^S |\hat{\mu}_2(p)|\,\mathrm{d}p = 0$ holds if μ_2 is continuous. In that case, it follows that

$$
\limsup_{T\to\infty} \sup_{\omega\in\mathbb{R}} \left| \frac{1}{T} \int_0^T \mathrm{e}^{\mathrm{i}\omega t} f_2(\Phi_t x)\mathrm{d}t \right|^2 = \lim_{S\to\infty} \limsup_{T\to\infty} \sup_{\omega\in\mathbb{R}} \left| \frac{1}{T} \int_0^T \mathrm{e}^{\mathrm{i}\omega t} f_2(\Phi_t x)\mathrm{d}t \right|^2
$$

$$
\leq \lim_{S\to\infty} \frac{4}{S\mu(X)} \int_0^S |\hat{\mu}_2(p)|\, \mathrm{d}p = 0.
$$

Hence if μ_2 is continuous, $\lim_{T\to\infty} \sup_{\omega\in\mathbb{R}} \left| 1/T \cdot \int_0^T \mathrm{e}^{\mathrm{i}\omega t} f_2(\Phi_t x)\mathrm{d}t \right|^2 = 0$. So it suffices to show continuity of μ_2.

We show that $\mu_2(\{\tau\}) = 0$ for all $\tau \in \mathbb{R}$. By Lemma 2.3.24, it holds that $\mu_2(\{\tau\}) = \lim_{T\to\infty} 1/T \cdot \int_0^T \hat{\mu}_2(\xi)\mathrm{e}^{\mathrm{i}\xi\tau}\mathrm{d}\xi$. By the construction of μ_2, it follows that

$$
\mu_2(\{\tau\}) = \lim_{T\to\infty} \frac{1}{T} \int_0^T a(\xi)\mathrm{e}^{\mathrm{i}\xi\tau}\mathrm{d}\xi = \lim_{T\to\infty} \frac{1}{T} \int_0^T \langle f_2, U_\xi f_2\rangle \mathrm{e}^{\mathrm{i}\xi\tau}\mathrm{d}\xi.
$$

In order to show that this equals zero, it suffices to show that

$$
p := \lim_{T\to\infty} \frac{1}{T} \int_0^T U_\xi f_2 \mathrm{e}^{\mathrm{i}\xi\tau}\mathrm{d}\xi = 0
$$

almost everywhere.

Define the semi-group of linear operators $\tilde{U}_t : L^2 \to L^2$ by $\tilde{U}_t f := \mathrm{e}^{\mathrm{i}t\tau} U_t f$, $t \geq 0$. It holds that $\|\tilde{U}_t\| = 1$ for all $t \geq 0$, because

$$
\|\tilde{U}_t f\|^2 = \|\mathrm{e}^{\mathrm{i}t\tau} U_t f\|^2 = \|U_t f\|^2 = \int |U_t f|^2 \mathrm{d}\mu = \int |f \circ \Phi_t|^2 \mathrm{d}\mu = \int |f|^2 \mathrm{d}\mu = \|f\|^2
$$

by invariance of μ. So the Mean Ergodic Theorem (Theorem 2.1.10) implies that p is the orthogonal projection of f_2 onto $\{g \in L^2 \mid \forall t \geq 0 : \tilde{U}_t f = f\} = \{g \in L^2 \mid \forall t \geq 0 : U_t f = \mathrm{e}^{-\mathrm{i}t\tau} f\} \subset \mathcal{K}$. As $f_2 \in \mathcal{K}^\perp$, it follows that $p = 0$. $\qquad\square$

2.3.4 Asymptotics

Knowledge of the asymptotic behaviour of the system can help in the analysis of harmonic limits. If, e.g., the trajectory starting in $x \in X$ converges to a fixed point $x_0 \in X$, then $f_\omega^*(x) = f(x_0)$ for continuous f. Similarly, if a trajectory asymptotically approaches a periodic orbit, one can apply results from the following Subsection 2.3.5.

Lemma 2.3.25. Let $T_1 \geq T_0$, and $k, l \in \mathbb{N}$. Let $g : [T_0, \infty) \to \mathbb{C}^{k \times l}$ and $h : [T_1, \infty) \to \mathbb{C}^{k \times l}$ be locally integrable. If $g(t) - h(t) \to 0$ for $t \to \infty$, then it holds for every $\omega \in \mathbb{R}$, that

$$\lim_{T \to \infty} \frac{1}{T} \int_{T_0}^{T} e^{i\omega t} g(t) dt = \lim_{T \to \infty} \frac{1}{T} \int_{T_1}^{T} e^{i\omega t} h(t) dt,$$

provided that one of the limits exists.

Proof. Due to the convergence $g(t) - h(t) \to 0$, for every $\varepsilon > 0$, there is $T_\varepsilon \geq T_1$ such that, for all $t \geq T_\varepsilon$, it holds that $|g(t) - h(t)| < \varepsilon$. So for any $\varepsilon > 0$ and all $T \geq T_\varepsilon$, it follows that

$$\left| \frac{1}{T} \int_{T_\varepsilon}^{T} e^{i\omega t} g(t) dt - \frac{1}{T} \int_{T_\varepsilon}^{T} e^{i\omega t} h(t) dt \right| \leq \frac{1}{T} \int_{T_\varepsilon}^{T} \underbrace{|g(t) - h(t)|}_{< \varepsilon} dt$$

$$< \frac{T - T_\varepsilon}{T} \varepsilon,$$

which tends to ε for $T \to \infty$. This means,

$$\left| \lim_{T \to \infty} \frac{1}{T} \int_{T_\varepsilon}^{T} e^{i\omega t} g(t) dt - \lim_{T \to \infty} \frac{1}{T} \int_{T_\varepsilon}^{T} e^{i\omega t} h(t) dt \right| < \varepsilon, \qquad (2.3.16)$$

if one of the limits exists.

As both g and h are locally integrable, it holds for every $\varepsilon > 0$, that

$$\lim_{T \to \infty} \frac{1}{T} \int_{T_0}^{T_\varepsilon} e^{i\omega t} g(t) dt = 0$$

and

$$\lim_{T \to \infty} \frac{1}{T} \int_{T_1}^{T_\varepsilon} e^{i\omega t} h(t) dt = 0.$$

So (2.3.16) implies

$$\left| \lim_{T \to \infty} \frac{1}{T} \int_{T_0}^{T} e^{i\omega t} g(t) dt - \lim_{T \to \infty} \frac{1}{T} \int_{T_1}^{T} e^{i\omega t} h(t) dt \right| < \varepsilon, \qquad (2.3.17)$$

provided that one of the limits exists. As this is true for every $\varepsilon > 0$, it holds that

$$\lim_{T \to \infty} \frac{1}{T} \int_{T_0}^{T} \mathrm{e}^{\mathrm{i}\omega t} g(t) \mathrm{d}t = \lim_{T \to \infty} \frac{1}{T} \int_{T_1}^{T} \mathrm{e}^{\mathrm{i}\omega t} h(t) \mathrm{d}t,$$

if one of the limits exists. □

Proposition 2.3.26. *Let X be a metric space, $x \in X$ and $T_0 \geq 0$. Let $g : [T_0, \infty) \to \mathbb{C}$ be locally integrable. If $g(t) - f(\Phi_t x) \to 0$ for $t \to \infty$, then it holds for every $\omega \in \mathbb{R}$, that*

$$f_\omega^*(x) = \lim_{T \to \infty} \frac{1}{T} \int_{T_0}^{T} \mathrm{e}^{\mathrm{i}\omega t} g(t) \mathrm{d}t,$$

provided that one of the limits exists.

Proof. This follows from Lemma 2.3.25 by setting $T_1 := 0$, $k = l = 1$ and $h(t) := f(\Phi_t x)$. □

Corollary 2.3.27. *Let X be a metric space and $f : X \to \mathbb{C}$. Assume that f is continuous, and let $x \in X$. If there is $y \in X$ such that $d(\Phi_t x, \Phi_t y) \to 0$ for $t \to \infty$, then $f_\omega^*(x) = f_\omega^*(y)$ for every $\omega \in \mathbb{R}$, provided that one of the limits exists.*

Proof. Let $x, y \in X$ such that $d(\Phi_t x, \Phi_t y) \to 0$. Then by continuity, $f(\Phi_t x) - f(\Phi_t y) \to 0$. So Proposition 2.3.26 with $g(t) := f(\Phi_t y)$ and $T_0 := 0$ implies the assertion. □

2.3.5 Periodicity

For almost periodic and especially quasi-periodic and periodic integrands, one can make use of the following results. Quasi-periodic integrands appear, e. g., in the computation of the harmonic limit for a linear autonomous ordinary differential equation projected onto the unit sphere. This example will be discussed in detail in Subsection 2.5.2. The definitions of almost periodicity and quasi-periodicity, which we use here, can be found in Section 1.1.

First, we give two existence results. Proposition 2.3.28 is concerned with periodic points, where the harmonic limit always exists and can be given without a limit. For almost periodic points, Proposition 2.3.29 also shows existence of the harmonic limit. Note that we have already used the following proposition in Lemma 2.1.8.

Proposition 2.3.28. *If $x \in X$ is such that $\Phi_\tau x = x$ for some $\tau > 0$, then $f_{2k\pi/\tau}^*(x) = 1/\tau \cdot \int_0^\tau \mathrm{e}^{\mathrm{i}t2k\pi/\tau} f(\Phi_t x) \mathrm{d}t$ for every $f : X \to \mathbb{C}$ and all $k \in \mathbb{Z}$. Particularly, $f_{2k\pi/\tau}^*(x)$ exists.*

Proof. Let $x \in X$ and $\tau > 0$ be such that $\Phi_\tau x = x$. Let $k \in \mathbb{Z}$ and $T > 0$, and define $\tilde{T} := \lfloor T/\tau \rfloor \cdot \tau$. Then it holds that

$$
\begin{aligned}
f_{2k\pi/\tau}^T(x) &= \frac{1}{T} \int_0^T e^{\mathrm{i}t2k\pi/\tau} f(\Phi_t x) \mathrm{d}t \\
&= \frac{1}{T} \int_0^{\tilde{T}} e^{\mathrm{i}t2k\pi/\tau} f(\Phi_t x) \mathrm{d}t + \frac{1}{T} \int_{\tilde{T}}^T e^{\mathrm{i}t2k\pi/\tau} f(\Phi_t x) \mathrm{d}t \qquad (2.3.18) \\
&= \frac{1}{T} \sum_{l=0}^{\lfloor T/\tau \rfloor - 1} \int_{l\tau}^{(l+1)\tau} e^{\mathrm{i}t2k\pi/\tau} f(\Phi_t x) \mathrm{d}t + \frac{1}{T} \int_{\tilde{T}}^T e^{\mathrm{i}t2k\pi/\tau} f(\Phi_t x) \mathrm{d}t \\
&\overset{(*)}{=} \frac{1}{T} \sum_{l=0}^{\lfloor T/\tau \rfloor - 1} \int_0^\tau e^{\mathrm{i}t2k\pi/\tau} f(\Phi_t x) \mathrm{d}t + \frac{1}{T} \int_{\tilde{T}}^T e^{\mathrm{i}t2k\pi/\tau} f(\Phi_t x) \mathrm{d}t \\
&= \frac{\lfloor T/\tau \rfloor}{T} \int_0^\tau e^{\mathrm{i}t2k\pi/\tau} f(\Phi_t x) \mathrm{d}t + \frac{1}{T} \int_{\tilde{T}}^T e^{\mathrm{i}t2k\pi/\tau} f(\Phi_t x) \mathrm{d}t,
\end{aligned}
$$

where the equation marked with $(*)$ holds tue to periodicity. Note that

$$
\frac{\lfloor T/\tau \rfloor}{T} = \frac{T/\tau - r}{T} = \frac{1}{\tau} - \frac{r}{T}
$$

for some $r \in [0, 1)$, and hence

$$
\lim_{T \to \infty} \frac{\lfloor T/\tau \rfloor}{T} = \frac{1}{\tau}. \qquad (2.3.19)
$$

Furthermore,

$$
\lim_{T \to \infty} \frac{1}{T} \int_{\tilde{T}}^T e^{\mathrm{i}t2k\pi/\tau} f(\Phi_t x) \mathrm{d}t = 0 \qquad (2.3.20)
$$

due to local integrability. Equations (2.3.18), (2.3.19) and (2.3.20) together imply that $f_{2k\pi/\tau}^*(x) = 1/\tau \int_0^\tau e^{\mathrm{i}t2k\pi/\tau} f(\Phi_t x) \mathrm{d}t$. This particularly means that the harmonic limit exists. $\qquad\square$

Along almost periodic orbits, the harmonic limit also always exists.

Proposition 2.3.29.

1. *If $f : X \to \mathbb{C}$ is continuous, and $x \in X$ is such that $\Phi_t x$ is almost periodic in t, then $f_\omega^*(x)$ exists for all $\omega \in \mathbb{R}$.*

2. *If $x \in X$ and $f : X \to \mathbb{C}$ are such that $f(\Phi_t x)$ is almost periodic in t, then $f_\omega^*(x)$ exists for all $\omega \in \mathbb{R}$.*

Proof. If $\Phi_t x$ is almost periodic, and f is continuous, then $f(\Phi_t x)$ is almost periodic by Lemma 1.1.2. So it suffices to show part 2. of this proposition.

Let $\omega \in \mathbb{R}$. By Theorem IV in [Boh47, p. 38], the map $t \to e^{i\omega t} f(\Phi_t x)$ is almost periodic. So part 2. follows from the Mean Value Theorem for almost periodic functions [Boh47, Section 50]. □

For almost periodic integrands, one can show how fast the limit converges.

Proposition 2.3.30. *Let $x \in X$ and $f : X \to \mathbb{C}$ be such that $f(\Phi_t x)$ is almost periodic in t, and let $\omega \in \mathbb{R}$. Define $P := \sup_{t \in \mathbb{R}} |f(\Phi_t x)| < \infty$, and denote by $L_\varepsilon := L(\varepsilon/2)$ the interval length to $\varepsilon/2$ to the almost periodic function $e^{i\omega t} f(\Phi_t x)$ (compare Definition 1.1.1 and note that L_ε is not uniquely determined). Then $|f_\omega^T(x) - f_\omega^\star(x)| < \varepsilon$ for all $T > 4PL_\varepsilon/\varepsilon$.*

Proof. Compare Equation 4 in [Boh47, p. 42]. □

Recall Proposition 1.3.21, where we have shown a connection between the harmonic limit in discrete time and certain Fourier coefficients. Something similar to the type of harmonic limits we analyze in this chapter already occured there. In fact, the following similar result holds for continuous-time systems.

Proposition 2.3.31. *Let $x \in X$ and $f : X \to \mathbb{C}$ be such that $f(\Phi_t x)$ is almost periodic in t. Let $\alpha_k \in \mathbb{R}$ be the Fourier exponents of $f(\Phi_t x)$, and let $c_k \in \mathbb{C}$ be its Fourier coefficients, $k \in \mathbb{N}_0$, i.e., $f(\Phi_t x) = \sum_{k \in \mathbb{N}_0} c_k e^{i\alpha_k t}$. Assume that the exponents α_k are pairwise different. Then for every $\omega \in \mathbb{R}$,*

$$f_\omega^\star(x) = \begin{cases} c_k & \text{if } k \in \mathbb{N}_0 \text{ is such that } \alpha_k = -\omega, \\ 0 & \text{if there is no such } k. \end{cases}$$

Proof. For every trigonometric polynomial $P(t) = \sum_{k=0}^m b_k e^{i\beta_k t}$ with coefficients $b_k \in \mathbb{C}$, (pairwise different) exponents $\beta_k \in \mathbb{R}$, $k = 0, \ldots, m$, and $\beta_0 = 0$, we have

$$\lim_{T \to \infty} \frac{1}{T} \int_0^T P(t) \mathrm{d}t = b_0, \tag{2.3.21}$$

because

$$\frac{1}{T} \int_0^T e^{i\beta t} \mathrm{d}t = \begin{cases} 1 & \text{if } \beta = 0, \\ \frac{e^{i\beta T} - 1}{Ti\beta} & \text{otherwise,} \end{cases}$$

and hence

$$\lim_{T \to \infty} \frac{1}{T} \int_0^T e^{i\beta t} \mathrm{d}t = \begin{cases} 1 & \text{if } \beta = 0, \\ 0 & \text{otherwise.} \end{cases}$$

We may assume, that $\alpha_0 = -\omega$, by reordering the Fourier exponents and coefficients, and by setting $c_0 = 0$ if necessary. So it suffices to show that $f_\omega^\star(x) = c_0$.

As $e^{it\omega}f(\Phi_t x)$ is almost periodic with Fourier exponents α_k in $\mathbb{R}$, it can be approximated by trigonometric polynomials $P_m(t) = \sum_{k=1}^m b_{k,m}e^{i\alpha_k t}$, $m \in \mathbb{N}$, $b_{k,m} \in \mathbb{C}$, such that $\sup_{t \in \mathbb{R}}|e^{it\omega}f(\Phi_t x) - P_m(t)| \to 0$ and $b_{k,m} \to c_k$ for $m \to \infty$, see [Zha03, Theorem 4.3]. So for every $\varepsilon > 0$, there is $m_\varepsilon \in \mathbb{N}$ such that $\sup_{t \in \mathbb{R}}|e^{it\omega}f(\Phi_t x) - P_{m_\varepsilon}(t)| \le \varepsilon/2$ and $|b_{0,m_\varepsilon} - c_0| \le \varepsilon/2$.

By (2.3.21), we have that

$$
\begin{aligned}
\left|\lim_{T\to\infty}\frac{1}{T}\int_0^T e^{i\omega t}f(\Phi_t x)\mathrm{d}t - c_0\right| &\le \left|\lim_{T\to\infty}\frac{1}{T}\int_0^T e^{i\omega t}f(\Phi_t x)\mathrm{d}t - \lim_{T\to\infty}\frac{1}{T}\int_0^T P_{m_\varepsilon}(t)\mathrm{d}t\right| \\
&\quad + \left|\lim_{T\to\infty}\frac{1}{T}\int_0^T P_{m_\varepsilon}(t)\mathrm{d}t - c_0\right| \\
&\le \lim_{T\to\infty}\frac{1}{T}\left|\int_0^T e^{i\omega t}f(\Phi_t x) - P_{m_\varepsilon}(t)\right|\mathrm{d}t \\
&\quad + |b_{0,m_\varepsilon} - c_0| \\
&\le \frac{\varepsilon}{2} + |b_{0,m_\varepsilon} - c_0| \\
&\le \varepsilon.
\end{aligned}
$$

Hence $f_\omega^*(x) = c_0$. $\qquad\square$

Now we turn to the question, which frequencies can occur at periodic and quasi-periodic orbits. With *(quasi-)periodic orbit* we mean an orbit such that $f(\Phi_t x)$ is (quasi-)periodic (see Definition 1.1.3). In the case, where $f(\Phi_t x)$ is quasi-periodic with periods $\tau_1, \ldots, \tau_n$, the following proposition shows that $f_\omega^*(x) = 0$ if the numbers $\omega, 2\pi/\tau_1, \ldots, 2\pi/\tau_n$ are rationally independent, i.e., if there is no nonzero $(c_0, c_1, \ldots, c_n) \in \mathbb{Q}^{n+1}$ such that $c_0\omega + c_1 2\pi/\tau_1 + \cdots + c_n 2\pi/\tau_n = 0$. To put it the other way round, $f_\omega^*(x) \ne 0$ can only hold if the numbers $\omega, 2\pi/\tau_1, \ldots, 2\pi/\tau_n$ are rationally dependent, i.e., if either $2\pi/\tau_1, \ldots, 2\pi/\tau_n$ are rationally dependent or $\omega = c_1 2\pi/\tau_1 + \cdots + c_n 2\pi/\tau_n$ for some $c_1, \ldots, c_n \in \mathbb{Q}$.

Proposition 2.3.32. *Let $x \in X$ and $f : X \to \mathbb{C}$ be such that $f(\Phi_t x)$ is quasi-periodic in t with a locally Lebesgue integrable generating function and periods τ_j, $j = 1, \ldots, n$, (see Definition 1.1.3). If $\omega \in \mathbb{R}$ is such that the numbers $\omega, 2\pi/\tau_1, \ldots, 2\pi/\tau_n$ are rationally independent, then $f_\omega^*(x) = 0$.*

Proof. Define the flow $\rho_t = (\rho_t^1, \rho_t^2)$ on $[0, 2\pi) \times [0, 2\pi)^n$ by $\rho_t^1(\alpha) = \alpha + \omega t \bmod 2\pi$ and $\rho_t^2(\beta_1, \ldots, \beta_n) = (\beta_1 + 2\pi/\tau_1 \cdot t \bmod 2\pi, \ldots, \beta_n + 2\pi/\tau_n \cdot t \bmod 2\pi)$, which preserves the Lebesgue measure. Note that $[0, 2\pi) \times [0, 2\pi)^n$ can be interpreted as an $(n+1)$-dimensional torus. This flow is ergodic, as $\omega, 2\pi/\tau_1, \ldots, 2\pi/\tau_n$ are rationally independent, compare [LM95, Example 7.7.1]. Note that it even is uniquely ergodic, compare [KH06, Proposition 4.2.3].

Let $Q : \mathbb{R}^n \to \mathbb{C}$ be a locally Lebesgue integrable generating function for $f(\Phi_t x)$, i.e., τ_j-periodic in its j-th component, $j = 1, \ldots, n$, and locally Lebesgue integrable with $Q(t, \ldots, t) = f(\Phi_t x)$. Let $g : [0, 2\pi) \times [0, 2\pi)^n \to \mathbb{C}$ be given by

$$(\alpha, \beta_1, \ldots, \beta_n) \mapsto e^{i\alpha} Q\left(\frac{\tau_1}{2\pi}\beta_1, \ldots, \frac{\tau_n}{2\pi}\beta_n\right).$$

In the following, if $\beta \in \mathbb{R}^n$, interpret $Q(\tau/2\pi \cdot \beta)$ as $Q(\tau_1/2\pi \cdot \beta_1, \ldots, \tau_n/2\pi \cdot \beta_n)$.

The map g is Lebesgue integrable, as it has compact domain, and as Q is locally integrable. Thus by the Ergodic Theorem [CFS82, Theorem 1.2.1 and Lemma 1.2.2] and by Fubini's Theorem [Bau92, Korollar 23.7], it holds for almost all $(\alpha^0, \beta^0) \in [0, 2\pi) \times [0, 2\pi)^n$, that

$$
\begin{aligned}
\lim_{T \to \infty} \frac{1}{T} \int_0^T g\big(\rho_t(\alpha^0, \beta^0)\big) \mathrm{d}t &= \frac{1}{\mu\big([0, 2\pi) \times [0, 2\pi)^n\big)} \int_{[0,2\pi) \times [0,2\pi)^n} g(\alpha, \beta) \mathrm{d}(\alpha, \beta) \\
&= \frac{1}{(2\pi)^{1+n}} \int_{[0,2\pi) \times [0,2\pi)^n} e^{i\alpha} Q\left(\frac{\tau}{2\pi}\beta\right) \mathrm{d}(\alpha, \beta) \\
&= \frac{1}{(2\pi)^{1+n}} \int_{[0,2\pi)^n} \int_0^{2\pi} e^{i\alpha} Q\left(\frac{\tau}{2\pi}\beta\right) \mathrm{d}\alpha \mathrm{d}\beta \\
&= \frac{1}{(2\pi)^{1+n}} \int_{[0,2\pi)^n} \left(\int_0^{2\pi} e^{i\alpha} \mathrm{d}\alpha\right) \cdot Q\left(\frac{\tau}{2\pi}\beta\right) \mathrm{d}\beta = 0, \quad (2.3.22)
\end{aligned}
$$

as $\int_0^{2\pi} e^{i\alpha} \mathrm{d}\alpha = 0$.

As ρ is uniquely ergodic with respect to a Borel measure, (2.3.22) actually holds everywhere, see Theorem 2.1.9. Note that $g\big(\rho_t(0,0)\big) = e^{i\omega t} Q(t, \ldots, t) = e^{i\omega t} f(\Phi_t x)$. So in particular, $f_\omega^*(x) = 0$. $\qquad \square$

In the periodic case, we get a stronger result, i.e., the harmonic limit can only be different from zero if the period is a multiple of $2\pi/\omega$. If, e.g., $x \in X$ and $f : X \to \mathbb{C}$ are such that $f(\Phi_t x)$ is 2π-periodic, then $f_\omega^*(x) \neq 0$ implies $2\pi = k \cdot 2\pi/\omega$ for some $k \in \mathbb{Z}$, i.e., $\omega \in \mathbb{Z}$.

Proposition 2.3.33. *Let $x \in X$, $f : X \to \mathbb{C}$, and $\tau > 0$ be such that $f(\Phi_t x)$ is τ-periodic. If $\omega \in \mathbb{R}$ is such that $\omega\tau/2\pi \notin \mathbb{Z}$, then $f_\omega^*(x) = 0$.*

Proof. By Proposition 2.3.32, $f_\omega^*(x) = 0$ if $\omega, 2\pi/\tau$ are rationally independent, i.e., if $\omega\tau/2\pi \notin \mathbb{Q}$.

Assume that $\omega\tau/2\pi \in \mathbb{Q} \setminus \mathbb{Z}$. Then there are $p \in \mathbb{Z} \setminus \{0\}$, $q \in \mathbb{N} \setminus \{1\}$ coprime such that $q\tau = p \cdot 2\pi/\omega =: \sigma > 0$. Clearly, both $e^{i\omega t}$ and $P(t)$ are σ-periodic in t. Thus

$$
\begin{aligned}
f_\omega^*(x) &= \lim_{T \to \infty} \frac{1}{T} \int_0^T e^{i\omega t} f(\Phi_t x) \mathrm{d}t \\
&= \frac{1}{\sigma} \int_0^\sigma e^{i\omega t} f(\Phi_t x) \mathrm{d}t
\end{aligned}
$$

$$= \frac{1}{\sigma} \sum_{j=0}^{q-1} \int_{j\tau}^{(j+1)\tau} e^{i\omega t} f(\Phi_t x) dt$$

$$= \frac{1}{\sigma} \sum_{j=0}^{q-1} \int_0^\tau e^{i\omega(t+j\tau)} f(\Phi_t x) dt$$

$$= \frac{1}{\sigma} \sum_{j=0}^{q-1} e^{i\omega j\tau} \cdot \int_0^\tau e^{i\omega t} f(\Phi_t x) dt.$$

As $q > 1$, $p \neq 0$, p, q coprime, and $\omega\tau = 2\pi p/q$, it holds that $e^{i\omega\tau} \neq 1$. Then

$$\sum_{j=0}^{q-1} e^{i\omega j\tau} = \frac{e^{i\omega q\tau} - 1}{e^{i\omega\tau} - 1} = \frac{e^{i2\pi p} - 1}{e^{i\omega\tau} - 1} = 0. \qquad \square$$

To illustrate Proposition 2.3.33, consider the following simple examples.

Example 2.3.34. Consider the differential equation

$$\dot{x} = y$$
$$\dot{y} = -x$$

in $\mathbb{R}^2$. Its solution is given by $\Phi_t(x, y) = (x \cos t + y \sin t, y \cos t - x \sin t)$.

Let $f(x, y) := x + 2$. Then $f(\Phi_t(0, 1)) = \sin t + 2$, which is 2π-periodic. It holds that

$$\frac{1}{T} \int_0^T e^{i0t}(\sin t + 2) dt = \frac{1}{T} \int_0^T (\sin t + 2) dt$$

$$= \frac{1}{T}(-\cos t + 2t)|_{t=0}^T$$

$$= \frac{1}{T}(-\cos T + 1) + 2 \to 2$$

for $T \to \infty$, and

$$\frac{1}{T} \int_0^T e^{i(\pm 1)t}(\sin t + 2) dt = \frac{1}{T} \int_0^T e^{i(\pm 1)t} \sin t dt + \frac{2}{T} \int_0^T e^{i(\pm 1)t} dt \to \pm\frac{1}{2}i$$

for $T \to \infty$, by Lemma 2.1.8.

For $\omega \notin \{0, \pm 1\}$, it holds that

$$\frac{1}{T} \int_0^T e^{i\omega t}(\sin t + 2) dt = \frac{1}{T} \int_0^T e^{i\omega t} \sin t dt + \frac{2}{T} \int_0^T e^{i\omega t} dt \to 0$$

for $T \to \infty$, by Lemma 2.1.7.

So in this example, $f_\omega^*(0, 1) \neq 0$ if and only if $\omega\tau/2\pi = \omega \in \{0, \pm 1\}$. ⌟

Example 2.3.35. Consider the differential equation

$$\dot{x}_1 = x_2$$
$$\dot{x}_2 = -x_1$$
$$\dot{y}_1 = y_2$$
$$\dot{y}_2 = -4y_1$$

in $\mathbb{R}^4$. Its solution is given by $\Phi_t(x_1, x_2, y_1, y_2) = \big(x_1 \cos t + x_2 \sin t, x_2 \cos t - x_1 \sin t, \frac{1}{2}y_2 \sin(2t) + y_1 \cos(2t), y_2 \cos(2t) - 2y_1 \sin(2t)\big)$.

Let $f(x_1, x_2, y_1, y_2) := x_1 + y_1$. Then $f\big(\Phi_t(0, 1, 0, 2)\big) = \sin t + \sin(2t)$, which is 2π-periodic. Then it holds that $f_\omega^* \neq 0$ if and only if $\omega\tau/2\pi = \omega \in \{\pm 1, \pm 2\}$.

To see this, first note that, for $\omega \neq \pm 1$, it holds by Lemma 2.1.7 that

$$\lim_{T \to \infty} \frac{1}{T} \int_0^T e^{i\omega t} \sin t \, dt = 0, \qquad (2.3.23)$$

and that, for $\omega \neq \pm 2$, it holds by Lemma 2.1.7 that

$$\lim_{T \to \infty} \frac{1}{T} \int_0^T e^{i\omega t} \sin 2t \, dt = 0 \qquad (2.3.24)$$

for $T \to \infty$. So it is clear, that $f_\omega^* \neq 0$ can only be true for $\omega \in \{\pm 1, \pm 2\}$.

On the other hand, by Lemma 2.1.8 and equations (2.3.23) and (2.3.24), it follows that $f_\omega^* \neq 0$ for $\omega \in \{\pm 1, \pm 2\}$. ⌋

The harmonic limit can be interpreted as a generalized inverse Fourier transform. In fact, for periodic integrands $f(\Phi_t x)$, the harmonic limits are the Fourier coefficients of $t \mapsto f(\Phi_t x)$, compare [Kat04, Section I.1]. And for almost periodic integrands $f(\Phi_t x)$, the harmonic limit coincides with the inverse Fourier transform, compare [Kat04, Section VI.5]. From this theory, we get the following result.

Lemma 2.3.36. *Let $x \in X$ and $f : X \to \mathbb{C}$ be such that $f(\Phi_t x)$ is almost periodic. Then it holds that $\lim_{T \to \infty} 1/T \cdot \int_0^T |f(\Phi_t x)|^2 dt = \sum_{\omega \in \mathbb{R}} |f_\omega^*(x)|^2$.*

Proof. This is Parseval's Equation, compare [Boh47, Sections 71 and 83; BN66, Theorem 9.11; Wer05, Satz V.4.9]. □

Remark 2.3.37. If $f(\Phi_t x)$ is not almost periodic but merely bounded and measurable, only Bessel's Inequality $\lim_{T \to \infty} 1/T \cdot \int_0^T |f(\Phi_t x)|^2 dt \geq \sum_{\omega \in \mathbb{R}} |f_\omega^*(x)|^2$ holds, compare (2.3.9) and see [BN66, Theorem 9.3 (1)]. ⌋

Proposition 2.3.32 and Proposition 2.3.33 only gave necessary criteria for the harmonic limit $f_\omega^*(x)$ to be different from zero. In Example 2.3.34 and Example 2.3.35, we have seen, that, even in the periodic case, the criterion is not sufficient. The following theorem will fill this gap for periodic points. Under an additional condition, at a τ-periodic point $x \in X$, for every frequency where a non-vanishing harmonic limit can possibly exist by Proposition 2.3.33, i.e., for every ω that is a multiple of $2\pi/\tau$, one can find a continuous function $f : X \to \mathbb{C}$ such that $f_\omega^*(x) \neq 0$.

Theorem 2.3.38. *Let $\omega > 0$. Let X be a metric space, and $p : \mathbb{R} \to X$ a continuous $2\pi/\omega$-periodic function. If there is a nontrivial open interval $J \subset \left[0, 2\pi/\omega\right)$ such that p is injective on J, then for every $k \in \mathbb{Z}$, there is $f : X \to \mathbb{C}$ continuous, such that $\int_0^{2\pi/\omega} \mathrm{e}^{\mathrm{i}k\omega t} f\left(p(t)\right) \mathrm{d}t \neq 0$.*

In particular, if $\Phi_t x = p(t)$ for some $x \in X$, then for every $k \in \mathbb{Z}$, there is $f : X \to \mathbb{C}$ continuous, such that $f^*_{k\omega}(x) \neq 0$.

Remark 2.3.39. The assumption in this theorem, that there is a nontrivial open interval $J \subset \left[0, 2\pi/\omega\right)$ such that p is injective on J, is not automatically satisfied. A trivial counterexample is any constant map p, for which Theorem 2.3.38 clearly does not hold. For a nontrivial counterexample, consider the Moore curve (see [Sag94, Section 2.7]), which is a space-filling curve. ⌟

Proof of Theorem 2.3.38. Let $[a, b] \subset J$ and $\varepsilon > 0$. Then it holds that $p^{-1}\left(p([a + \varepsilon, b - \varepsilon])\right) = [a + \varepsilon, b - \varepsilon] \cup A_\varepsilon$ for closed sets $A_\varepsilon \subset \left[0, 2\pi/\omega\right)$ with $A_\varepsilon \cap [a + \varepsilon, b - \varepsilon] = \emptyset$. To see this, first note that $[a + \varepsilon, b - \varepsilon] \subset p^{-1}\left(p([a + \varepsilon, b - \varepsilon])\right)$. So define $A_\varepsilon := p^{-1}\left(p([a + \varepsilon, b - \varepsilon])\right) \setminus [a + \varepsilon, b - \varepsilon]$, which clearly does not intersect $[a + \varepsilon, b - \varepsilon]$. We have to show that A_ε is closed. Note that $J \cap A_\varepsilon = \emptyset$. Otherwise there would be $t \in J \setminus [a + \varepsilon, b - \varepsilon]$ with $p(t) \in p([a + \varepsilon, b - \varepsilon])$, which would contradict injectivity of p on J. So $A_\varepsilon = p^{-1}\left(p([a + \varepsilon, b - \varepsilon])\right) \setminus J$. Note that $p^{-1}\left(p([a + \varepsilon, b - \varepsilon])\right)$ is closed, as p is continuous. So openness of J implies closedness of A_ε.

Let $k \in \mathbb{Z}$ and $H_\varepsilon := p([a + \varepsilon, b - \varepsilon]) \cup (X \setminus p([a, b]))$. Define a map $h_\varepsilon : H_\varepsilon \to \mathbb{C}$ by

$$h_\varepsilon(x) := \begin{cases} \mathrm{e}^{-\mathrm{i}k\omega t} & \text{if } x = p(t) \text{ for } t \in [a + \varepsilon, b - \varepsilon], \\ 0 & \text{otherwise.} \end{cases}$$

This map is well-defined due to injectivity of p on $[a, b]$, and because the sets $p([a + \varepsilon, b - \varepsilon])$ and $X \setminus p([a, b])$ are disjoint.

The map h_ε is continuous, because $p([a + \varepsilon, b - \varepsilon])$ and $(X \setminus p([a, b]))$ are disjoint, and because $p|_{[a + \varepsilon, b - \varepsilon]}$ is a homeomorphism. So by the Tietze extension theorem [Kön00, p. 24], there is $f_\varepsilon : X \to \mathbb{C}$ continuous, such that $f_\varepsilon|_{H_\varepsilon} = h_\varepsilon$. We may assume that $|f_\varepsilon(x)| \leq 1$ for all $x \in X$, because $|h_\varepsilon(x)| \leq 1$.

Note that $f_\varepsilon\left(p(t)\right) = h_\varepsilon\left(p(t)\right) = \mathrm{e}^{-\mathrm{i}k\omega t}$ for $t \in [a + \varepsilon, b - \varepsilon]$. So for $\varepsilon < {}^{b-a}/_2$,

$$\int_0^{2\pi/\omega} \mathrm{e}^{\mathrm{i}k\omega t} f_\varepsilon\left(p(t)\right) \mathrm{d}t = \int_{[0,a) \cup (b, 2\pi/\omega)} \mathrm{e}^{\mathrm{i}k\omega t} f_\varepsilon\left(p(t)\right) \mathrm{d}t + \int_{[a+\varepsilon, b-\varepsilon]} \mathrm{e}^{\mathrm{i}k\omega t} f_\varepsilon\left(p(t)\right) \mathrm{d}t$$
$$+ \int_{[a,a+\varepsilon) \cup (b-\varepsilon, b]} \mathrm{e}^{\mathrm{i}k\omega t} f_\varepsilon\left(p(t)\right) \mathrm{d}t$$
$$= 0 + \int_{a+\varepsilon}^{b-\varepsilon} \mathrm{e}^{\mathrm{i}k\omega t} \mathrm{e}^{-\mathrm{i}k\omega t} \mathrm{d}t + \int_{[a,a+\varepsilon) \cup (b-\varepsilon, b]} \mathrm{e}^{\mathrm{i}k\omega t} f_\varepsilon\left(p(t)\right) \mathrm{d}t$$

$$= b - a - 2\varepsilon + \underbrace{\int_{[a,a+\varepsilon)\cup(b-\varepsilon,b]} e^{ik\omega t} f_\varepsilon\big(p(t)\big) dt}_{=:I(k,\varepsilon)}.$$

So if $I(k,\varepsilon)$ is small enough, then $\int_0^{2\pi/\omega} e^{ik\omega t} f_\varepsilon\big(p(t)\big) dt \neq 0$.
It holds that

$$
\begin{aligned}
|I(k,\varepsilon)| &= \left| \int_{[a,a+\varepsilon)\cup(b-\varepsilon,b]} e^{ik\omega t} f_\varepsilon\big(p(t)\big) dt \right| \\
&\leq \int_{[a,a+\varepsilon)\cup(b-\varepsilon,b]} \big| f_\varepsilon\big(p(t)\big) \big| dt \\
&\leq \int_{[a,a+\varepsilon)\cup(b-\varepsilon,b]} 1 dt \\
&= 2\varepsilon.
\end{aligned}
$$

So $I(k,\varepsilon)$ can be made arbitrarily small by choosing ε small enough. This proves the first part of the theorem.

The second part follows with Proposition 2.3.28. $\qquad\square$

The following corollary shows, that Theorem 2.3.38 in particular applies to non-constant differentiable maps p, and hence to solutions of differential equations.

Corollary 2.3.40. *Let X be an n-dimensional normed linear space over $\mathbb{K} \in \{\mathbb{R}, \mathbb{C}\}$, $n \in \mathbb{N}$. Let $p : \mathbb{R} \to X$ be $2\pi/\omega$-periodic, $\omega > 0$. If there is a nontrivial open interval $I \subset \big[0, 2\pi/\omega\big)$ such that $p|_I$ is continuously $\mathbb{K}$-differentiable and not constant, then for every $k \in \mathbb{Z}$, there is $f : X \to \mathbb{C}$ continuous, such that $\int_0^{2\pi/\omega} e^{ik\omega t} f\big(p(t)\big) dt \neq 0$.*

Proof. As p is not constant on I, there are $t_0 \in I$ and $j \in \{1,\ldots,n\}$ such that $p_j'(t_0) \neq 0$. So by the inverse function theorem (see [Kön00, Section 3.3]), there is an open interval $J \subset \big[0, 2\pi/\omega\big)$ containing t_0 such that p is injective on J. Hence Theorem 2.3.38 implies the assertion. $\qquad\square$

2.4 The Koopman Operator

Rotational factor maps can also be interpreted as eigenfunctions of the Koopman operator. A system has a rotational factor map to the frequency $\omega/2\pi$, $\omega \in \mathbb{R}$, if and only if $e^{-it\omega}$ is an eigenvalue of the Koopman operator (Proposition 2.4.4). For square-integrable functions f, we will be able to show that the operator $f \mapsto f_\omega^*$ is the orthogonal projection onto the space of eigenfunctions of the Koopman operator associated with eigenvalue $e^{-it\omega}$ (Proposition 2.4.5).

First define the Koopman operator.

Definition 2.4.1 (Koopman operator). Assume that there is a finite invariant measure μ on X. For every $t \geq 0$ and all $1 \leq p \leq \infty$, let $U_t : L^p(\mu, \mathbb{C}) \to L^p(\mu, \mathbb{C})$ be the *Koopman operator associated with* Φ_t given by $U_t f := f \circ \Phi_t$ for all $f \in L^p(\mu, \mathbb{C})$. ⌟

Remark 2.4.2. If μ is not invariant, the Koopman operator does not map L^p into itself for finite p and is thus defined on L^∞ only. But in the case of an invariant measure, the Koopman operator maps every L^p space, $1 \leq p \leq \infty$, into itself: Let $1 \leq p \leq \infty$, and let μ be an invariant measure on X. In general (also without invariance), for every $t \geq 0$ and $f \in L^p(\mu, \mathbb{C})$, it holds that $\int |U_t f|^p \mathrm{d}\mu = \int |f|^p \mathrm{d}\Phi_t(\mu)$, where $\Phi_t(\mu) := \mu \circ \Phi_t^{-1}$. As μ is invariant, $\Phi_t(\mu)(A) = \mu(\Phi_t^{-1} A) = \mu(A)$ for every measurable $A \subset X$ and thus $\Phi_t(\mu) = \mu$. Hence $\int |U_t f|^p \mathrm{d}\mu = \int |f|^p \mathrm{d}\mu$, which implies $U_t f \in L^p(\mu, \mathbb{C})$. ⌟

A result we have already used in the proof of Theorem 2.3.18 is that all eigenvalues of the Koopman operator have modulus 1.

Lemma 2.4.3. *Let* $t \geq 0$ *and* $1 \leq p < \infty$. *The Koopman operator* $U_t : L^p \to L^p$ *is an isometry, and the eigenvalues of the the Koopman operator* U_t *have modulus 1.*

Proof. Let $t \geq 0$ and $1 \leq p < \infty$. By invariance of μ, it holds for every $f \in L^p$, that

$$\|U_t f\|^p = \int |U_t f|^p \mathrm{d}\mu = \int |f \circ \Phi_t|^p \mathrm{d}\mu = \int |f|^p \mathrm{d}\Phi_t(\mu) = \int |f|^p \mathrm{d}\mu = \|f\|^p.$$

So U_t is an isometry. If $f \in L^p$ is an eigenfunction associated to the eigenvalue $\lambda \in \mathbb{C}$, then $\|f\| = \|U_t f\| = \|\lambda f\| = |\lambda|\|f\|$. Hence $|\lambda| = 1$. □

The eigenfunctions of the Koopman operator are related to rotational factor maps in the following way.

Proposition 2.4.4. *Let* $1 \leq p \leq \infty$ *and* $\omega \in \mathbb{R}$. *If* $F \in L^p(\mu, \mathbb{C})$ *is a rotational factor map for* Φ *with frequency* $\omega/2\pi$, *then* F *is an eigenfunction of* U_t *to the eigenvalue* $\mathrm{e}^{\mathrm{i}\omega t}$ *for all* $t \geq 0$. *Conversely, every map* $g \in L^p(\mu, \mathbb{C})$ *that is an eigenfunction of* U_t^p *for all* $t \geq 0$, *associated to the eigenvalue* $\mathrm{e}^{\mathrm{i}\omega t} \in \mathbb{C}$, $\omega \in \mathbb{R}$, *is a rotational factor map for* Φ *to frequency* $\omega/2\pi$.

Proof. Let $1 \leq p \leq \infty$ and $\omega \in \mathbb{R}$. Assume that $F \in L^p(\mu, \mathbb{C})$ is a rotational factor map for Φ with frequency $\omega/2\pi$. Then $U_t F = F \circ \Phi_t = \mathrm{e}^{\mathrm{i}\omega t} \cdot F$ for all $t \geq 0$.

Let $g \in L^p(\mu, \mathbb{C})$, and assume that, for all $t \geq 0$, the map g is an eigenfunction of U_t associated to the eigenvalue $\mathrm{e}^{\mathrm{i}\omega t}$. Then $g \circ \Phi_t = U_t g = \mathrm{e}^{\mathrm{i}\omega t} \cdot g$ for all $t \geq 0$. □

The Mean Ergodic Theorem (see Theorem 2.1.10) gives the following result on the operator $f \mapsto f_\omega^*$ on L^2.

Proposition 2.4.5. *Assume that there is a finite invariant measure* μ *on* X. *The map* $P_\omega : L^2(\mu, \mathbb{C}) \to L^2(\mu, \mathbb{C})$ *given by* $P_\omega f := f_\omega^*$, $\omega \in \mathbb{R}$, *is the orthogonal projection onto* $\{g \in L^2 \mid \forall t \geq 0 : U_t g = \mathrm{e}^{-\mathrm{i}\omega t} g\}$, *i. e., onto the set of functions that are eigenfunctions of* U_t *to the eigenvalue* $\mathrm{e}^{\mathrm{i}\omega t}$ *for all* $t \geq 0$.

Proof. This follows from the Mean Ergodic Theorem, see Theorem 2.1.10. More precisely, we apply this theorem to the Hilbert space L^2 and the semi-group of linear operators $e^{i\omega t}U_t$. Then the theorem implies that P_ω is the orthogonal projection onto

$$\{g \in L^2 \mid \forall t \geq 0 : e^{i\omega t}U_t g = g\} = \{g \in L^2 \mid \forall t \geq 0 : U_t g = e^{-i\omega t}g\}. \qquad \square$$

2.5 Linear ordinary differential equations

A very important class of dynamical systems in continuous time is given by solutions of linear ordinary differential equations $\dot{x} = A(t)x$ in $\mathbb{C}^n$. For these systems, there is a close connection between the occuring frequencies and the system matrix A. For autonomous equations $\dot{x} = Ax$, the imaginary parts of the eigenvalues play an important role here, and in the nonautonomous periodic case, the imaginary parts of the Floquet exponents have to be considered.

Note that, throughout this section, we consider systems given by the projection of linear differential equations onto the unit sphere, i.e., for a solution $\phi(t, x)$ of a linear differential equation, we consider $\Phi_t x := {\phi(t,x)}/{\|\phi(t,x)\|}$, where $\|\cdot\|$ denotes the Euclidean norm. This has the advantage that, with S^{n-1}, we now have a compact state space. Note that, by the following proposition, every rotational factor map for the projected system also is a rotational factor map for the original system.

Proposition 2.5.1. *If $F : S^{n-1} \to \mathbb{C}$, $F \not\equiv 0$, satisfies $F(\Phi_t s) = e^{it\omega}F(s)$ for all $t \geq 0$ and all $s \in S^{n-1}$, then there is a map $G : \mathbb{C}^n \to \mathbb{C}$, $G \not\equiv 0$, that satisfies $G(\phi(t,x)) = e^{it\omega}G(x)$ for all $t \geq 0$ and $x \in \mathbb{C}^n$.*

Proof. As $\phi(t, x)$ is the solution of a linear differential equation, it holds that $\Phi_t x = \Phi_t\left({x}/{\|x\|}\right)$ for all $x \neq 0$.

Define $G(x) := F\left({x}/{\|x\|}\right)$ for $x \neq 0$, and $G(0) := 0$. Then for $x \neq 0$, it holds that $G(\phi(t,x)) = F(\Phi_t x) = F\left(\Phi_t\left({x}/{\|x\|}\right)\right) = e^{it\omega}F\left({x}/{\|x\|}\right) = e^{it\omega}G(x)$. For $x = 0$, it holds that $G(\phi(t,0)) = G(0) = 0 = e^{it\omega}G(0)$. $\qquad \square$

Before we start to analyze the rotational behaviour of Φ, we introduce a special kind of basis of $\mathbb{C}^n$ in Subsection 2.5.1, which we call Jordan chain basis. Using such a basis, we sometimes can avoid having to bring the system matrix into Jordan normal form in examples. Subsections 2.5.2 and 2.5.3 contain the results that we obtained for autonomous and periodic equations, respectively. In both subsections, we first show that the harmonic limit for a continuous function $f : S^{n-1} \to \mathbb{C}$ can be written as a harmonic limit with a quasi-periodic integrand (Proposition 2.5.16 and Proposition 2.5.31). From this, we derive how the occuring frequencies depend on the imaginary parts of the eigenvalues of A in the autonomous case and on the imaginary parts of the Floquet exponents of $A(t)$ in the periodic case.

Notation and Symbols

Let us fix some additional notation that we will use in this section. Note that, in Definition 2.5.7 and Definition 2.5.9, we will introduce some more notation. Remember that there is a general list of notation and symbols on page 197.

E_λ For a matrix $M \in \mathbb{C}^{n \times n}$ and an eigenvalue $\lambda \in \operatorname{spec} M$, we denote by $E_\lambda \subset \mathbb{C}^n$ the generalized eigenspace of M to λ.

$\iota(x)$ For a matrix $M \subset \mathbb{C}^{n \times n}$ and a generalized eigenvector $x \in \mathbb{C}^n$ to the eigenvalue $\lambda \in \operatorname{spec} M$, we denote by $\iota(x) := \min\{k \in \mathbb{N} \mid (M - \lambda I)^k x = 0\}$ the *nilpotency index* of x.

$J(x)$ For a basis $\{x_1, \ldots, x_n\}$ of $\mathbb{C}^n$ and a point $x \in \mathbb{C}^n$, we denote by $J(x) \subset \{1, \ldots, n\}$ the set of indices such that there exist $\alpha_j \in \mathbb{C} \setminus \{0\}$, $j \in J(x)$, with $x = \sum_{j \in J(x)} \alpha_j x_j$.

$J(x, \lambda)$ For a basis $\{x_1, \ldots, x_n\}$ of $\mathbb{C}^n$ consisting of generalized eigenvectors to the eigenvalues $\lambda_j \in \mathbb{C}$, for a point $x \in \mathbb{C}^n$, and for an eigenvalue $\lambda \in \operatorname{spec} M$, we denote by $J(x, \lambda) \subset \{1, \ldots, n\}$ the set of indices such that $\lambda_j = \lambda$ for all $j \in J(x, \lambda)$ and such that there exist $\alpha_j \in \mathbb{C} \setminus \{0\}$, $j \in J(x)$, with $x = \sum_{j \in J(x)} \alpha_j x_j$, i.e., the set $\{j \in J(x) \mid \lambda_j = \lambda\}$.

Π_{λ_0} For a matrix $M \in \mathbb{C}^{n \times n}$ and an eigenvalue $\lambda_0 \in \operatorname{spec} M$, we denote by $\Pi_{\lambda_0} : \mathbb{C}^n \to \mathbb{C}^n$ the projection onto E_{λ_0} along $\sum_{\lambda \neq \lambda_0} E_\lambda$.

$\operatorname{spec}_J M$ For a matrix $M \subset \mathbb{C}^{n \times n}$ with eigenvalues $\lambda_j \in \mathbb{C}$, $j = 1, \ldots, n$, and a set $J \subset \{1, \ldots, n\}$, define $\operatorname{spec}_J M := \{\lambda_j \mid j \in J\}$.

2.5.1 Jordan chain bases

We will use the concept of a Jordan chain basis, which is a generalization of the kind of basis used to construct a Jordan normal form, see [KM03, Section 6.3]. Using such a basis, we sometimes can avoid having to bring the system matrix into Jordan normal form in examples.

Definition 2.5.2. Let $M \subset \mathbb{C}^{n \times n}$ and $x \in \mathbb{C}^n$ be given. A basis $\{x_1, \ldots, x_n\}$ of $\mathbb{C}^n$ is called *Jordan chain basis relative to x*, if

    1. every x_j, $j \in J(x)$, is a generalized eigenvector of M to some eigenvalue $\lambda_j \in \operatorname{spec} M$, and

2. for every $\lambda \in \operatorname{spec}_{J(x)} M$ and all $k \in \mathbb{N}$, it holds that all nonvanishing vectors $(M - \lambda I)^k x_j$, $j \in J(x, \lambda)$, are linearly independent, i.e., the set $\{(M - \lambda I)^k x_j \mid j \in J(x, \lambda)\} \setminus \{0\}$ is linearly independent. ⌟

To illustrate this concept, consider the following example.

Example 2.5.3. Let

$$M := \begin{pmatrix} 1 & 1 & 1 \\ 0 & 1 & 0 \\ 0 & 0 & 1 \end{pmatrix},$$

and

$$x := \begin{pmatrix} 0 \\ 1 \\ 0 \end{pmatrix} = e_2.$$

Then $B_1 := \{x_1 := e_1, x_2 := e_2, x_3 := e_3\}$ is a Jordan chain basis relative to x. To see this, note that $J(x) = \{2\}$, and that e_2 is a generalized eigenvector of M to the eigenvalue 1. ⌟

Jordan chain bases exist for all matrices and all points.

Lemma 2.5.4. *For every matrix $M \subset \mathbb{C}^{n \times n}$ and all points $x \in \mathbb{C}^n$, there is a Jordan chain basis.*

Proof. Let $\{x_1, \ldots, x_n\}$ be a basis of $\mathbb{C}^n$ such that M is in Jordan normal form with respect to this basis. Then $\{x_1, \ldots, x_n\}$ is a Jordan chain basis, compare [KM03, Satz 6.3.11]. To be precise, properties 1. and 2. of Jordan chain bases follow from the construction of the basis in the proof of [KM03, Satz 6.3.11 (d)]. □

In the following, we usually assume that a Jordan chain basis relative to some point is given. By the proof of Lemma 2.5.4, we could assume that the matrix M is given in Jordan normal form instead, and use the standard basis. But when we deal with examples, it can be a nuisance to have to bring the matrix in normal form first. Using Jordan chain bases, we often can avoid this.

The solution of a differential equation $\dot{x} = Mx$, $x(0) = x_0$, is given by $e^{Mt}x_0$. The exponential term can be represented as follows.

Lemma 2.5.5. *Let $M \subset \mathbb{C}^{n \times n}$, and $x \in \mathbb{C}^n$. Let $\{x_1, \ldots, x_n\}$ be a Jordan chain basis relative to x. Denote the eigenvalue associated with x_k by λ_k, $k = 1, \ldots, n$. Then for every eigenvalue $\lambda \in \operatorname{spec}_{J(x)} M$, there are nonzero polynomial vectors $\pi^\lambda \in \mathbb{C}[t]^n$ with*

$$e^{Mt}x = \sum_{\lambda \in \operatorname{spec}_{J(x)} M} e^{\lambda t} \pi^\lambda(t)$$

such that

$$\deg \pi^\lambda = \max_{k \in J(x, \lambda)} \iota(x_k) - 1 \tag{2.5.1}$$

for every $\lambda \in \operatorname{spec}_{J(x)} M$.

Proof. Let $x = \sum_{j \in J(x)} \alpha_j x_j$ with $\alpha_j \in \mathbb{C} \setminus \{0\}$. For $j = 1, \ldots, n$, let $p_j \in \mathbb{C}[t]^n$ be given by

$$p_j(t) := \sum_{k=0}^{\iota(x_j)-1} \frac{1}{k!}(M - \lambda_j I)^k t^k x_j.$$

For every $\lambda \in \mathrm{spec}_{J(x)} M$, let $\pi^\lambda \in \mathbb{C}[t]^n$ be given by $\pi^\lambda := \sum_{k \in J(x,\lambda)} \alpha_k p_k$.
As x_j are generalized eigenvectors, it holds that

$$\mathrm{e}^{Mt} x = \sum_{j \in J(x)} \mathrm{e}^{\lambda_j t} \alpha_j p_j(t) = \sum_{\lambda \in \mathrm{spec}_{J(x)} M} \mathrm{e}^{\lambda t} \pi^\lambda(t).$$

To show (2.5.1), let $\lambda \in \mathrm{spec}_{J(x)} M$ and $\iota_0 := \max_{k \in J(x,\lambda)} \iota(x_k) - 1$. Then

$$\pi^\lambda(t) = \sum_{k \in J(x,\lambda)} \alpha_k \sum_{l=0}^{\iota(x_k)-1} \frac{1}{l!}(M - \lambda_k I)^l t^l x_k$$

$$= \sum_{k \in J(x,\lambda)} \alpha_k \sum_{l=0}^{\iota_0} \frac{1}{l!}(M - \lambda I)^l t^l x_k$$

$$= \sum_{l=0}^{\iota_0} \frac{1}{l!} t^l (M - \lambda I)^l \sum_{k \in J(x,\lambda)} \alpha_k x_k.$$

As $\{x_1, \ldots, x_n\}$ is a Jordan chain basis, it holds that

$$(M - \lambda I)^{\iota_0} \sum_{k \in J(x,\lambda)} \alpha_k x_k \neq 0 \qquad (2.5.2)$$

for every $\lambda \in \mathrm{spec}_{J(x)} M$, which proves (2.5.1). To be precise, by definition of ι_0, there is at least one $k \in J(x, \lambda)$ such that $(M - \lambda I)^{\iota_0} x_k \neq 0$. As $\{x_1, \ldots, x_n\}$ is a Jordan chain basis, the non-vanishing $(M - \lambda I)^{\iota_0} x_k$, $k \in J(x, \lambda)$, are linearly independent. Together, this implies (2.5.2). $\qquad \square$

To illustrate this lemma, look at the following example.

Example 2.5.6. Consider the matrix

$$M := \begin{pmatrix} 1+\mathrm{i} & 2 & 0 & 0 & 0 \\ 0 & 1+\mathrm{i} & 0 & 0 & 0 \\ 0 & 0 & 1 & 0 & 0 \\ 0 & 0 & 0 & 1 & 2 \\ 0 & 0 & 0 & 1 & 1 \end{pmatrix}$$

and the vector

$$x := \begin{pmatrix} 1 \\ 2 \\ 3 \\ 0 \\ 0 \end{pmatrix}.$$

A Jordan chain basis to x is given by the standard basis $\{e_1, \dots, e_n\}$. Then $J(x) = \{1, 2, 3\}$, and $\mathrm{spec}_{J(x)} M = \{1+\mathrm{i}, 1\}$. Furthermore, $J(x, 1+\mathrm{i}) = \{1, 2\}$, and $J(x, 1) = \{3\}$. The relevant nilpotency indices are $\iota(e_1) = 1$, $\iota(e_2) = 2$ and $\iota(e_3) = 1$. It holds that $e^{Mt} x = e^{(1+\mathrm{i})t} \pi^{1+\mathrm{i}}(t) + e^t \pi^1(t)$, where $\pi^{1+\mathrm{i}}(t) = (1+4t)e_1 + 2e_2$ and $\pi^1(t) = 3e_3$. So $\deg \pi^{1+\mathrm{i}} = 1 = \max_{k \in \{1,2\}} \iota(e_k) - 1$, and $\deg \pi^1 = 0 = \iota(e_3) - 1$. ⌟

Let us fix some more notation.

Definition 2.5.7. Let a matrix $M \in \mathbb{C}^{n \times n}$ with eigenvalues $\{\lambda_1, \dots, \lambda_n\}$, a point $x \in \mathbb{C}^n \setminus \{0\}$, and a Jordan chain basis $\{x_1, \dots, x_n\}$ relative to x be given, such that every x_j is a generalized eigenvector of M to the eigenvalue λ_j. Then define

$$\lambda_{\max}(x) := \max\{\Re\lambda \mid \lambda \in \mathrm{spec}_{J(x)} M\},$$

$$J_{\max}(x) := \{j \in J(x) \mid \Re\lambda_j = \lambda_{\max}(x)\},$$

and

$$k_0(x) := \max\{\iota(x_j) \mid j \in J_{\max}(x)\} - 1.$$ ⌟

Proposition 2.5.8. *The quantities $\lambda_{\max}$ and k_0 from Definition 2.5.7 are independent of the choice of the Jordan chain basis $\{x_1, \dots, x_n\}$.*

Proof. Let $x \in \mathbb{C}^n \setminus \{0\}$. Let $\{x_1, \dots, x_n\}$ be a Jordan chain basis, and let $\alpha_j \in \mathbb{C}$, $j = 1, \dots, n$, be such that $x = \sum_{j=1}^n \alpha_j x_j$.

First of all, note that $\mathrm{spec}_{J(x)} M$ is independent of the choice of the basis, as generalized eigenvectors to different eigenvalues are linearly independent. Thus also $\lambda_{\max}(x)$ does not depend on the choice.

Clearly, $J_{\max}(x)$ depends on the order of the basis vectors. But $\mathrm{spec}_{J_{\max}(x)} M$ is independent of the basis, as it equals $\{\lambda \in \mathrm{spec}_{J(x)} M \mid \Re\lambda = \lambda_{\max}(x)\}$, which does not depend on the basis.

To show the independence of $k_0(x)$, note that

$$k_0(x) + 1 = \max\{\max\{\iota(x_j) \mid j \in J(x, \lambda)\} \mid \lambda \in \mathrm{spec}_{J_{\max}(x)} M\}. \qquad (2.5.3)$$

Further note that
$$\max\{\iota(x_j) \mid j \in J(x, \lambda)\} = \iota(\Pi_\lambda x) \qquad (2.5.4)$$
for every $\lambda \in \mathrm{spec}_{J_{\max}(x)} M$. To verify (2.5.4), note that $\Pi_\lambda x = \sum_{j \in J(x,\lambda)} \alpha_j x_j$. Thus $(M - \lambda I)^k \Pi_\lambda x = \sum_{j \in J(x,\lambda)} \alpha_j (M - \lambda I)^k x_j$, which equals zero if and only if all $(M - \lambda I)^k x_j$ equal zero, as the nonvanishing $(M - \lambda I)^k x_j$ are linearly independent. Equations (2.5.3) and (2.5.4) together imply $k_0(x) = \max\{\iota(\Pi_\lambda x) \mid \lambda \in \mathrm{spec}_{J_{\max}(x)} M\} - 1$, which does not depend on the basis. □

101

The map q given in the following definition will be essential in the analysis of harmonic limits, as it can be used to bring the harmonic limit into a simpler form.

Definition 2.5.9. For a matrix $M \in \mathbb{C}^{n \times n}$, a vector $x \in \mathbb{C}^n \setminus \{0\}$, and $t \in \mathbb{R}$, define $q(M, x, t) \in \mathbb{C}^n$ by

$$q(M, x, t) := \sum_{\lambda \in \operatorname{spec}_{J_{\max}(x)} M} \mathrm{e}^{\mathrm{i} \Im \lambda t} (M - \lambda I)^{k_0(x)} \Pi_\lambda x. \qquad \lrcorner$$

Lemma 2.5.10. *If $\{x_1, \ldots, x_n\}$ is a Jordan chain basis to x such that each x_j is a generalized eigenvector of M to the eigenvalue λ_j, then it holds that*

$$q(M, x, t) = \sum_{j \in J_{\max}(x)} \alpha_j \mathrm{e}^{\mathrm{i} \Im \lambda_j t} (M - \lambda_j I)^{k_0(x)} s_j.$$

Consider the following example.

Example 2.5.11. Consider M and x from Example 2.5.6 again. Here $\lambda_{\max}(x) = 1$, $J_{\max}(x) = \{1, 2, 3\}$ and $k_0(x) = 1$. So

$$
\begin{aligned}
q(M, x, t) &= \mathrm{e}^{\mathrm{i}t} \big(M - (1+\mathrm{i})I \big) e_1 + 2\mathrm{e}^{\mathrm{i}t} \big(M - (1+\mathrm{i})I \big) e_2 + 3(M - I) e_3 \\
&= \mathrm{e}^{\mathrm{i}t} \big(M - (1+\mathrm{i})I \big) (e_1 + 2e_2) \\
&= \mathrm{e}^{\mathrm{i}t}
\begin{pmatrix}
0 & 2 & 0 & 0 & 0 \\
0 & 0 & 0 & 0 & 0 \\
0 & 0 & -\mathrm{i} & 0 & 0 \\
0 & 0 & 0 & -\mathrm{i} & 2 \\
0 & 0 & 0 & 1 & -\mathrm{i}
\end{pmatrix}
\begin{pmatrix}
1 \\ 2 \\ 0 \\ 0 \\ 0
\end{pmatrix} \\
&= 4\mathrm{e}^{\mathrm{i}t} e_1.
\end{aligned}
$$

$\lrcorner$

As we will consider $q(M,x,t)/\|q(M,x,t)\|$ later, the following lemma is useful.

Lemma 2.5.12. *For every $M \in \mathbb{C}^{n \times n}$, every $x \in \mathbb{C}^n \setminus \{0\}$, and all $t \in \mathbb{R}$, it holds that $q(M, x, t) \neq 0$.*

Proof. Let $\{x_1, \ldots, x_n\}$ be some Jordan chain basis to x. Then all nonvanishing $(M - \lambda_j)^{k_0(x)} x_j$, $j \in J_{\max}(x)$, are linearly independent. By the definition of k_0, there is at least one $j_0 \in J_{\max}(x)$ such that $(M - \lambda_{j_0})^{k_0(x)} x_{j_0} \neq 0$. Thus $q(M, x, t) \neq 0$. $\qquad \square$

2.5.2 Autonomous equations

We first examine linear *autonomous* ordinary differential equations in $\mathbb{C}^n$ projected onto the complex unit sphere S^{n-1}. Consider the linear autonomous equation

$$\dot{x} = Ax \qquad (2.5.5)$$

in $\mathbb{C}^n$ for some $A \in \mathbb{C}^{n \times n}$. The solution $\phi(t, x)$ of (2.5.5) with initial value $\phi(0, x) = x$ is given by $e^{At}x$. As discussed at the beginning of this section, we project the system onto the unit sphere S^{n-1}, i.e., we consider the flow Φ on S^{n-1} given by

$$\Phi_t s := \frac{e^{At}s}{\|e^{At}s\|} \tag{2.5.6}$$

for $s \in S^{n-1}$. This flow is the solution of the differential equation

$$\dot{s} = (A - s^T As \cdot I)s, \tag{2.5.7}$$

which can easily be shown by differentiating (2.5.6), compare [CK00, p. 142].

Naturally, the question arises, if there is a connection between the frequencies of the rotational factor maps of this system and the imaginary parts of the eigenvalues of the matrix A. We will analyze this question with harmonic limits f_ω^* for continuous functions $f : S^{n-1} \to \mathbb{C}$.

This harmonic limit always exists, as will be shown in Proposition 2.5.16. So by part 3. of Proposition 2.2.12, we can as well let f take values in $\mathbb{C}^n$, $\mathbb{R}$, or $\mathbb{R}^n$, as long as we are only interested in the question if f_ω^* vanishes.

Now we show the connection between $\Phi_t s$ and the map q introduced in Definition 2.5.9.

Lemma 2.5.13. *Let $A \in \mathbb{C}^{n \times n}$, and $x \in \mathbb{C}^n \setminus \{0\}$. Then*

$$\frac{e^{At}x}{\|e^{At}x\|} - \frac{q(A, x, t)}{\|q(A, x, t)\|} \to 0 \tag{2.5.8}$$

for $t \to \infty$.

Proof. Let λ_j, $j = 1, \ldots, n$, be the eigenvalues of A, and let $x_j \in \mathbb{C}^n$, $j = 1, \ldots, n$, be a Jordan chain basis relative to x, such that every x_j is a generalized eigenvector of A to the eigenvalue λ_j. Let $\alpha_j \in \mathbb{C} \setminus \{0\}$ be such that $x = \sum_{j \in J(x)} \alpha_j x_j \in \mathbb{C}^n \setminus \{0\}$, and write $J := J(x)$. Let $\omega \in \mathbb{R}$.

By Lemma 2.5.5, there are nonzero $\pi^\lambda \in \mathbb{C}[t]^n$, $\lambda \in \operatorname{spec}_J A$, such that

$$e^{At}x = \sum_{\lambda \in \operatorname{spec}_J A} e^{\lambda t} \pi^\lambda(t),$$

and $\deg \pi^\lambda = \max_{k \in J(x, \lambda)} \iota(x_k) - 1$. It is clear that $k_0 := k_0(x) = \max\{\deg \pi^\lambda \mid \lambda \in \operatorname{spec}_{J_{\max}(x)} A\}$.

It holds that

$$\lim_{t \to \infty} \frac{\|e^{At}x\| k_0!}{e^{\lambda_{\max}(x)t} \|q(A, x, t)\| t^{k_0}} = 1. \tag{2.5.9}$$

This can be seen by writing

$$
\begin{aligned}
\mathrm{e}^{At}x &= \mathrm{e}^{At}\sum_{j\in J_{\max}(x)}\alpha_j x_j + \mathrm{e}^{At}\sum_{j\in J\setminus J_{\max}(x)}\alpha_j x_j \\
&= \sum_{j\in J_{\max}(x)}\alpha_j \mathrm{e}^{\lambda_j t}\sum_{k=0}^{k_0}\frac{1}{k!}(A-\lambda_j I)^k t^k x_j + \mathrm{e}^{At}\sum_{j\in J\setminus J_{\max}(x)}\alpha_j x_j \\
&= \sum_{j\in J_{\max}(x)}\alpha_j \mathrm{e}^{\lambda_j t}\frac{1}{k_0!}(A-\lambda_j I)^{k_0} t^{k_0} x_j \\
&\quad + \sum_{j\in J_{\max}(x)}\alpha_j \mathrm{e}^{\lambda_j t}\sum_{k=0}^{k_0-1}\frac{1}{k!}(A-\lambda_j I)^k t^k x_j \\
&\quad + \mathrm{e}^{At}\sum_{j\in J\setminus J_{\max}(x)}\alpha_j x_j \\
&= \mathrm{e}^{\lambda_{\max}(x)t}\cdot \underbrace{\sum_{j\in J_{\max}(x)}\alpha_j \mathrm{e}^{\mathrm{i}\Im\lambda_j t}\frac{1}{k_0!}(A-\lambda_j I)^{k_0} t^{k_0} x_j}_{=:E_1} \\
&\quad + \mathrm{e}^{\lambda_{\max}(x)t}\cdot \underbrace{\sum_{j\in J_{\max}(x)}\alpha_j \mathrm{e}^{\mathrm{i}\Im\lambda_j t}\sum_{k=0}^{k_0-1}\frac{1}{k!}(A-\lambda_j I)^k t^k x_j}_{=:E_2} \\
&\quad + \underbrace{\mathrm{e}^{At}\sum_{j\in J\setminus J_{\max}(x)}\alpha_j x_j}_{=:E_3},
\end{aligned}
\tag{2.5.10}
$$

i. e.,

$$
\mathrm{e}^{At}x = \mathrm{e}^{\lambda_{\max}(x)t}\cdot E_1 + \mathrm{e}^{\lambda_{\max}(x)t}\cdot E_2 + E_3.
\tag{2.5.11}
$$

Clearly, $E_1 k_0! = q(A,x,t)t^{k_0}$ and thus

$$
\frac{\mathrm{e}^{\lambda_{\max}(x)t}\|E_1\|k_0!}{\mathrm{e}^{\lambda_{\max}(x)t}\|q(A,x,t)\|t^{k_0}} = 1.
$$

Furthermore,

$$
\frac{E_2 k_0!}{\|q(A,x,t)\|t^{k_0}} = \sum_{j\in J_{\max}(x)}\sum_{k=-k_0}^{-1} b_{j,k}(t)t^k,
$$

where

$$
b_{j,k}(t) := \alpha_j \frac{k_0!}{(k+k_0)!}(A-\lambda_j I)^{k+k_0}x_j \cdot \mathrm{e}^{\Im\lambda_j t}\cdot\frac{1}{\|q(A,x,t)\|}.
$$

The terms $b_{j,k}(t)$ are bounded, thus

$$\lim_{t \to \infty} \frac{\mathrm{e}^{\lambda_{\max}(x)t}\|E_2\|k_0!}{\mathrm{e}^{\lambda_{\max}(x)t}\|q(A,x,t)\|t^{k_0}} = 0. \tag{2.5.12}$$

Finally,

$$\lim_{t \to \infty} \frac{\|E_3\|k_0!}{\mathrm{e}^{\lambda_{\max}(x)t}\|q(A,x,t)\|t^{k_0}} = 0, \tag{2.5.13}$$

as $\Re\lambda_j < \lambda_{\max}(x)$ for $j \in J \setminus J_{\max}(x)$. As $\mathrm{e}^{At}x = \mathrm{e}^{\lambda_{\max}(x)t} \cdot E_1 + \mathrm{e}^{\lambda_{\max}(x)t} \cdot E_2 + E_3$, this implies (2.5.9).

So because $\mathrm{e}^{At}x/\|\mathrm{e}^{At}x\|$ is bounded, it holds that

$$\frac{\mathrm{e}^{At}x}{\|\mathrm{e}^{At}x\|} - \frac{\mathrm{e}^{At}xk_0!}{\mathrm{e}^{\lambda_{\max}(x)t}\|q(A,x,t)\|t^{k_0}} \to 0 \tag{2.5.14}$$

for $t \to \infty$. To see this, note that

$$\frac{\mathrm{e}^{At}xk_0!}{\mathrm{e}^{\lambda_{\max}(x)t}\|q(A,x,t)\|t^{k_0}} = \frac{\mathrm{e}^{At}x}{\|\mathrm{e}^{At}x\|} \cdot \frac{\|\mathrm{e}^{At}x\|k_0!}{\mathrm{e}^{\lambda_{\max}(x)t}\|q(A,x,t)\|t^{k_0}}.$$

As by (2.5.11)

$$
\begin{aligned}
\frac{\mathrm{e}^{At}xk_0!}{\mathrm{e}^{\lambda_{\max}(x)t}\|q(A,x,t)\|t^{k_0}} &= \frac{\mathrm{e}^{\lambda_{\max}(x)t}E_1k_0! + \mathrm{e}^{\lambda_{\max}(x)t}E_2k_0! + E_3k_0!}{\mathrm{e}^{\lambda_{\max}(x)t}\|q(A,x,t)\|t^{k_0}} \\
&= \frac{\mathrm{e}^{\lambda_{\max}(x)t}q(A,x,t)t^{k_0} + \mathrm{e}^{\lambda_{\max}(x)t}E_2k_0! + E_3k_0!}{\mathrm{e}^{\lambda_{\max}(x)t}\|q(A,x,t)\|t^{k_0}}
\end{aligned}
$$

holds, equations (2.5.12) and (2.5.13) imply that

$$
\begin{aligned}
&\frac{\mathrm{e}^{At}xk_0!}{\mathrm{e}^{\lambda_{\max}(x)t}\|q(A,x,t)\|t^{k_0}} - \frac{q(A,x,t)}{\|q(A,x,t)\|} \\
&= \frac{\mathrm{e}^{At}xk_0!}{\mathrm{e}^{\lambda_{\max}(x)t}\|q(A,x,t)\|t^{k_0}} - \frac{\mathrm{e}^{\lambda_{\max}(x)t}q(A,x,t)t^{k_0}}{\mathrm{e}^{\lambda_{\max}(x)t}\|q(A,x,t)\|t^{k_0}} \\
&\to 0
\end{aligned}
\tag{2.5.15}
$$

for $t \to \infty$. Finally, equations (2.5.15) and (2.5.14) imply (2.5.8). $\qquad\square$

The term $q(A,x,t)/\|q(A,x,t)\|$, which appears in Lemma 2.5.13, is in fact a solution of (2.5.7). In general, $q(A,t,x)$ is not a solution of (2.5.5), but $\mathrm{e}^{\lambda_{\max}(x)t}q(A,t,x)$ is.

Lemma 2.5.14. *Let $A \in \mathbb{C}^{n \times n}$ and $x \in \mathbb{C}^n \setminus \{0\}$. Then $\mathrm{e}^{\lambda_{\max}(x)t}q(A,x,t)$ is a solution of (2.5.5).*

Proof. Note that

$$e^{\lambda_{\max}(x)t}q(A,x,t) = e^{\lambda_{\max}(x)t}\sum_{\lambda\in\mathrm{spec}_{J_{\max}(x)}A} e^{i\Im\lambda t}(A-\lambda I)^{k_0(x)}\Pi_\lambda x$$

$$= \sum_{\lambda\in\mathrm{spec}_{J_{\max}(x)}A} e^{\lambda t}(A-\lambda I)^{k_0(x)}\Pi_\lambda x.$$

Let $\tilde{q}(A,x,t) := e^{\lambda_{\max}(x)t}q(A,x,t)$. Then

$$\frac{d}{dt}\tilde{q}(A,x,t) = \sum_{\lambda\in\mathrm{spec}_{J_{\max}(x)}A} \lambda e^{\lambda t}(A-\lambda I)^{k_0(x)}\Pi_\lambda x$$

$$= A\tilde{q}(A,x,t) - A\tilde{q}(A,x,t) + \sum_{\lambda\in\mathrm{spec}_{J_{\max}(x)}A} e^{\lambda t}\lambda I(A-\lambda I)^{k_0(x)}\Pi_\lambda x$$

$$= A\tilde{q}(A,x,t) - \sum_{\lambda\in\mathrm{spec}_{J_{\max}(x)}A} e^{\lambda t}(A-\lambda I)(A-\lambda I)^{k_0(x)}\Pi_\lambda x$$

$$= A\tilde{q}(A,x,t) - \sum_{\lambda\in\mathrm{spec}_{J_{\max}(x)}A} e^{\lambda t}(A-\lambda I)^{k_0(x)+1}\Pi_\lambda x$$

$$= A\tilde{q}(A,x,t),$$

i.e., $\tilde{q}(A,x,t)$ is a solution of (2.5.5). □

Lemma 2.5.15. *Let $A\in\mathbb{C}^{n\times n}$ and $x\in\mathbb{C}^n\setminus\{0\}$. Then*

$$\frac{q(A,x,t)}{\|q(A,x,t)\|}$$

is a solution of (2.5.7).

Proof. By Lemma 2.5.14, $\tilde{q}(A,x,t) := e^{\lambda_{\max}(x)t}q(A,x,t)$ is a solution of (2.5.5). So by [CK00, p. 142], it follows that

$$\frac{\tilde{q}(A,x,t)}{\|\tilde{q}(A,x,t)\|} = \frac{e^{\lambda_{\max}(x)t}q(A,x,t)}{\|e^{\lambda_{\max}(x)t}q(A,x,t)\|} = \frac{q(A,x,t)}{\|q(A,x,t)\|}$$

is a solution of (2.5.7). □

With Lemma 2.5.13, we can prove the following proposition, which shows that the harmonic limit to a continuous function $f : S^{n-1} \to \mathbb{C}$ reduces to a harmonic limit with a quasi-periodic integrand.

Proposition 2.5.16. *Let $A\in\mathbb{C}^{n\times n}$ with eigenvalues λ_j, $j=1,\dots,n$. Then for every $s\in S^{n-1}$, every continuous $f : S^{n-1} \to \mathbb{C}$, and all $\omega\in\mathbb{R}$, the harmonic limit $f_\omega^*(s)$ exists, and it holds that*

$$f_\omega^*(s) = \lim_{T\to\infty}\frac{1}{T}\int_0^T e^{i\omega t}f\left(\frac{q(A,s,t)}{\|q(A,s,t)\|}\right)dt. \qquad (2.5.16)$$

Proof. By Lemma 2.5.13, it holds that

$$\frac{e^{At}s}{\|e^{At}s\|} - \frac{q(A,s,t)}{\|q(A,s,t)\|} \to 0$$

for $t \to \infty$. By continuity of f, and because $e^{At}s/\|e^{At}s\|$ and $q(A,s,t)/\|q(A,s,t)\|$ are bounded, it follows that

$$f\left(\frac{e^{At}s}{\|e^{At}s\|}\right) - f\left(\frac{q(A,s,t)}{\|q(A,s,t)\|}\right) \to 0 \tag{2.5.17}$$

for $t \to \infty$.

As $f\big(q(A,s,t)/\|q(A,s,t)\|\big)$ is quasi-periodic and hence almost periodic (see [IJ90, Section IX.7]), the limit

$$\lim_{T \to \infty} \frac{1}{T} \int_0^T e^{i\omega t} f\left(\frac{q(A,s,t)}{\|q(A,s,t)\|}\right) \mathrm{d}t$$

exists due to Proposition 2.3.29. So by Lemma 2.3.25 and Equation (2.5.17), Equation (2.5.16) and existence of $f_\omega^*(s)$ follow. $\qquad\square$

As by the preceding Theorem, $f_\omega^*(s)$ exists for all continuous $f : S^{n-1} \to \mathbb{C}$, it follows by Proposition 2.2.12 that it does not matter if we let f have values in $\mathbb{C}$, $\mathbb{R}$, $\mathbb{C}^n$ or $\mathbb{R}^n$ for the purpose of determining if $f_\omega^*(s) \neq 0$. In particular, we will consider id_ω^* later, for $\mathrm{id} : S^{n-1} \to S^{n-1}$.

The quasi-periodicity of the integrand in (2.5.16) not only proves existence of the limit, but can also be used to show that f_ω^* vanishes, if ω is rationally independent of the imaginary parts of all eigenvalues that appear in q.

Corollary 2.5.17. *Let $s \in S^{n-1}$ and $\omega \in \mathbb{R}$. Let $\mathrm{spec}_{J_{\max}(s)} A = \{\lambda_1, \ldots, \lambda_m\}$. Let $Z := \{\omega, \Im\lambda_1, \ldots, \Im\lambda_m\} \setminus \{0\}$. If the numbers $1/z$, $z \in Z$, are rationally independent, then $f_\omega^*(s) = 0$ holds for all continuous f.*

Proof. By Proposition 2.5.16, it holds that

$$f_\omega^*(s) = \lim_{T \to \infty} \frac{1}{T} \int_0^T e^{i\omega t} f\left(\frac{q(A,s,t)}{\|q(A,s,t)\|}\right) \mathrm{d}t,$$

where

$$q(A,s,t) = \sum_{j=1}^m e^{i\Im\lambda_j t}(A - \lambda_j I)^{k_0(s)} \Pi_{\lambda_j} s.$$

Assume that the eigenvalues λ_j, $j = 1, \ldots, m$, are ordered such that those with zero imaginary part (if any) appear at the end of the list. Let $\tilde{m} \in \{0, \ldots, m\}$ be such that $\lambda_1, \ldots, \lambda_{\tilde{m}}$ are the eigenvalues with nonzero imaginary part. Then there is $c \in \mathbb{C}^n$ independent of t, such that

$$q(A, s, t) = \sum_{j=1}^{\tilde{m}} e^{i\Im\lambda_j t} (A - \lambda_j I)^{k_0(s)} \Pi_{\lambda_j} s + c.$$

Define $\tilde{q} : \mathbb{R}^{\tilde{m}} \to \mathbb{C}^n$ by

$$\tilde{q}(t_1, \ldots, t_{\tilde{m}}) := \sum_{j=1}^{\tilde{m}} e^{i\Im\lambda_j t_j} (A - \lambda_j I)^{k_0(s)} \Pi_{\lambda_j} s + c,$$

and $Q : \mathbb{R}^{\tilde{m}} \to \mathbb{C}$ by

$$Q(t_1, \ldots, t_{\tilde{m}}) := f\left(\frac{\tilde{q}(t_1, \ldots, t_{\tilde{m}})}{\|\tilde{q}(t_1, \ldots, t_{\tilde{m}})\|} \right).$$

Note that $\tilde{q}(t_1, \ldots, t_{\tilde{m}}) \neq 0$ for all $t_1, \ldots, t_{\tilde{m}} \in \mathbb{R}$, which can be shown analogously to Lemma 2.5.12: Let $\{s_1, \ldots, s_n\}$ be some Jordan chain basis to s. Then all nonvanishing $(A - \lambda_j)^{k_0(s)} s_j$, $j \in J_{\max}(s)$, are linearly independent. By the definition of k_0, there is at least one $j_0 \in J_{\max}(s)$ such that $(A - \lambda_{j_0})^{k_0(s)} s_{j_0} \neq 0$. Thus $\tilde{q}(t_1, \ldots, t_{\tilde{m}}) \neq 0$.

Then Q is continuous and $2\pi/\Im\lambda_j$-periodic in its j-th component, $j = 1, \ldots, \tilde{m}$. Assume that $\omega \neq 0$. By Proposition 2.3.32, it holds that

$$\lim_{T \to \infty} \frac{1}{T} \int_0^T e^{i\omega t} Q(t, \ldots, t) dt = 0$$

if the numbers $2\pi/\omega, 2\pi/\Im\lambda_1, \ldots, 2\pi/\Im\lambda_{\tilde{m}}$ are rationally independent, i.e., if the numbers $1/\omega, 1/\Im\lambda_1, \ldots, 1/\Im\lambda_{\tilde{m}}$ are rationally independent. Similarly, if $\omega = 0$, by Proposition 2.3.32, it holds that

$$\lim_{T \to \infty} \frac{1}{T} \int_0^T e^{i\omega t} Q(t, \ldots, t) dt = \lim_{T \to \infty} \frac{1}{T} \int_0^T Q(t, \ldots, t) dt = 0$$

if the numbers $2\pi/\Im\lambda_1, \ldots, 2\pi/\Im\lambda_{\tilde{m}}$ are rationally independent, i.e., if $1/\Im\lambda_1, \ldots, 1/\Im\lambda_{\tilde{m}}$ are rationally independent. As $\tilde{q}(t, \ldots, t) = q(A, s, t)$, and thus

$$Q(t, \ldots, t) = f\left(\frac{q(A, s, t)}{\|q(A, s, t)\|} \right).$$

This completes the proof. $\qquad\square$

If q is periodic in t, one can show an even stronger result.

Theorem 2.5.18. *Let $s \in S^{n-1}$, and $\omega \in \mathbb{R}$. For a $\tau > 0$, assume that $q(A, s, t)$ is τ-periodic in t, i.e., that $\tau/2\pi \Im \lambda \in \mathbb{Z}$ for all $\lambda \in \operatorname{spec}_{J_{\max}(s)} A$. Further assume that $q(A, s, t)$ is not constant in t, i.e., that there is at least one $\lambda \in \operatorname{spec}_{J_{\max}(s)} A$ with $\Im \lambda \neq 0$. If $\omega \tau/2\pi \notin \mathbb{Z}$, then $f_\omega^\star(s) = 0$ for all continuous $f : S^{n-1} \to \mathbb{C}$. On the other hand, for every $k \in \mathbb{Z}$, there is a continuous function $f : S^{n-1} \to \mathbb{C}$ such that $f_{k\omega}^\star(s) \neq 0$.*

Proof. The first assertion follows from Proposition 2.5.16 and Proposition 2.3.33. The second assertion follows from Proposition 2.5.16 and Theorem 2.3.38. Note that $p(t) := q(A,s,t)/\|q(A,s,t)\|$ is solution of an ordinary autonomous differential equation with locally Lipschitz continuous right-hand side (compare Lemma 2.5.15), and hence is continuously differentiable. Thus, by Corollary 2.3.40, p satisfies the assumptions in Theorem 2.3.38. □

Example 2.5.19. Let

$$A := \begin{pmatrix} -1 & 1 \\ -2 & 1 \end{pmatrix} \text{ and } s := \begin{pmatrix} 1 \\ 2 \end{pmatrix}.$$

The vector s is the sum of the eigenvectors

$$s_\pm := \begin{pmatrix} \frac{1}{2} \mp \frac{1}{2}\mathrm{i} \\ 1 \end{pmatrix}$$

of A to the eigenvalues $\pm\mathrm{i}$, respectively. Note that $k_0(s) = 0$. Thus

$$q(A, s, t) = \mathrm{e}^{\mathrm{i}t} s_+ + \mathrm{e}^{-\mathrm{i}t} s_- = \mathrm{e}^{\mathrm{i}t} s_+ + \overline{\mathrm{e}^{\mathrm{i}t} s_+} = 2\Re(\mathrm{e}^{\mathrm{i}t} s_+)$$

$$= 2\cos t \Re s_+ - 2\sin t \Im s_+ = \cos t \begin{pmatrix} 1 \\ 2 \end{pmatrix} + \sin t \begin{pmatrix} 1 \\ 0 \end{pmatrix},$$

which is 2π-periodic. By Theorem 2.5.18, the harmonic limit $f_\omega^\star(s)$ equals zero, if $\omega \notin \mathbb{Z}$. It holds that $\|q(A, s, t)\|^2 = 5\cos^2 t + 2\cos t \sin t + \sin^2 t = 4\cos^2 t + 2\cos t \sin t + 1 = 2\cos 2t + \sin 2t + 3$. So for $k \in \mathbb{Z}$ and $f = \mathrm{id}$, it holds that

$$\mathrm{id}_k^\star(s) = \frac{1}{2\pi} \int_0^{2\pi} \mathrm{e}^{\mathrm{i}kt} (2\cos 2t + \sin 2t + 3)^{-1/2} \left[\cos t \begin{pmatrix} 1 \\ 2 \end{pmatrix} + \sin t \begin{pmatrix} 1 \\ 0 \end{pmatrix} \right] \mathrm{d}t.$$

This integral can be evaluated numerically, and one finds that not only $\mathrm{id}_{\pm 1}^\star(s) \neq 0$ but, e.g., also $\mathrm{id}_3^\star(s)$. Table 2.2 shows the results for $k = 0, \ldots, 5$. ⌟

We now introduce the concept of monic vectors. Those are vectors, for which the map $q(A, x, t)$ is particularly simple. This is the case if there is only one $\lambda \in \operatorname{spec}_{J(x)} A$ with maximal nilpotency index.

Definition 2.5.20 (Monicity). Let a matrix $M \in \mathbb{C}^{n \times n}$ be given. A vector $x \in \mathbb{C}^n \setminus \{0\}$ is called *monic* if there is exactly one $\lambda \in \operatorname{spec} M$ such that $\iota(\Pi_\lambda x) - 1 = k_0(x)$. The eigenvalue λ is called *monic eigenvalue* of x, and $\Pi_\lambda x$ is called *monic eigenvector* of x. Monic points with monic eigenvalue λ will also be called λ-monic. ⌟

k	$(\mathrm{id}_k^*(s))_1$	$(\mathrm{id}_k^*(s))_2$
0	0	0
1	$0.2311 + 0.3685\mathrm{i}$	$0.5309 + 0.06868\mathrm{i}$
2	0	0
3	$0.002057 - 0.09783\mathrm{i}$	$-0.1062 - 0.02709\mathrm{i}$
4	0	0
5	$-0.01652 + 0.02856\mathrm{i}$	$0.02772 + 0.02244\mathrm{i}$

Table 2.2: Harmonic limits of the linear autonomous ODE from Example 2.5.19 for integer multiples of the imaginary part of an eigenvalue

Remark 2.5.21. This notion of monicity does not coincide with that of monic morphisms or monic polynomials. ⌟

Look at the following two examples for monicity.

Example 2.5.22. In Example 2.5.6, the vector s is $(1 + \mathrm{i})$-monic. To see this, note that $\operatorname{spec} M = \{1 + \mathrm{i}, 1, 1 \pm \sqrt{2}\}$. Furthermore, $\Pi_{1+\mathrm{i}}s = e_1 + 2e_2$, $\Pi_1 s = 3e_3$ and $\Pi_{1\pm\sqrt{2}}s = 0$. As $\iota(e_1) = \iota(e_3) = 1$ and $\iota(e_2) = 2$, this implies $\iota(\Pi_{1+\mathrm{i}}s) = 2$, $\iota(\Pi_1 s) = 1$, and $\iota(\Pi_{1\pm\sqrt{2}}s) = 0$. Thus $1 + \mathrm{i}$ is the only eigenvalue with maximal nilpotency index, which means $(1 + \mathrm{i})$-monicity. ⌟

Example 2.5.23. Generalized eigenvectors are monic. To see this, let $\lambda \in \operatorname{spec} M$ and $x \in E_\lambda$. Clearly, $\Pi_\lambda x = x$ and $\Pi_\mu x = 0$ for $\mu \neq \lambda$. Thus x is λ-monic. ⌟

Proposition 2.5.16 applied at a monic point yields the following result.

Proposition 2.5.24. *Let $A \in \mathbb{C}^{n \times n}$. For every monic $s \in S^{n-1}$ with monic eigenvalue $\lambda_\mathrm{m}(s)$, and all $\omega \in \mathbb{R}$, it holds that*

$$\mathrm{id}_\omega^*(s) \neq 0 \tag{2.5.18}$$

if and only if

$$\omega = -\Im\lambda_\mathrm{m}(s). \tag{2.5.19}$$

Loosely speaking, Proposition 2.5.24 states, that $\mathrm{id}_\omega^*(s)$ does not vanish if and only if $-\omega$ is the imaginary part of an eigenvalue with maximal real part and maximal index. In particular, rotational factor maps exist with frequencies related to these ω.

Proof of Proposition 2.5.24. Let $s \in S^{n-1}$ be monic with monic eigenvalue $\lambda_\mathrm{m} := \lambda_\mathrm{m}(s)$ and monic eigenvector s_m. Then $q(A, s, t) = \mathrm{e}^{\lambda_\mathrm{m} t}(A - \lambda_\mathrm{m} I)^{k_0(s)} s_\mathrm{m}$. Thus

$$\frac{q(A, s, t)}{\|q(A, s, t)\|} = \mathrm{e}^{\mathrm{i}\Im\lambda_\mathrm{m} t} r$$

with $r := {}^{(A-\lambda_{\mathrm{m}} I)^{k_0(s)} s_{\mathrm{m}}} / {\|(A-\lambda_{\mathrm{m}} I)^{k_0(s)} s_{\mathrm{m}}\|}$. It holds that $r \neq 0$ by definition of $k_0(s)$ and s_{m}.

So by Proposition 2.5.16,

$$
\begin{aligned}
\mathrm{id}_\omega^*(s) &= \lim_{T \to \infty} \frac{1}{T} \int_0^T \mathrm{e}^{\mathrm{i}\omega t} \frac{q(A, s, t)}{\|q(A, s, t)\|} \mathrm{d}t \\
&= \lim_{T \to \infty} \frac{1}{T} \int_0^T \mathrm{e}^{\mathrm{i}(\omega + \Im \lambda_{\mathrm{m}})t} \mathrm{d}t \cdot r.
\end{aligned}
$$

If $\omega + \Im\lambda_{\mathrm{m}} = 0$, this implies $\mathrm{id}_\omega^*(s) = r \neq 0$. If $\omega + \Im\lambda_{\mathrm{m}} \neq 0$, it holds that

$$
\int_0^T \mathrm{e}^{\mathrm{i}(\omega + \Im\lambda_{\mathrm{m}})t} \mathrm{d}t = \frac{1}{\mathrm{i}(\omega + \Im\lambda_{\mathrm{m}})} \left(\mathrm{e}^{\mathrm{i}(\omega + \Im\lambda_{\mathrm{m}})T} - 1 \right),
$$

which is bounded in T. So in this case, $\mathrm{id}_\omega^*(s) = 0$. $\qquad\square$

2.5.3 Periodic equations

In this subsection, we consider continuous linear periodic ordinary differential equations, i. e., systems of the form

$$
\dot{x} = A(t)x \tag{2.5.20}
$$

in $\mathbb{C}^n$, where $A : \mathbb{R} \to \mathbb{C}^{n \times n}$ is continuous and τ-periodic, $\tau > 0$.

Let us recall some basic definitions and results from Floquet theory, which gives some insight into the structure of the fundamental solutions of (2.5.20).

Definition 2.5.25 (Characteristic multipliers and exponents). Let $\lambda \in \mathbb{C}$. If there are $P : \mathbb{R} \to \mathbb{C}^{n \times n}$ nonsingular, differentiable and τ-periodic, and $R \in \mathbb{C}^{n \times n}$ such that $P(t)\mathrm{e}^{Rt}$ is a fundamental solution of (2.5.20), then the eigenvalues of $\mathrm{e}^{R\tau}$ are called *characteristic multipliers*, and the eigenvalues of R are called the *characteristic exponents* of the system. ⌟

By the following theorem, a fundamental solution of (2.5.20) can always be given by $P(t)\mathrm{e}^{Rt}$, where $P : \mathbb{R} \to \mathbb{C}^{n \times n}$ is nonsingular, differentiable and τ-periodic, and $R \in \mathbb{C}^{n \times n}$. Note that P and R are not uniquely determined.

Theorem 2.5.26 (Floquet). *Consider the τ-periodic linear differential equation* (2.5.20).

1. *For every fundamental solution $Y(t)$ of (2.5.20), there is a (nonunique) Floquet representation $Y(t) = P(t)e^{Rt}$ with some $P : \mathbb{R} \to \mathbb{C}^{n \times n}$ nonsingular, differentiable and τ-periodic, and $R \in \mathbb{C}^{n \times n}$.*

2. *The characteristic multipliers do not depend on the choice of a fundamental solution or on the Floquet representation.*

3. *The characteristic exponents of (2.5.20) do not depend on the choice of a fundamental solution and are unique up to addition of integer multiples of $2\pi i/\tau$.*

4. *For the fundamental solution $Y_0(t)$ that satisfies $Y_0(0) = I$, $\{ 1/\tau \cdot \log \lambda \mid \lambda \in \operatorname{spec} Y_0(\tau)\}$ are characteristic exponents of (2.5.20).*

Proof. This theorem can be derived from [Har82, Theorem 6.1] and [Hig08, Theorem 1.28]. For the sake of completeness, we provide a full proof

First, recall that every fundamental solution $Y(t)$ can be expressed by the (unique) fundamental solution $Y_0(t)$ that satisfies $Y_0(0) = I$, by $Y(t) = Y_0(t)C$ for some invertible $C \in \mathbb{C}^{n \times n}$, compare [Har82, p. 47]. In fact, $C = Y(0)$. Further note that, if $Y_0(t)$ has a Floquet representation $Y_0(t) = P(t)e^{Rt}$, then it holds for every invertible $C \in \mathbb{C}^{n \times n}$, that $Y_0(t)C = P(t)CC^{-1}e^{Rt}C = P(t)Ce^{C^{-1}RCt}$. As $\operatorname{spec} C^{-1}e^{Rt}C = \operatorname{spec} e^{Rt}$, and $\operatorname{spec} C^{-1}RC = \operatorname{spec} R$, the characteristic multipliers and exponents do not depend on the choice of a fundamental solution. Furthermore, it follows that, in order to prove 1., it suffices to show this assertion for $Y_0(t)$.

Let $M := Y_0(\tau)$. This matrix is called the *monodromy matrix*. Then there is a (nonunique) matrix $\log M \in \mathbb{C}^{n \times n}$ such that

$$e^{\log M} = M, \qquad (2.5.21)$$

compare [Hig08, Theorem 1.27]. Let $R := 1/\tau \log M$, and define $P(t) := Y_0(t)e^{-Rt}$. Then clearly, $P(t)e^{Rt} = Y_0(t)e^{-Rt}e^{Rt} = Y_0(t)$. Furthermore, P is nonsingular and differentiable, as Y_0 and $t \mapsto e^{-Rt}$ are. It remains to show that P is τ-periodic. Note that

$$\frac{\mathrm{d}}{\mathrm{d}t}Y_0(t + \tau) = A(t + \tau)Y_0(t + \tau) = A(t)Y_0(t + \tau).$$

So $Y_0(t + \tau)$ is the unique solution of (2.5.20) with initial value $Y_0(\tau)$, and thus equals $Y_0(t)Y_0(\tau)$. Hence P is τ-periodic, because

$$P(t + \tau) = Y_0(t + \tau)e^{-R(t+\tau)} = Y_0(t)Y_0(\tau)e^{-R\tau}e^{-Rt}$$
$$= Y_0(t)e^{R\tau}e^{-R\tau}e^{-Rt} = Y_0(t)e^{-Rt} = P(t).$$

This proves 1.

For 2., we already have shown that the characteristic multipliers do not depend on the choice of the fundamental solution. Let a Floquet representation $Y_0(t) = P(t)e^{Rt}$ be given. Setting $t = 0$, one gets $I = Y_0(0) = P(0)$. Hence with $t = \tau$, one gets

$$M = Y_0(\tau) = P(\tau)e^{R\tau} = P(0)e^{R\tau} = e^{R\tau}. \tag{2.5.22}$$

So the characteristic multipliers are always given by $\operatorname{spec} M$.

For 3., we already have shown that the characteristic exponents do not depend on the choice of the fundamental solution. So it remains to be shown that they are unique up to addition of integer multiples of $2\pi i/\tau$. By (2.5.22), it holds that $e^{R\tau} = M$ for every Floquet representation by P and R, i.e., the matrix $R\tau$ is a logarithm of M, and conversely, every logarithm $\log M$ gives rise to a Floquet representation with $R := 1/\tau \log M$. So it suffices to show that the eigenvalues of logarithms of M are unique up to addition of integer multiples of $2\pi i$.

Let $\Lambda(M)$ denote the set of all complex $n \times n$ matrices $\log M$ with $e^{\log M} = M$, i.e., the set of all logarithms of M. By [Hig08, Theorem 1.28], $\Lambda(M)$ is given either by the union of all matrices ZDZ^{-1}, $D \in \mathcal{D}_1$, or by the union of all matrices $ZUDU^{-1}Z^{-1}$, $D \in \mathcal{D}_2$, $U \in \mathcal{U}$, depending on the Jordan structure of M, where $Z \in \mathbb{C}^{n \times n}$ is invertible, $\mathcal{U} \subset \mathbb{C}^{n \times n}$ is some set of invertible matrices, and $\mathcal{D}_1$, $\mathcal{D}_2$ are sets of certain block diagonal matrices. It further holds by [Hig08], that the set of all spectra of matrices in $\mathcal{D}_1$ coincides with the set of all spectra of matrices in $\mathcal{D}_2$, and is given by

$$\bigl\{\{\log \lambda_1 + 2\pi i j_1, \ldots, \log \lambda_n + 2\pi i j_n\} \in \mathbb{C}^n \ \big| \ j_1, \ldots, j_n \in \mathbb{Z}\bigr\}, \tag{2.5.23}$$

where $\lambda_1, \ldots, \lambda_n$ are the eigenvalues of M. Hence the set of all spectra of matrices in $\Lambda(M)$ is also given by that set, i.e., the eigenvalues of logarithms of M are unique up to addition of integer multiples of $2\pi i$.

Choosing $j_k = 0$, $k = 1, \ldots, n$, in (2.5.23), proves part 4. $\qquad\square$

Consider the following example for nonuniqueness of the Floquet representation.

Example 2.5.27. As a very simple example for the nonuniqueness of P and R, consider $A(t) := \sin t$, i.e., consider the differential equation $\dot{x} = \sin t \cdot x$ in $\mathbb{C}$. A fundamental solution is given by $e^{-\cos t}$. Hence, possible Floquet representations are $P(t) = e^{-\cos t - k i}$, $R = e^{k i}$, $k \in \mathbb{Z}$, because $e^{-\cos t - k i}e^{k i} = e^{-\cos t}$. $\qquad\lrcorner$

As in the previous section, we consider the flow projected onto the unit sphere, i.e., the flow

$$\Phi_t s := \frac{P(t)e^{Rt}s}{\|P(t)e^{Rt}s\|}. \tag{2.5.24}$$

Again, this projected flow is the solution of a differential equation, i.e., of

$$\dot{s} = \bigl(A(t) - s^T A(t)s \cdot I\bigr)s, \tag{2.5.25}$$

which can be shown analogously to the autonomous case, compare [CK00, p. 142], by replacing A with $A(t)$.

Now we show the connection between $\Phi_t s$ and the map q introduced in Definition 2.5.9.

Lemma 2.5.28. *Let* $A : \mathbb{R} \to \mathbb{C}^{n \times n}$ *be continuous and periodic, and consider the flow* Φ *from* (2.5.24). *Let* $x \in \mathbb{C}^n \setminus \{0\}$. *Then for any Floquet representation by* P *and* R, *it holds that*

$$\frac{P(t)\mathrm{e}^{Rt}x}{\|P(t)\mathrm{e}^{Rt}x\|} - \frac{P(t)q(R,x,t)}{\|P(t)q(R,x,t)\|} \to 0 \qquad (2.5.26)$$

for $t \to \infty$.

Proof. This can be shown analogously to Lemma 2.5.13:

Let λ_j, $j = 1, \ldots, n$, be the eigenvalues of R, and let $x_j \in \mathbb{C}^n$, $j = 1, \ldots, n$, be a Jordan chain basis to x, such that every x_j is a generalized eigenvector of R to the eigenvalue λ_j. Let $\alpha_j \in \mathbb{C} \setminus \{0\}$ be such that $x = \sum_{j \in J(x)} \alpha_j x_j$, and write $J := J(x)$. Let $\omega \in \mathbb{R}$.

By Lemma 2.5.5, there are nonzero $\pi^\lambda \in \mathbb{C}[t]^n$, $\lambda \in \mathrm{spec}_J R$, such that $\mathrm{e}^{Rt}x = \sum_{\lambda \in \mathrm{spec}_J R} \mathrm{e}^{\lambda t} \pi^\lambda(t)$ and $\deg \pi^\lambda = \max_{k \in J(x,\lambda)} \iota(x_k) - 1$. It is clear that $k_0 := k_0(x) = \max\{\deg \pi^\lambda \mid \lambda \in \mathrm{spec}_{J_{\max}(x)} R\}$. It holds that

$$\lim_{t \to \infty} \frac{\|P(t)\mathrm{e}^{Rt}x\| k_0!}{\mathrm{e}^{\lambda_{\max}(x)t}\|P(t)q(R,x,t)\| t^{k_0}} = 1. \qquad (2.5.27)$$

This can be seen by writing

$$\mathrm{e}^{Rt}x = \mathrm{e}^{\lambda_{\max}(x)t} \cdot E_1 + \mathrm{e}^{\lambda_{\max}(x)t} \cdot E_2 + E_3, \qquad (2.5.28)$$

where

$$E_1 := \sum_{j \in J_{\max}(x)} \alpha_j \mathrm{e}^{\mathrm{i}\Im\lambda_j t} \frac{1}{k_0!}(R - \lambda_j I)^{k_0} t^{k_0} x_j,$$

$$E_2 := \sum_{j \in J_{\max}(x)} \alpha_j \mathrm{e}^{\mathrm{i}\Im\lambda_j t} \sum_{k=0}^{k_0-1} \frac{1}{k!}(R - \lambda_j I)^k t^k x_j,$$

and

$$E_3 := \mathrm{e}^{Rt} \sum_{j \in J \setminus J_{\max}(x)} \alpha_j x_j,$$

compare (2.5.10).

Clearly, $E_1 k_0! = q(R,x,t) t^{k_0}$, and thus

$$\frac{\mathrm{e}^{\lambda_{\max}(x)t}\|P(t)E_1\| k_0!}{\mathrm{e}^{\lambda_{\max}(x)t}\|P(t)q(R,x,t)\| t^{k_0}} = 1.$$

Furthermore,

$$\frac{P(t)E_2k_0!}{\|P(t)q(R,x,t)\|t^{k_0}} = \sum_{j\in J_{\max}(x)} \sum_{k=-k_0}^{-1} b_{j,k}(t)t^k,$$

where

$$b_{j,k}(t) := \alpha_j \frac{k_0!}{(k+k_0)!} \frac{P(t)(R-\lambda_j I)^{k+k_0}x_j \cdot e^{\Im \lambda_j t}}{\|P(t)q(R,x,t)\|}.$$

The terms $b_{j,k}(t)$ are bounded, thus

$$\lim_{t\to\infty} \frac{e^{\lambda_{\max}(x)t}\|P(t)E_2\|k_0!}{e^{\lambda_{\max}(x)t}\|P(t)q(R,x,t)\|t^{k_0}} = 0. \tag{2.5.29}$$

Last,

$$\lim_{t\to\infty} \frac{\|P(t)E_3\|k_0!}{e^{\lambda_{\max}(x)t}\|P(t)q(R,x,t)\|t^{k_0}} = 0, \tag{2.5.30}$$

as $\Re\lambda_j < \lambda_{\max}(x)$ for $j \in J \setminus J_{\max}(x)$. By (2.5.28), this implies (2.5.27).

So because $e^{Rt}x/\|e^{Rt}x\|$ is bounded, it holds that

$$\frac{P(t)e^{Rt}x}{\|P(t)e^{Rt}x\|} - \frac{P(t)e^{Rt}xk_0!}{e^{\lambda_{\max}(x)t}\|P(t)q(R,x,t)\|t^{k_0}} \to 0 \tag{2.5.31}$$

for $t \to \infty$. To see this, note that

$$\frac{P(t)e^{Rt}xk_0!}{e^{\lambda_{\max}(x)t}\|P(t)q(R,x,t)\|t^{k_0}} = \frac{P(t)e^{Rt}x}{\|P(t)e^{Rt}x\|} \cdot \frac{\|P(t)e^{Rt}x\|k_0!}{e^{\lambda_{\max}(x)t}\|P(t)q(R,x,t)\|t^{k_0}}.$$

As by (2.5.28),

$$\frac{P(t)e^{Rt}xk_0!}{e^{\lambda_{\max}(x)t}\|P(t)q(R,x,t)\|t^{k_0}}$$
$$= \frac{e^{\lambda_{\max}(x)t}P(t)E_1k_0! + e^{\lambda_{\max}(x)t}P(t)E_2k_0! + P(t)E_3k_0!}{e^{\lambda_{\max}(x)t}\|P(t)q(R,x,t)\|t^{k_0}}$$
$$= \frac{e^{\lambda_{\max}(x)t}P(t)q(R,x,t)t^{k_0} + e^{\lambda_{\max}(x)t}P(t)E_2k_0! + P(t)E_3k_0!}{e^{\lambda_{\max}(x)t}\|P(t)q(R,x,t)\|t^{k_0}}$$

holds, Equations (2.5.29) and (2.5.30) imply that

$$\frac{P(t)e^{Rt}xk_0!}{e^{\lambda_{\max}(x)t}\|P(t)q(R,x,t)\|t^{k_0}} - \frac{P(t)q(R,x,t)}{\|P(t)q(R,x,t)\|}$$
$$= \frac{P(t)e^{Rt}xk_0!}{e^{\lambda_{\max}(x)t}\|P(t)q(R,x,t)\|t^{k_0}} - \frac{e^{\lambda_{\max}(x)t}P(t)q(R,x,t)t^{k_0}}{e^{\lambda_{\max}(x)t}\|P(t)q(R,x,t)\|t^{k_0}} \tag{2.5.32}$$
$$\to 0$$

for $t \to \infty$. Finally, Equations (2.5.32) and (2.5.31) imply (2.5.26). $\qquad\square$

The term $P(t)q(R,x,t)/\|P(t)q(R,x,t)\|$, which appears in Lemma 2.5.28, is in fact a solution of (2.5.25). In general, $P(t)q(R,x,t)$ is not a solution of (2.5.20), but $\mathrm{e}^{\lambda_{\max}(x)t}P(t)q(R,x,t)$ is.

Lemma 2.5.29. *Let* $A : \mathbb{R} \to \mathbb{C}^{n \times n}$ *continuous and periodic, and let a Floquet representation by* P *and* R *be given. Then* $\mathrm{e}^{\lambda_{\max}(x)t}P(t)q(R,x,t)$ *is a solution of* (2.5.20).

Proof. As $P(t)\mathrm{e}^{Rt}$ is a fundamental solution of (2.5.20), the assertion is true if

$$\mathrm{e}^{\lambda_{\max}(x)t}P(t)q(R,x,t) = P(t)\mathrm{e}^{Rt} \cdot \mathrm{e}^{\lambda_{\max}(x)0}P(0)q(R,x,0). \tag{2.5.33}$$

Note that $P(t)$ is invertible and that $P(0) = I$. So (2.5.33) is equivalent to

$$\mathrm{e}^{\lambda_{\max}(x)t}q(R,x,t) = \mathrm{e}^{Rt}q(R,x,0). \tag{2.5.34}$$

Clearly, Equation (2.5.34) is true for $t = 0$. So it suffices to show that the terms on both sides of (2.5.34) solve the differential equation $\dot{x} = Rx$. This is clearly true for the term on the right-hand side. For the other term, this follows from Lemma 2.5.14. $\qquad\square$

Lemma 2.5.30. *Let* $A : \mathbb{R} \to \mathbb{C}^{n \times n}$ *continuous and periodic, and let a Floquet representation by* P *and* R *be given. Then*

$$\frac{P(t)q(R,x,t)}{\|P(t)q(R,x,t)\|}$$

is a solution of (2.5.25).

Proof. By Lemma 2.5.29, $\tilde{q}(R,x,t) := \mathrm{e}^{\lambda_{\max}(x)t}P(t)q(R,x,t)$ is a solution of (2.5.5). So with an argument analogous to [CK00, p. 142], it follows that

$$\frac{\tilde{q}(R,x,t)}{\|\tilde{q}(R,x,t)\|} = \frac{\mathrm{e}^{\lambda_{\max}(x)t}P(t)q(R,x,t)}{\|\mathrm{e}^{\lambda_{\max}(x)t}P(t)q(R,x,t)\|} = \frac{P(t)q(R,x,t)}{\|P(t)q(R,x,t)\|}$$

is a solution of (2.5.25). $\qquad\square$

With Lemma 2.5.28, we can show the following proposition, which shows that the harmonic limit to a continuous function $f : S^{n-1} \to \mathbb{C}$ reduces to a harmonic limit with a quasi-periodic integrand.

Proposition 2.5.31. *Let* $A : \mathbb{R} \to \mathbb{C}^{n \times n}$ *be continuous and* τ-*periodic,* $\tau > 0$, *and consider system* (2.5.20). *Let* $s \in S^{n-1}$ *and* $\omega \in \mathbb{R}$. *Then* $f_\omega^*(s)$ *exists for every continuous* $f : S^{n-1} \to \mathbb{C}$, *and for any Floquet representation by* P *and* R, *it holds that*

$$f_\omega^*(s) = \lim_{T \to \infty} \frac{1}{T} \int_0^T \mathrm{e}^{\mathrm{i}\omega t} f\left(\frac{P(t)q(R,s,t)}{\|P(t)q(R,s,t)\|}\right) \mathrm{d}t. \tag{2.5.35}$$

Proof. By Lemma 2.5.28, it holds that

$$\frac{P(t)\mathrm{e}^{Rt}s}{\|P(t)\mathrm{e}^{Rt}s\|} - \frac{P(t)q(R,s,t)}{\|P(t)q(R,s,t)\|} \to 0$$

for $t \to \infty$. By continuity of f, and because $P(t)\mathrm{e}^{Rt}s/\|P(t)\mathrm{e}^{Rt}s\|$ and $P(t)q(R,s,t)/\|P(t)q(R,s,t)\|$ are bounded, it follows that

$$f\left(\frac{P(t)\mathrm{e}^{Rt}s}{\|P(t)\mathrm{e}^{Rt}s\|}\right) - f\left(\frac{P(t)q(R,s,t)}{\|P(t)q(R,s,t)\|}\right) \to 0 \qquad (2.5.36)$$

for $t \to \infty$.

As $f\left(P(t)q(R,s,t)/\|P(t)q(R,s,t)\|\right)$ is quasi-periodic and hence almost periodic (see [IJ90, Section IX.7]), the limit $\lim_{T\to\infty} 1/T \cdot \int_0^T \mathrm{e}^{\mathrm{i}\omega t}f\left(P(t)q(R,s,t)\|P(t)q(R,s,t)\|^{-1}\right)\mathrm{d}t$ exists due to Proposition 2.3.29. So by Lemma 2.3.25 and Equation (2.5.36), Equation (2.5.35) and existence of $f_\omega^*(s)$ follow. $\qquad\square$

The quasi-periodicity of the integrand in (2.5.35) not only proves existence of the limit, but can also be used to show that f_ω^* vanishes, if ω is rationally independent of the imaginary parts of all eigenvalues that appear in q, and of the period τ in the following sense.

Corollary 2.5.32. *Let a Floquet representation by P and R be given. Let $s \in S^{n-1}$ and $\omega \in \mathbb{R}$. Let $\mathrm{spec}_{J_{\max}(s)}R = \{\lambda_1,\ldots,\lambda_m\}$. Let $Z := \left\{\omega, 2\pi/\tau, \Im\lambda_1,\ldots,\Im\lambda_m\right\} \setminus \{0\}$. If the numbers $1/z$, $z \in Z$, are rationally independent, then $f_\omega^*(s) = 0$ holds for all continuous f.*

Proof. By Proposition 2.5.31, it holds that

$$f_\omega^*(s) = \lim_{T\to\infty}\frac{1}{T}\int_0^T \mathrm{e}^{\mathrm{i}\omega t}f\left(\frac{P(t)q(R,s,t)}{\|P(t)q(R,s,t)\|}\right)\mathrm{d}t,$$

where $q(R,s,t) = \sum_{j=1}^m \mathrm{e}^{\mathrm{i}\Im\lambda_j t}(R - \lambda_j I)^{k_0(s)}\Pi_{\lambda_j}s$. Assume that the eigenvalues λ_j, $j = 1,\ldots,m$, are ordered such that those with zero imaginary part (if any) appear at the end of the list. Let $\tilde{m} \in \{0,\ldots,m\}$ be such that $\lambda_1,\ldots,\lambda_{\tilde{m}}$ are the eigenvalues with nonzero imaginary part. Then there is $c \in \mathbb{C}^n$ independent of t, such that

$$q(R,s,t) = \sum_{j=1}^{\tilde{m}} \mathrm{e}^{\mathrm{i}\Im\lambda_j t}(R - \lambda_j I)^{k_0(s)}\Pi_{\lambda_j}s + c.$$

Define $\tilde{q}: \mathbb{R}^{\tilde{m}} \to \mathbb{C}^n$ by

$$\tilde{q}(t_1,\ldots,t_{\tilde{m}}) := \sum_{j=1}^{\tilde{m}} \mathrm{e}^{\mathrm{i}\Im\lambda_j t_j}(R - \lambda_j I)^{k_0(s)}\Pi_{\lambda_j}s + c,$$

and $Q : \mathbb{R}^{\tilde{m}+1} \to \mathbb{C}$ by

$$Q(t_0,\dots,t_{\tilde{m}}) := f\left(\frac{P(t_0)\tilde{q}(t_1,\dots,t_{\tilde{m}})}{\|P(t_0)\tilde{q}(t_1,\dots,t_{\tilde{m}})\|}\right).$$

Note that $\tilde{q}(t_1,\dots,t_{\tilde{m}}) \neq 0$ for all $t_1,\dots,t_{\tilde{m}} \in \mathbb{R}$, which can be shown analogously to Lemma 2.5.12: Let $\{s_1,\dots,s_n\}$ be some Jordan chain basis to s. Then all nonvanishing $(R-\lambda_j)^{k_0(s)}s_j$, $j \in J_{\max}(s)$, are linearly independent. By the definition of k_0, there is at least one $j_0 \in J_{\max}(s)$ such that $(R-\lambda_{j_0})^{k_0(s)}s_{j_0} \neq 0$. Thus $\tilde{q}(t_1,\dots,t_{\tilde{m}}) \neq 0$.

Then Q is continuous, τ-periodic in its first component and $2\pi/\Im\lambda_j$-periodic in its $(j+1)$-st component, $j = 1,\dots,\tilde{m}$. Assume that $\omega \neq 0$. By Proposition 2.3.32, it holds that if the numbers $2\pi/\omega, \tau, 2\pi/\Im\lambda_1, \dots, 2\pi/\Im\lambda_{\tilde{m}}$ are rationally independent, i.e., if the numbers $1/\omega, \tau/2\pi, 1/\Im\lambda_1, \dots, 1/\Im\lambda_{\tilde{m}}$ are rationally independent. Similarly, if $\omega = 0$, by Proposition 2.3.32, it holds that

$$\lim_{T\to\infty}\frac{1}{T}\int_0^T e^{i\omega t}Q(t,\dots,t)\mathrm{d}t = \lim_{T\to\infty}\frac{1}{T}\int_0^T Q(t,\dots,t)\mathrm{d}t = 0$$

if $\tau, 2\pi/\Im\lambda_1, \dots, 2\pi/\Im\lambda_{\tilde{m}}$ are rationally independent, i.e., if $\tau/2\pi, 1/\Im\lambda_1, \dots, 1/\Im\lambda_{\tilde{m}}$ are rationally independent. As $\tilde{q}(t,\dots,t) = q(R,s,t)$, and thus

$$Q(t,\dots,t) = f\left(\frac{P(t)q(A,s,t)}{\|P(t)q(A,s,t)\|}\right),$$

this completes the proof. $\qquad\square$

This result can be used to show the following corollary:

Corollary 2.5.33. *Let $A : \mathbb{R} \to \mathbb{C}^{n\times n}$ be continuous and periodic, and let a Floquet representation by P and R be given. Assume that $q(R,s,t)$ is σ-periodic for some $\sigma \neq 0$ such that σ and τ are rationally dependent, i.e., $\sigma = a/b \cdot \tau$ for coprime $a,b \in \mathbb{N}$. Further assume that $q(R,s,t)$ is not constant in t. Then for every $k \in \mathbb{Z}$, there is a continuous map $f : S^{n-1} \to \mathbb{C}$ such that $f^*_{2k\pi/(a\tau)}(s) \neq 0$.*

Proof. If $q(R,s,t)$ is σ-periodic, and $\sigma = a/b\tau$ for coprime $a,b \in \mathbb{N}$, then it holds that $P(t)q(R,s,t)/\|P(t)q(R,s,t)\|$ is $a\tau$-periodic. Hence by Lemma 2.5.30 and Proposition 2.5.31, this lemma follows from Corollary 2.3.40. $\qquad\square$

Remark 2.5.34. If $s \in S^{n-1}$ is λ-monic with respect to R for some $\lambda \in \mathbb{C}$, then $q(R,s,t)$ is $2\pi/\Im\lambda$-periodic. So Corollary 2.5.33 particularly applies to λ-monic points with

$$\Im\lambda = \frac{2\pi b}{a\tau}$$

for $a,b \in \mathbb{N}$ coprime. $\qquad\lrcorner$

Theorem 2.5.35. *Let $A : \mathbb{R} \to \mathbb{C}^{n \times n}$ be continuous and τ-periodic, $\tau > 0$, and consider system (2.5.20). Let a fundamental solution be given by $P(t)e^{Rt}$, where $P : \mathbb{R} \to \mathbb{C}^{n \times n}$ is τ-periodic and $R \in \mathbb{C}^{n \times n}$. Let $s \in S^{n-1}$, and assume that, for every $\lambda \in \mathrm{spec}_{J_{\max}(s)} R \setminus \mathbb{R}$, there are coprime numbers $a_\lambda, b_\lambda \in \mathbb{N}$ such that*

$$a_\lambda \tau = b_\lambda \frac{2\pi}{\Im \lambda} \tag{2.5.37}$$

Let $a := \mathrm{lcm}\{a_\lambda \mid \lambda \in \mathrm{spec}_{J_{\max}(s)} R \setminus \mathbb{R}\}$, and $\sigma := a\tau$. Then for every $k \in \mathbb{Z}$, there is a continuous map $f : S^{n-1} \to \mathbb{C}$ such that $f^*_{2k\pi/\sigma}(s) \neq 0$.

Proof. By Definition 2.5.9, Equation (2.5.37) implies that $q(R, s, t)$ is σ-periodic. Hence, this theorem follows from Corollary 2.5.33. $\qquad\square$

For monic points, the following holds.

Proposition 2.5.36. *Let $A : \mathbb{R} \to \mathbb{C}^{n \times n}$ be continuous and τ-periodic, $\tau > 0$, and consider system (2.5.20). Let a fundamental solution be given by $P(t)e^{Rt}$, where $P : \mathbb{R} \to \mathbb{C}^{n \times n}$ is τ-periodic, and $R \in \mathbb{C}^{n \times n}$. Let $s \in S^{n-1}$ be λ-monic with respect to R. Then it holds that*

$$\mathrm{id}^*_{-\Im\lambda}(s) \neq 0 \tag{2.5.38}$$

if and only if

$$\int_0^\tau \frac{P(t)\Pi_\lambda s}{\|P(t)\Pi_\lambda s\|} \mathrm{d}t \neq 0. \tag{2.5.39}$$

Furthermore, $\mathrm{id}^*_\omega(s) = 0$ for all $\omega \in \mathbb{R}$ with $(\omega+\Im\lambda)\tau/2\pi \notin \mathbb{Z}$.

Proof. Let s be λ-monic with respect to R. Then $P(t)q(R, s, t) = e^{\lambda t}P(t)\Pi_\lambda s$, and thus

$$
\begin{aligned}
\mathrm{id}^*_\omega(s) &= \lim_{T \to \infty} \frac{1}{T} \int_0^T e^{i\omega t} \frac{e^{\lambda t}P(t)\Pi_\lambda s}{\|e^{\lambda t}P(t)\Pi_\lambda s\|} \\
&= \lim_{T \to \infty} \frac{1}{T} \int_0^T e^{i(\omega+\Im\lambda)t} \frac{P(t)\Pi_\lambda s}{\|P(t)\Pi_\lambda s\|}.
\end{aligned}
$$

So

$$
\begin{aligned}
\mathrm{id}^*_{-\Im\lambda}(s) &= \lim_{T \to \infty} \frac{1}{T} \int_0^T \frac{P(t)\Pi_\lambda s}{\|P(t)\Pi_\lambda s\|} \mathrm{d}t \\
&= \frac{1}{\tau} \int_0^\tau P(t)\Pi_\lambda s \|P(t)\Pi_\lambda s\|^{-1}\mathrm{d}t,
\end{aligned}
$$

which implies that (2.5.38) and (2.5.39) are equivalent. Furthermore, the map $t \mapsto P(t)\Pi_\lambda s/\|P(t)\Pi_\lambda s\|$ is τ-periodic, and thus Proposition 2.3.33 gives the second assertion. $\qquad\square$

2.6 Spectra and growth rates

In this section, we will introduce the harmonic growth spectrum, which contains all possible limit points of the harmonic average, and discuss its connection to the uniform growth spectrum and the Morse spectrum. This is based on methods originally presented in [CFJ07] and refined in [Ste09].

In Subsection 2.6.1, we will first define the harmonic growth spectrum and a kind of growth rate similar to the growth rates indroduced in [Ste09, Definition 2.1.1]. Then we discuss some properties of these objects. In order to be able to apply results from [Ste09], we identify $\mathbb{C}$ with $\mathbb{R}^2$, and prove some technical results. It turns out that the harmonic growth spectrum is in fact a uniform growth spectrum, as defined in [Ste09, Definition 2.1.9], which is a generalization of the uniform exponential spectrum that was introduced in [Grü00].

In Subsection 2.6.2, we introduce the Morse spectrum and show its connection to the harmonic growth spectrum in Proposition 2.6.19. The Morse spectrum is easier to compute numerically, because it deals with (ε, T)-chains, and due to Proposition 2.6.22, which we will show in Subsection 2.6.3. The Morse spectrum has a particularly simple structure. It is the closed disc around the origin with a radius given by the maximum of $|f_\omega^\star(x)|$ for all x, see Theorem 2.6.25. Finally, we discuss the relation to the harmonic spectrum and the existence of rotational factor maps in Subsection 2.6.4.

In this section, we let Φ be a semi-flow, fix a bounded function $f : X \to \mathbb{C}$, and assume that $t \mapsto f(\Phi_t x)$ is locally integrable.

2.6.1 Harmonic spectra and growth rates

Define the harmonic growth spectrum as the set of all limit points of the harmonic average of f. This spectrum particularly contains the harmonic limit, if it exists.

Definition 2.6.1. Let $M \subset X$ be compact and invariant. For $\omega \in \mathbb{R}$, the *harmonic growth spectrum over M to frequency* $\omega/2\pi$ is defined by

$$\Sigma_{\mathrm{HG}}^\omega(M) := \left\{ \lambda \in \mathbb{C} \ \middle| \ \begin{array}{c} \text{there are } t_k \to \infty \text{ and } x_k \in M \\ \text{such that } f_\omega^{t_k}(x_k) \to \lambda \end{array} \right\}. \qquad \lrcorner$$

In [CFJ07; Ste09], the concept of a growth rate for a semiflow is introduced (compare [Ste09, Definition 2.1.1]). We want to interpret the harmonic average f_ω^t as a growth rate in this sense. For this, we have to extend our view to the semiflow Ψ^ω on $S^1 \times X$ given by $(z, x) \mapsto (e^{it\omega} z, \Phi_t x)$ for $\omega \in \mathbb{R}$. Note that in [Ste09], continuity of the semiflow is required, but Ψ is not continuous in general. But all results and ideas we use here, also work with discontinuous semiflows.

For $s_1, s_2 \in S^1$, we choose $d(s_1, s_2) := \min\{(\arg s_1 - \arg s_2 \mod 2\pi), (\arg s_2 - \arg s_1 \mod 2\pi)\}$ as the metric on S^1, and endow $S^1 \times X$ with the ∞-product metric, i.e., $d\big((s_1,x_1),(s_2,x_2)\big) := \max\{d(s_1,s_2),d(x_1,x_2)\}$ for $(s_1,x_1),(s_2,x_2) \in S^1 \times X$. Note that we do not use different symbols for the metrics on S^1, X and $S^1 \times X$, but always write d.

We now define the growth rate ρ_ω, which is a growth rate in the sense of [Ste09, Definition 2.1.1] with the difference that we do not assume continuity of ρ_ω.

Definition 2.6.2. For every $\omega \in \mathbb{R}$, let $\rho_\omega : \mathbb{R}^+ \times S^1 \times X \to \mathbb{C}$ be given by

$$(t,z,x) \mapsto \rho_\omega^t(z,x) := z f_\omega^t(x). \qquad \lrcorner$$

Let us look at some properties of the growth rate and the harmonic growth spectrum. First of all, ρ_ω is bounded, continuous in t, and has the property that essentially defines growth rates in the sense of Stender.

Lemma 2.6.3. *For every $\omega \in \mathbb{R}$, the map ρ_ω is bounded, the map $t \to \rho_\omega^t(z,x)$ is continuous for all $(z,x) \in S^1 \times X$, and*

$$t_2 \rho_\omega^{t_2}\big(\Psi_{t_1}^\omega(z,x)\big) = (t_1 + t_2)\rho_\omega^{t_1+t_2}(z,x) - t_1 \rho_\omega^{t_1}(z,x) \qquad (2.6.1)$$

holds for all $t_1, t_2 \geq 0$ and $(z,x) \in S^1 \times X$.

Proof. Boundedness follows from Proposition 2.3.4 and boundedness of f. Continuity is clear from the Fundamental Theorem of Calculus (compare [Kön04, p. 200]). Equation (2.6.1) is straightforward to show:

$$t_2 \rho_\omega^{t_2}\big(\Psi_{t_1}^\omega(z,x)\big) = t_2 \rho_\omega^{t_2}(\mathrm{e}^{\mathrm{i}t_1\omega}z, \Phi_{t_1}x) = t_2 \mathrm{e}^{\mathrm{i}t_1\omega} z f_\omega^{t_2}(\Phi_{t_1}x)$$

$$= z \int_0^{t_2} \mathrm{e}^{\mathrm{i}(t_1+t)\omega} f(\Phi_{t_1+t}x)\mathrm{d}t = z \int_{t_1}^{t_1+t_2} \mathrm{e}^{\mathrm{i}t\omega} f(\Phi_t x)\mathrm{d}t$$

$$= z \int_0^{t_1+t_2} \mathrm{e}^{\mathrm{i}t\omega} f(\Phi_t x)\mathrm{d}t - z \int_0^{t_1} \mathrm{e}^{\mathrm{i}t\omega} f(\Phi_t x)\mathrm{d}t$$

$$= (t_1 + t_2)z f_\omega^{t_1+t_2}(x) - t_1 z f_\omega^{t_1}(x) = (t_1 + t_2)\rho_\omega^{t_1+t_2}(z,x) - t_1 \rho_\omega^{t_1}(z,x) \qquad \square$$

Remark 2.6.4. By letting $t \to \infty$, one gets that the harmonic limit f_ω^* is a Lyapunov growth rate in the sense of [Ste09, Definition 2.1.4], if ρ_ω is continuous. $\qquad \lrcorner$

Next, we show that limit points of ρ remain limit points along trajectories in the following sense.

Lemma 2.6.5. *Let $t \geq 0$, $\omega \in \mathbb{R}$ and $(z,x) \in S^1 \times X$. Let $(t_k)_{k\in\mathbb{N}} \subset [t,\infty)$ be a sequence with $t_k \to \infty$ as $k \to \infty$. If $\lim_{k\to\infty} \rho_\omega^{t_k}(z,x) = \lambda$, then also*

$$\lim_{k\to\infty} \rho_\omega^{t_k-t}\big(\Psi_t^\omega(z,x)\big) = \lambda.$$

Proof. By Lemma 2.6.3, for all $t_1 \geq 0$ and $t_2 > 0$, it holds that

$$\rho_\omega^{t_2}\big(\Psi_{t_1}^\omega(z,x)\big) = \frac{t_1 + t_2}{t_2}\rho_\omega^{t_1+t_2}(z,x) + \frac{t_1}{t_2}\rho_\omega^{t_1}(z,x).$$

Setting $t_1 := t$ and $t_2 := t_k - t$, this equation becomes

$$\rho_\omega^{t_k-t}\big(\Psi_t^\omega(z,x)\big) = \frac{t_k}{t_k - t}\rho_\omega^{t_k}(z,x) + \frac{t}{t_k - t}\rho_\omega^t(z,x).$$

So

$$\lim_{k\to\infty}\rho_\omega^{t_k-t}\big(\Psi_t^\omega(z,x)\big) = 1\cdot\lambda + 0\cdot\rho_\omega^t(z,x) = \lambda. \qquad \square$$

The harmonic growth spectrum is compact.

Lemma 2.6.6. *Let $M \subset X$ be compact and invariant, and let $\omega \in \mathbb{R}$. Then $\Sigma_{\mathrm{HG}}^\omega(M)$ is compact.*

Proof. This follows from [Ste09, Lemma 2.1.10]. Note that continuity of Ψ and ρ is required there. But Stender's proof also works without these assumptions. $\qquad \square$

The harmonic growth spectrum is rotationally invariant.

Lemma 2.6.7. *Let $M \subset X$ be compact and invariant, and let $\omega \in \mathbb{R}$. For every $s \in S^1$, it holds that $s\Sigma_{\mathrm{HG}}^\omega(M) = \Sigma_{\mathrm{HG}}^\omega(M)$, i. e., the harmonic spectrum is invariant under rotation around the origin.*

Proof. Fix $\omega \in \mathbb{R}$ and $s \in S^1$. Let $\lambda \in \Sigma_{\mathrm{HG}}^\omega(M)$. Let $t_k \to \infty$ and $x_k \in M$ be such that $f_\omega^{t_k}(x_k) \to \lambda$. Let $t \geq 0$ be such that $e^{it\omega} = s$. Then by Corollary 2.3.5, it holds that $\lim_{k\to\infty} f_\omega^{t_k-t}(\Phi_t x_k) = e^{-it\omega}\lim_{k\to\infty} f_\omega^{t_k}(x_k)$, which implies $\lambda = s\lim_{k\to\infty} f_\omega^{t_k-t}(\Phi_t x_k) \in s\Sigma_{\mathrm{HG}}^\omega(M)$. Thus $s\Sigma_{\mathrm{HG}}^\omega(M) \supset \Sigma_{\mathrm{HG}}^\omega(M)$.

The other inclusion can be shown analogously. $\qquad \square$

In order to apply more results from [Ste09], we need to identify $\mathbb{C}$ with $\mathbb{R}^2$. To this end, we define the following scalar product on $\mathbb{C}$.

Definition 2.6.8. For two complex numbers $a, b \in \mathbb{C}$, define

$$\langle a, b\rangle := \Re(\overline{a}b) = \Re(a\overline{b}) = \langle b, a\rangle.$$

This product coincides with the Euclidean scalar product on $\mathbb{R}^2$, as the following lemma shows.

Lemma 2.6.9. *If we identify $\mathbb{C}$ with $\mathbb{R}^2$, then $\langle\cdot,\cdot\rangle$ coincides with the Euclidean scalar product, i. e., $\langle a, b\rangle = \Re a\Re b + \Im a\Im b$ for all $a, b \in \mathbb{C}$. In particular, $\langle\cdot,\cdot\rangle$ is $\mathbb{R}$-linear in each argument, and the Cauchy-Schwarz inequality*

$$|\langle a, b\rangle| \leq |a|\cdot|b| \tag{2.6.2}$$

holds.

Proof. Let $a, b \in \mathbb{C}$. Then it holds that

$$\langle a, b \rangle = \Re(\overline{a}b) = \Re[(\Re a - i\Im a)(\Re b + i\Im b)]$$
$$= \Re[\Re a \Re b + \Im a \Im b + i(\Re a \Im b - \Im a \Re b)] = \Re a \Re b + \Im a \Im b,$$

i. e., $\langle \cdot, \cdot \rangle$ is the Euclidean scalar product in $\mathbb{R}^2$. Hence it is $\mathbb{R}$-linear in each argument, and (2.6.2) holds. $\qquad\square$

Now we prove some technical results, which use this scalar product.

Lemma 2.6.10. *Let $M \subset X$ be compact and invariant. Then for every $z \in S^1$, it holds that $\langle \Sigma_{\mathrm{HG}}(M), z \rangle$ is a compact interval.*

Proof. By Lemma 2.6.6, $\Sigma_{\mathrm{HG}}(M)$ is compact. So there is $\lambda_0 \in \Sigma_{\mathrm{HG}}(M)$ such that $|\lambda_0| = \max_{\lambda \in \Sigma_{\mathrm{HG}}(M)} |\lambda|$. Clearly, $\langle \Sigma_{\mathrm{HG}}(M), z \rangle \subset [-|\lambda_0|, |\lambda_0|]$ for every $z \in S^1$, because for every $\lambda \in \Sigma_{\mathrm{HG}}(M)$, it holds that $|\langle \lambda, z \rangle| = |\Re[\overline{z}\lambda]| \le |\lambda| \le |\lambda_0|$. To show the other inclusion

$$\Re[\overline{z} \cdot \Sigma_{\mathrm{HG}}(M)] \supset [-|\lambda_0|, |\lambda_0|], \qquad (2.6.3)$$

fix $z \in S^1$. By Lemma 2.6.7,

$$z e^{it}|\lambda_0| = \left(z e^{it} \frac{\overline{\lambda_0}}{|\lambda_0|} \right) \lambda_0 \in \Sigma_{\mathrm{HG}}(M)$$

for all $t \in \mathbb{R}$. So $\langle \Sigma_{\mathrm{HG}}(M), z \rangle \ni \langle z e^{it}|\lambda_0|, z \rangle = \Re[\overline{z} \cdot z e^{it}|\lambda_0|] = \Re[e^{it}|\lambda_0|] = \sin(t)|\lambda_0|$ for all $t \in \mathbb{R}$. This implies (2.6.3). $\qquad\square$

Lemma 2.6.11. *Let $\xi_0 \in S^1 \times X$, $z \in S^1$, $t \ge 0$, and $\omega \in \mathbb{R}$. Define*

$$P := \sup\{|\rho_\omega^t(\xi)| \mid t \ge 0, \xi \in X\}.$$

Then for every $\varepsilon \in (0, 2P)$, there is a time $t_1 \le {}^{[(2P-\varepsilon)t]}/(2P)$ such that

$$\langle \rho_\omega^s(\Psi_{t_1}\xi_0), z \rangle \ge \langle \rho_\omega^t(\xi_0), z \rangle - \varepsilon \qquad (2.6.4)$$

for all $s \in (0, t - t_1]$ and $t - t_1 \ge {}^{\varepsilon t}/(2P)$.

Proof. This can be shown analogously to [Ste09, Lemma 2.1.7]. But as Stender assumes continuity of Ψ, we provide a full proof here.

First note that P is finite, as ρ_ω is bounded according to Lemma 2.6.3. Abbreviate $\sigma := \langle \rho_\omega^t(\xi_0), z \rangle$, and let $\varepsilon \in (0, 2P)$. Define $\beta := \sup_{s \in (0,t]} \langle \rho_\omega^s(\xi_0), z \rangle$. If $\beta \ge \sigma - \varepsilon$, the assertion clearly holds with $t_1 = 0$. So assume that $\beta < \sigma - \varepsilon$, and let

$$t_1 := \sup\{s \in (0, t] \mid \langle \rho_\omega^s(\xi_0), z \rangle \le \sigma - \varepsilon\}.$$

By continuity of ρ in t, and because $\beta < \sigma - \varepsilon$ and $\langle \rho_\omega^t(\xi_0), z \rangle = \sigma > \sigma - \varepsilon$, the intermediate value theorem implies existence of a time $t_1 \in (0, t)$, such that $\langle \rho_\omega^{t_1}(\xi_0), z \rangle = \sigma - \varepsilon$.

Let $t_2 := t - t_1$. It holds that

$$
\begin{aligned}
\varepsilon &= |\sigma - (\sigma - \varepsilon)| \\
&= |\langle \rho_\omega^t(\xi_0), z \rangle - \langle \rho_\omega^{t_1}(\xi_0), z \rangle| \\
&= |\langle \rho_\omega^{t_1+t_2}(\xi_0), z \rangle - \langle \rho_\omega^{t_1}(\xi_0), z \rangle| \\
&\overset{(*)}{=} |\langle \rho_\omega^{t_1+t_2}(\xi_0) - \rho_\omega^{t_1}(\xi_0), z \rangle| \\
&\overset{(*)}{\leq} |\rho_\omega^{t_1+t_2}(\xi_0) - \rho_\omega^{t_1}(\xi_0)| \cdot |z| \\
&= |\rho_\omega^{t_1+t_2}(\xi_0) - \rho_\omega^{t_1}(\xi_0)| \cdot |z| \\
&\overset{(**)}{=} \left| \frac{1}{t_1 + t_2} [t_2 \rho_\omega^{t_2}(\Psi_{t_1}\xi_0) + t_1 \rho_\omega^{t_1}(\xi_0)] - \rho_\omega^{t_1}(\xi_0) \right| \\
&\leq \frac{t_2}{t_1 + t_2} |\rho_\omega^{t_2}(\Psi_{t_1}\xi_0)| + \left| \frac{t_1}{t_1 + t_2} - 1 \right| |\rho_\omega^{t_1}(\xi_0)| \\
&\leq \frac{t_2}{t_1 + t_2} P + \frac{t_2}{t_1 + t_2} P \\
&= \frac{2 P t_2}{t},
\end{aligned}
\tag{2.6.5}
$$

where the steps marked with $(*)$ follow from Lemma 2.6.9, and $(**)$ follows from Lemma 2.6.3. So $t_2 \geq {}^{\varepsilon t}/_{2P}$, which implies $t_1 \leq {}^{(2P-\varepsilon)t}/_{2P}$.

It remains to show that t_1 satisfies (2.6.4). Let $s \in (0, t - t_1]$, i.e., $t_1 + s \in (t_1, t]$. Then $\langle \rho_\omega^{t_1+s}(\xi_0), z \rangle > \sigma - \varepsilon$ by the definition of t_1. Hence by Lemmas 2.6.3 and 2.6.9, it follows that

$$
\begin{aligned}
\langle \rho_\omega^s(\Psi_{t_1}\xi_0), z \rangle &= \left\langle \frac{1}{s}[(t_1 + s)\rho_\omega^{t_1+s}(\xi_0) - t_1 \rho_\omega^{t_1}(\xi_0)], z \right\rangle \\
&= \frac{1}{s}[(t_1 + s)\langle \rho_\omega^{t_1+s}(\xi_0), z \rangle - t_1 \langle \rho_\omega^{t_1}(\xi_0), z \rangle] \\
&\geq \frac{1}{s}[(t_1 + s)(\sigma - \varepsilon) - t_1(\sigma - \varepsilon)] \\
&= \sigma - \varepsilon. \qquad \square
\end{aligned}
$$

By Lemma 2.6.10, $\langle \Sigma_{\mathrm{HG}}^\omega(M), z \rangle$ is a compact interval. Its endpoints are actual limits of ρ_ω^t, not only limit points.

Proposition 2.6.12. *Let $M \subset X$ be compact and invariant. Let $\omega \in \mathbb{R}$ and $z \in S^1$. Assume that ρ is continuous. Then there is $(s^*, x^*) \in S^1 \times M$ such that*

$$
\lim_{t \to \infty} \rho_\omega^t(s^*, x^*) = \max \langle \Sigma_{\mathrm{HG}}^\omega(M), z \rangle.
$$

Proof. This follows from [Ste09, Theorem 2.1.12], but as Stender assumes connectedness of M and continuity of Ψ, we provide a complete proof here.

By Lemma 2.6.10, $\langle \Sigma^\omega_{\mathrm{HG}}(M), z \rangle$ is a compact interval. Let $\rho^* := \max\langle \Sigma^\omega_{\mathrm{HG}}(M), z \rangle$. Then there is $p^* \in \Sigma^\omega_{\mathrm{HG}}(M)$ such that $\langle p^*, z \rangle = \rho^*$. So there are sequences $(s_k, x_k) \in S^1 \times M$, and $t_k \to \infty$, such that $\rho^{t_k}_\omega(s_k, x_k) \to p^*$, by definition of Σ_{HG}. Hence $\rho^{t_k}_\omega(x_k), z \rangle \to \rho^*$ for $k \to \infty$, which implies that there is a sequence $\varepsilon_k \to 0$ with $\langle \rho^{t_k}_\omega(x_k), z \rangle > \rho^* - \varepsilon_k$ for all k. Let $\tilde{\varepsilon}_k := 1/\sqrt{t_k} \to 0$ and $P := \sup\{|\rho^t_\omega(x)| \mid t \geq 0, x \in X\}$. Note that P is finite, as ρ_ω is bounded according to Lemma 2.6.3. Then by Lemma 2.6.11, there are times t^*_k such that

$$\langle \rho^s_\omega\big(\Psi_{t^*_k}(s_k, x_k)\big), z \rangle \geq \rho^* - \varepsilon_k - \tilde{\varepsilon}_k \tag{2.6.6}$$

for all $s \in (0, t_k - t^*_k]$ and $t_k - t^*_k \geq \tilde{\varepsilon}_k t_k/(2P) = \sqrt{t_k}/(2P)$.

Define $(\tilde{s}_k, \tilde{x}_k) := \Psi_{t^*_k}(s_k, x_k)$, and $\tilde{t}_k := t_k - t^*_k \to \infty$. Then by (2.6.6), it holds that

$$\langle \rho^s_\omega(\tilde{s}_k, \tilde{x}_k), z \rangle \geq \rho^* - \varepsilon_k - \tilde{\varepsilon}_k \tag{2.6.7}$$

for all $s \in (0, \tilde{t}_k]$. By compactness of $S^1 \times M$, we may assume that $(\tilde{s}_k, \tilde{x}_k)$ converges to some $(\tilde{s}, \tilde{x}) \in S^1 \times M$.

Fix $\varepsilon > 0$ and $t > 0$. Then it holds by (2.6.7), that

$$\langle \rho^t_\omega(\tilde{s}, \tilde{x}), z \rangle = \langle \rho^t_\omega(\tilde{s}_k, \tilde{x}_k), z \rangle + \langle \rho^t_\omega(\tilde{s}, \tilde{x}), z \rangle - \langle \rho^t_\omega(\tilde{s}_k, \tilde{x}_k), z \rangle$$
$$\geq \rho^* - \varepsilon_k - \tilde{\varepsilon}_k + \langle \rho^t_\omega(\tilde{s}, \tilde{x}), z \rangle - \langle \rho^t_\omega(\tilde{s}_k, \tilde{x}_k), z \rangle,$$

if k is large enough such that $t \leq \tilde{t}_k$. As ρ is continuous, letting $k \to \infty$ yields $\langle \rho^t_\omega(\tilde{s}, \tilde{x}), z \rangle - \langle \rho^t_\omega(\tilde{s}_k, \tilde{x}_k), z \rangle \to 0$, and so $\langle \rho^t_\omega(\tilde{s}, \tilde{x}), z \rangle \geq \rho^*$. As this holds for all $t > 0$, it follows that

$$\liminf_{t \to \infty} \langle \rho^t_\omega(\tilde{s}, \tilde{x}), z \rangle \geq \rho^*. \tag{2.6.8}$$

It remains to show that

$$\limsup_{t \to \infty} \langle \rho^t_\omega(\tilde{s}, \tilde{x}), z \rangle \leq \rho^*, \tag{2.6.9}$$

because this, together with (2.6.8) and the fact that the limes superior cannot be strictly smaller than the limes inferior, implies that $\lim_{t \to \infty} \langle \rho^t_\omega(\tilde{s}, \tilde{x}), z \rangle = \rho^*$. So assume that (2.6.9) is wrong. Then there is a sequence $t_k \to \infty$ such that $\lim_{k \to \infty} \langle \rho^{t_k}_\omega(\tilde{s}, \tilde{x}), z \rangle > \rho^* = \max\langle \Sigma^\omega_{\mathrm{HG}}(M), z \rangle$, which is a contradiction. $\square$

In [Ste09, Definition 2.1.9] the concept of the *uniform growth spectrum* was introduced. Our harmonic growth spectrum coincides with the uniform growth spectrum with respect to the growth rate ρ_ω.

Lemma 2.6.13. *Let $M \subset X$ be compact and invariant, and $\omega \in \mathbb{R}$. Then it holds that*

$$\Sigma^\omega_{\mathrm{HG}}(M) = \left\{ \lambda \in \mathbb{C} \;\middle|\; \begin{array}{l} \text{there are } t_k \to \infty, \; z_k \in S^1 \text{ and} \\ x_k \in M \text{ such that } \rho^{t_k}_\omega(z_k, x_k) \to \lambda \end{array} \right\}. \tag{2.6.10}$$

Proof. Fix $\omega \in \mathbb{R}$, and denote by $\Sigma_{\mathrm{UG}}(M)$ the set on the right-hand side of (2.6.10). The inclusion $\Sigma_{\mathrm{UG}}(M) \supset \Sigma_{\mathrm{HG}}^{\omega}(M)$ is clear, as for every $t_k \to \infty$ and $x_k \in M$ with $f_{\omega}^{t_k}(x_k) \to \lambda \in \Sigma_{\mathrm{HG}}^{\omega}(M)$, one has $\rho_{\omega}^{t_k}(1, x_k) = 1 \cdot f_{\omega}^{t_k}(x_k) \to \lambda$.

On the other hand, let $t_k \to \infty$, $z_k \in S^1$ and $x_k \in M$ such that $\rho_{\omega}^{t_k}(z_k, x_k) \to \lambda \in \Sigma_{\mathrm{UG}}(M)$. Without loss of generality, we can assume that $z_k \to z \in S^1$, as S^1 is compact. As z_k is bounded away from zero, and $\rho_{\omega}^{t_k}(z_k, x_k) = z_k f_{\omega}^{t_k}(x_k)$ converges to λ, it holds that $z f_{\omega}^{t_k}(x_k) \to \lambda$. Let $t \geq 0$ be such that $\mathrm{e}^{-\mathrm{i}t\omega} = z$. We may assume that $t_k \geq 0$ for all k. Then by Lemma 2.6.5, $\lim_{k \to \infty} f_{\omega}^{t_k - t}(\Phi_t x_k) = \lim_{k \to \infty} \rho_{\omega}^{t_k - t}(1, \Phi_t x_k) = \lim_{k \to \infty} \rho_{\omega}^{t_k - t}(\Psi(z, x_k)) = \lim_{k \to \infty} \rho_{\omega}^{t_k}(z, x_k) = \lim_{k \to \infty} z f_{\omega}^{t_k}(x_k) = \lambda$. $\square$

2.6.2 Morse spectrum

Now consider the Morse spectrum over compact chain transitive sets. The Morse spectrum is similar to the harmonic growth spectrum, but the Morse spectrum contains limit points of ρ evaluated along chains. For an (ε, T)-chain ζ in $S^1 \times X$, given by $n \in \mathbb{N}$, points $(z_0, x_0), \ldots, (z_n, x_n) \in S^1 \times X$, and times $T_0, \ldots, T_{n-1} > T$ with $d\left(\Psi_{T_i}^{\omega}(z_i, x_i), (z_{i+1}, x_{i+1})\right) < \varepsilon$ for all $i = 0, \ldots, n-1$, define

$$\rho_{\omega}(\zeta) := \frac{1}{\sum_{i=0}^{n-1} T_i} \sum_{i=0}^{n-1} T_i \rho_{\omega}^{T_i}(z_i, x_i)$$

$$= \frac{1}{\sum_{i=0}^{n-1} T_i} \sum_{i=0}^{n-1} z_i \int_0^{T_i} \mathrm{e}^{\mathrm{i}t\omega} f(\Phi_t x_i) \mathrm{d}t \in \mathbb{C}.$$

Definition 2.6.14. Let $M \subset X$ be compact and chain transitive. Then define the *Morse spectrum over M* by

$$\Sigma_{\mathrm{MO}}^{\omega}(M) := \left\{ \lambda \in \mathbb{C} \ \middle| \ \begin{array}{c} \text{there are } \varepsilon^k \to 0, \ T^k \to \infty \text{ and} \\ (\varepsilon^k, T^k)\text{-chains } \zeta^k \text{ in } S^1 \times M \text{ with } \rho_{\omega}(\zeta^k) \to \lambda \end{array} \right\}. \quad \lrcorner$$

We let M be a subset of X and not a subset of $S^1 \times X$, which would be analogous to the notation in [Ste09, Equation (2.5)]. This is no restriction, as every compact chain transitive subset of $S^1 \times X$ is of the form $S^1 \times M$ for some compact chain transitive $M \subset X$.

Lemma 2.6.15. *Let $\omega \in \mathbb{R}$. A set $M \subset X$ is chain transitive for Φ, if and only if $S^1 \times M$ is chain transitive for Ψ^{ω}.*

Proof. First assume that $S^1 \times M$ is chain transitive for Ψ^{ω}. Let $x, y \in M$, and let $\varepsilon, T > 0$. Then there is an (ε, T)-chain ξ from $(1, x)$ to $(1, y)$. Let ξ be given by the points $(s_0, x_0) = (1, x), (s_1, x_1), \ldots, (s_n, x_n)$, and times $T_0, \ldots, T_{n-1} \geq T$. Then the chain $\tilde{\xi}$ given by the points $x_0 = x, x_1, \ldots, x_n = y$, and times $T_0, \ldots, T_{n-1}$ is an (ε, T)-chain from x to y, because

$$\varepsilon > d\big(\Psi^\omega_{T_j}(s_j, x_j), (s_{j+1}, x_{j+1})\big) = d\big((e^{i\omega T_j} s_j, \Phi_{T_j} x_j), (s_{j+1}, x_{j+1}),\big)$$
$$= \max\{d(e^{i\omega T_j} s_j, s_{j+1}), d(\Phi_{T_j} x_j, x_{j+1})\} \geq d(\Phi_{T_j} x_j, x_{j+1}).$$

So M is chain transitive for Φ.

Now assume that M is chain transitive for Φ. Let $(r, x), (s, y) \in S^1 \times M$, and $\varepsilon, T > 0$. Then there is an (ε, T)-chain ξ_1 for Φ from x to y. Let ξ_1 be given by the points $x_0 = x, x_1, \ldots, x_n = y$, and times $T_0, \ldots, T_{n-1} \geq T$. Let $\delta := \big(\arg s - \arg r - \omega \sum_{j=0}^{n-1} T_j\big)$ mod 2π.

By chain transitivity of M, there also is an (ε, T)-chain ζ from x to x. Let ζ be given by the points $z_0, \ldots, z_m$, and times $U_0, \ldots, U_{m-1} \geq T$. Let $\gamma := \big(\omega \sum_{j=0}^{m-1} U_j\big)$ mod 2π.

Let $N := \lceil 4\pi/m\varepsilon \rceil$. Construct a chain ξ_2 for Φ by concatenating ζ and ξ_1 in the following way: $\xi_2 := \zeta^N \circ \xi_1$. Let $\beta := [(N\gamma + \delta) \mod 2\pi]/(Nm+n)$. Then

$$\begin{aligned}
\beta &= \frac{(N\gamma + \delta) \mod 2\pi}{\lceil \frac{4\pi}{m\varepsilon} \rceil m + n} \\
&\leq \frac{2\pi}{\frac{4\pi}{m\varepsilon} m} \\
&= \frac{1}{2}\varepsilon.
\end{aligned} \tag{2.6.11}$$

Construct a chain ξ_3 for the semiflow $s \mapsto e^{i\omega t} s$ on S^1 with the times

$$\underbrace{V_0 = U_0, \ldots, V_{m-1} = U_{m-1}, \ldots, V_{(N-1)m} = U_0, \ldots, V_{Nm-1} = U_{m-1},}_{U_0, \ldots, U_{m-1} \text{ repeated } N \text{ times}}$$

$$V_{Nm} = T_0, \ldots, V_{Nm+n-1} = T_{n-1},$$

and points $s_j = e^{i\omega \sum_{k=0}^{j-1} V_k} e^{ij\beta} r$, $j = 0, \ldots, Nm + n - 1$, and $s_{Nm+n} = s$.

Let $\xi := \xi_3 \times \xi_2$ (see Definition 2.1.4 for the definition of the product of chains). Then ξ is a chain from (r, x) to (s, y). So it remains to show, that it is an (ε, T)-chain. By Lemma 2.1.5, it suffices to show that ξ_2 and ξ_3 both are (ε, T)-chains. As ζ and ξ_1 are (ε, T)-chains, also ξ_2 is an (ε, T)-chain. In order to show that ξ_3 is an (ε, T)-chain, we have to prove that $d(e^{i\omega V_j} s_j, s_{j+1}) < \varepsilon$ for all $j = 0, \ldots, Nm+n-1$.

For $j = 0, \ldots, Nm + n - 2$, it holds that

$$e^{i\omega V_j} s_j = e^{i\omega V_j} e^{i\omega \sum_{k=0}^{j-1} V_k} e^{ij\beta} r = e^{i\omega \sum_{k=0}^{j} V_k} e^{ij\beta} r.$$

So $\arg e^{i\omega V_j} s_j = \omega \sum_{k=0}^{j} V_k + j\beta + \arg r$ mod 2π and $\arg s_{j+1} = \omega \sum_{k=0}^{j} V_k + (j+1)\beta + \arg r$ mod 2π. Hence $d(e^{i\omega V_j} s_j, s_{j+1}) = \min\{\pm\beta \mod 2\pi\} \leq \beta \leq 1/2\varepsilon$ due to (2.6.11). $\qquad\square$

Lemma 2.6.17, which states, that ρ_ω evaluated along a trivial chain ζ equals ρ_ω evaluated at the starting point of ζ, will be needed in the proofs of Proposition 2.6.19 and Theorem 2.6.29. In order to show this lemma, we need the following corollary to Lemma 2.6.3.

Corollary 2.6.16. *For all $t_0, \ldots, t_n \in \mathbb{R}_0^+$, and $(z, x) \in S^1 \times X$, it holds that*

$$\sum_{i=0}^{n} t_i \rho^{t_i} \left(\Psi^\omega_{\sum_{j=0}^{i-1} t_j}(z, x) \right) = \sum_{j=0}^{n} t_j \cdot \rho^{\sum_{j=0}^{n} t_j}(z, x).$$

Proof. This follows inductively from Lemma 2.6.3. $\qquad\square$

Lemma 2.6.17. *Let ζ be a trivial chain with total time T starting in (z_0, x_0), i. e., let ζ be given by $(z_0, x_0), \ldots, (z_n, x_n) \in S^1 \times X$, and $T_0, \ldots, T_{n-1} > 0$, with $\sum_{j=0}^{n-1} T_j = T$ and $\Psi^\omega_{T_j}(z_j, x_j) = (z_{j+1}, x_{j+1})$, $j = 0, \ldots, n-1$. Then the growth rate of ζ equals the growth rate of (z_0, x_0) at time T, i. e., $\rho_\omega(\zeta) = \rho_\omega^T(z_0, x_0)$.*

Proof. By assumption, it holds that $(z_i, x_i) = \Psi^\omega_{\sum_{j=0}^{i-1} T_j}(z_0, x_0)$ for every $i = 0, \ldots, n-1$. So

$$
\begin{aligned}
\rho_\omega(\zeta) &= \frac{1}{\sum_{i=0}^{n-1} T_i} \sum_{i=0}^{n-1} T_i \rho_\omega^{T_i}(z_i, x_i) \\
&= \frac{1}{T} \sum_{i=0}^{n-1} T_i \rho_\omega^{T_i} \left(\Psi^\omega_{\sum_{j=0}^{i-1} T_j}(z_0, x_0) \right) \\
&\overset{(*)}{=} \frac{1}{T} \cdot T \rho^T(z_0, x_0) \\
&= \rho^T(z_0, x_0),
\end{aligned}
$$

where the equation marked with $(*)$ holds due to Corollary 2.6.16. $\qquad\square$

In Proposition 2.6.19, we will show a relation between the harmonic growth spectrum and the Morse spectrum. We need the following lemma for the proof.

Lemma 2.6.18. *Let ζ be an (ε, T)-chain in $S^1 \times X$ with times $T_0, \ldots, T_{n-1} \geq T$, and points $(s_0, x_0), \ldots, (s_n, x_n) \in S^1 \times X$. Then it holds for every $\omega \in \mathbb{R}$ and $z \in \mathbb{C}$, that*

$$\min_{i=0,\ldots,n-1} \langle \rho_\omega^{T_i}, z \rangle \leq \langle \rho_\omega(\zeta), z \rangle \leq \max_{i=0,\ldots,n-1} \langle \rho_\omega^{T_i}, z \rangle.$$

Proof. See [Ste09, Lemma 2.2.8]. Note that Stender assumes continuity of ρ and Ψ there. But this is not needed in the proof. $\qquad\square$

Proposition 2.6.19. *$M \subset X$ be a compact and invariant set such that $S^1 \times M$ is chain transitive for $\Psi|_{S^1 \times M}$. Then it holds for all $\omega \in \mathbb{R}$ and $z \in \mathbb{C}$, that $\langle \Sigma^\omega_{\mathrm{MO}}(M), z \rangle = \langle \Sigma^\omega_{\mathrm{HG}}(M), z \rangle$.*

Proof. The inclusion $\langle \Sigma_{\mathrm{HG}}^{\omega}(M), z \rangle \subset \langle \Sigma_{\mathrm{MO}}^{\omega}(M), z \rangle$ is obvious, as by Lemma 2.6.17, $\rho(\zeta) = \rho^T(x)$ for every trivial chain ζ starting in x with total time T.

By Lemma 2.6.10, $\langle \Sigma_{\mathrm{HG}}^{\omega}(M), z \rangle$ is a compact interval. So in order to show the other inclusion $\langle \Sigma_{\mathrm{MO}}^{\omega}(M), z \rangle \subset \langle \Sigma_{\mathrm{HG}}^{\omega}(M), z \rangle$, it suffices to show that

$$\min\langle \Sigma_{\mathrm{HG}}^{\omega}(M), z \rangle \leq \min\langle \Sigma_{\mathrm{MO}}^{\omega}(M), z \rangle, \qquad (2.6.12)$$

and

$$\max\langle \Sigma_{\mathrm{HG}}^{\omega}(M), z \rangle \geq \max\langle \Sigma_{\mathrm{MO}}^{\omega}(M), z \rangle. \qquad (2.6.13)$$

Let $\lambda \in \Sigma_{\mathrm{MO}}^{\omega}(M)$ be such that $\langle \lambda, z \rangle = \min\langle \Sigma_{\mathrm{HG}}^{\omega}(M), z \rangle$. By definition of the Morse spectrum, there are $\varepsilon^k \to 0$, $T^k \to \infty$, and (ε^k, T^k)-chains ζ^k in $S^1 \times M$ with $\rho_\omega(\zeta^k) \to \lambda$. Hence $\langle \rho_\omega(\zeta^k), z \rangle \to \langle \lambda, z \rangle = \min\langle \Sigma_{\mathrm{HG}}^{\omega}(M), z \rangle$. By Lemma 2.6.18, there are $(z_k, x_k) \in S^1 \times M$ and $t_k \geq T^k$, such that $\langle \rho_\omega^{t_k}(x_k), z \rangle \leq \langle \rho_\omega(\zeta^k), z \rangle$. By boundedness, we can assume that $\lim_{k \to \infty} \langle \rho_\omega^{t_k}(x_k), z \rangle$ exists, and it follows that $\lim_{k \to \infty} \langle \rho_\omega^{t_k}(x_k), z \rangle \leq \min\langle \Sigma_{\mathrm{HG}}^{\omega}(M), z \rangle$. As $t_k \to \infty$, this implies (2.6.12).

Inequality (2.6.13) can be shown analogously. $\qquad \square$

2.6.3 Structure of Morse spectra

In this subsection, we will prove two results on the structure of the Morse spectrum. The first one, Proposition 2.6.22, shows that one only needs to consider periodic chains, if Φ is continuous. The second result, Theorem 2.6.30, states that the Morse spectrum is a closed disc, whose radius is related to harmonic limits.

We need the following technical result, which is taken from [CK00, Lemma B.2.23].

Lemma 2.6.20. *Let $M \subset X$ be compact and invariant, such that $\Phi|_M$ is chain transitive, and fix $\varepsilon, T > 0$. If Φ is continuous, then there is a time $\bar{T}(\varepsilon, T) > 0$ such that for all $x, y \in M$, there is an (ε, T)-chain in M from x to y with total length at most $\bar{T}(\varepsilon, T)$.*

Proof. By chain transitivity, for all $x, y \in M$ there is an $(\varepsilon/2, T)$-chain in M from x to y. Fix $z \in M$. By compactness of M and continuity of Φ, there are finitely many (ε, T)-chains connecting every $x \in M$ to z, if we do not distinguish between chains that only differ in their starting points. Similarly, there are (modulo their endpoints) finitely many (ε, T)-chains connecting z with arbitrary $y \in M$. Thus there are finitely many (ε, T)-chains connecting all points in M. The maximum of their total lengths is the desired upper bound $\bar{T}(\varepsilon, T)$. $\qquad \square$

We also need the following lemma, which tells us how to evaluate the growth rate ρ_ω for concatenated chains. It turns out that the growth rate of a concatenated chain is a convex combination of the growth rates of the original chains.

Lemma 2.6.21. *Let ξ and ζ be (ε, T)-chains in $S^1 \times X$ with total lengths σ and τ, respectively. Assume that the initial point of ζ coincides with the final point of ξ. Denote by $\zeta \circ \xi$ the concatenation of these chains. Then it holds that*

$$\rho_\omega(\zeta \circ \xi) = \frac{\sigma}{\sigma + \tau} \rho_\omega(\xi) + \frac{\tau}{\sigma + \tau} \rho_\omega(\zeta).$$

Proof. Let ξ be given by the points $(z_0, x_0), \ldots, (z_n, x_n) \in S^1 \times X$, and the times $T_0, \ldots, T_{n-1} > T$, and let ζ be given by the points $(z_n, x_n), \ldots, (z_{n+m}, x_{n+m}) \in S^1 \times X$, and times $T_n, \ldots, T_{n+m-1} > T$. Note that $\sum_{i=0}^{n-1} T_i = \sigma$ and $\sum_{i=n}^{n+m-1} T_i = \tau$. Then by definition,

$$\begin{aligned}
\rho_\omega(\zeta \circ \xi) &= \frac{1}{\sum_{i=0}^{n+m-1} T_i} \sum_{i=0}^{n+m-1} T_i \rho_\omega^{T_i}(z_i, x_i) \\
&= \frac{1}{\sigma + \tau} \sum_{i=0}^{n+m-1} T_i \rho_\omega^{T_i}(z_i, x_i) \\
&= \frac{1}{\sigma + \tau} \left[\sum_{i=0}^{n-1} T_i \rho_\omega^{T_i}(z_i, x_i) + \sum_{i=n}^{n+m-1} T_i \rho_\omega^{T_i}(z_i, x_i) \right] \\
&= \frac{1}{\sigma + \tau} \left[\sigma \rho_\omega(\xi) + \tau \rho_\omega(\zeta) \right] \\
&= \frac{\sigma}{\sigma + \tau} \rho_\omega(\xi) + \frac{\tau}{\sigma + \tau} \rho_\omega(\zeta).
\end{aligned}$$

$\square$

The following proposition is needed in the proof of Theorem 2.6.25, and will also simplify the numerical computation of the Morse spectrum, as it states that only *periodic* chains have to be considered, if we deal with a continuous semiflow.

Proposition 2.6.22. *Let $M \subset X$ be compact and invariant, such that $\Phi|_M$ is chain transitive. If Φ is continuous, it holds that*

$$\Sigma_{\mathrm{MO}}^\omega(M) = \left\{ \lambda \in \mathbb{C} \;\middle|\; \begin{array}{c} \text{there are } \varepsilon^k \to 0, \; T^k \to \infty, \text{ and periodic} \\ (\varepsilon^k, T^k)\text{-chains } \zeta^k \text{ in } S^1 \times M \text{ with } \rho_\omega(\zeta^k) \to \lambda \end{array} \right\}.$$

A proof of this proposition can be found in [CFJ07, Proposition 2.6; Ste09, Lemma 2.2.13]. Nevertheless, we will provide a complete proof here, because we do not require ρ_ω and Ψ to be continuous, and because we deal with semiflows, contrary to the cited sources. In fact, we will show that

$$\Sigma_{\mathrm{MO}}^\omega(M) = \left\{ \lambda \in \mathbb{C} \;\middle|\; \begin{array}{c} \text{for every } \varepsilon, T > 0, \text{ there are periodic} \\ (\varepsilon, T)\text{-chains } \zeta^k \text{ in } S^1 \times M \text{ with } \rho_\omega(\zeta^k) \to \lambda \end{array} \right\}.$$

Proof of Proposition 2.6.22. Clearly, the inclusion "⊃" holds. For the other inclusion, let $\omega \in \mathbb{R}$ and $\lambda \in \Sigma_{\mathrm{MO}}^{\omega}(M)$. Fix $\varepsilon, T > 0$. It suffices to show that, for every $\delta > 0$, there is a periodic (ε, T)-chain ζ' in $S^1 \times M$ with $|\lambda - \rho_{\omega}(\zeta')| < \delta$.

By Lemma 2.6.20, there exists $\bar{T}(\varepsilon, T) > 0$ such that for all $(z_1, x_1), (z_2, x_2) \in S^1 \times M$, there is an (ε, T)-chain in $S^1 \times M$ from (z_1, x_1) to (z_2, x_2) with total time less than $\bar{T}(\varepsilon, T)$. Note that chain transitivity of M implies chain transitivity of $S^1 \times M$, compare Lemma 2.6.15.

Let $S > T$, and choose an (ε, S)-chain ζ in $S^1 \times M$ with $|\lambda - \rho_{\omega}(\zeta)| < \delta/2$. Such a chain exists, as $\lambda \in \Sigma_{\mathrm{MO}}^{\omega}(M)$. Denote the total length of ζ by σ. By Lemma 2.6.20, there is an (ε, T)-chain ξ from the endpoint of ζ to its starting point with total length $\tau \leq \bar{T}(\varepsilon, T)$. Note that ξ depends on ζ, and thus particularly depends on S. Concatenation yields a periodic (ε, T)-chain $\zeta' = \zeta'(S) := \xi \circ \zeta$, which has the desired approximation property, as long as S is chosen big enough:

By Lemma 2.6.21, it holds that

$$\rho_{\omega}(\zeta') = \frac{\sigma}{\sigma + \tau} \rho_{\omega}(\zeta) + \frac{\tau}{\sigma + \tau} \rho_{\omega}(\xi).$$

So

$$
\begin{aligned}
|\lambda - \rho_{\omega}(\zeta')| &\leq |\lambda - \rho_{\omega}(\zeta)| + |\rho_{\omega}(\zeta) - \rho_{\omega}(\zeta')| \\
&< \frac{\delta}{2} + \left| \rho_{\omega}(\zeta) - \frac{\sigma}{\sigma + \tau} \rho_{\omega}(\zeta) - \frac{\tau}{\sigma + \tau} \rho_{\omega}(\xi) \right| \\
&\leq \frac{\delta}{2} + \left[1 - \frac{\sigma}{\sigma + \tau} \right] |\rho_{\omega}(\zeta)| + \frac{\tau}{\sigma + \tau} |\rho_{\omega}(\xi)| \\
&= \frac{\delta}{2} + \frac{\tau}{\sigma + \tau} \left[|\rho_{\omega}(\zeta)| + |\rho_{\omega}(\xi)| \right] \\
&\leq \frac{\delta}{2} + \frac{\bar{T}(\varepsilon, T)}{\sigma + \bar{T}(\varepsilon, T)} \left[|\rho_{\omega}(\zeta)| + |\rho_{\omega}(\xi)| \right] \\
&\leq \frac{\delta}{2} + \frac{\bar{T}(\varepsilon, T)}{\sigma + \bar{T}(\varepsilon, T)} \cdot 2 \sup_{x \in X} |f(x)|.
\end{aligned}
$$

For the last inequality, compare Lemma 2.6.28. Note that $\sigma \to \infty$ when $S \to \infty$. So $\lim_{S \to \infty} |\lambda - \rho_{\omega}(\zeta')| \leq \delta/2$. In particular, there is S_0 such that $|\lambda - \rho_{\omega}(\zeta'(S_0))| < \delta$. $\square$

As a final step towards Theorem 2.6.30, we show that every exposed point of the Morse spectrum is a limit of ρ_{ω}.

Lemma 2.6.23. *Assume that ρ_{ω} is continuous. Let $M \subset X$ be a compact and invariant set such that $S^1 \times M$ is chain transitive for $\Psi|_{S^1 \times M}$. Then for every $\omega \in \mathbb{R}$, the exposed points of the Morse spectrum $\Sigma_{\mathrm{MO}}^{\omega}(M)$ are limits of ρ_{ω}, i.e., for every exposed point $p \in \Sigma_{\mathrm{MO}}^{\omega}(M)$, there is $(s^*, x^*) \in S^1 \times M$ such that*

$$s^* f_{\omega}^*(x^*) = \lim_{t \to \infty} \rho_{\omega}^t(s^*, x^*) = p. \tag{2.6.14}$$

Remark 2.6.24. If M was connected, Lemma 2.6.23 would follow from [Ste09, Theorem 2.2.14]. But as we do not assume connectedness, we provide a proof here, which is based on Stender's proof of [Ste09, Theorem 2.2.14]. There, connectedness is only needed in order to show that $\langle \Sigma_{\mathrm{UG}}(M), z \rangle$ is an interval for every $z \in S^1$, compare [Ste09, Remark 2.2.10].

Proof of Lemma 2.6.23. Fix $\omega \in \mathbb{R}$, and let $p \in \Sigma_{\mathrm{MO}}^{\omega}(M)$ be an exposed point. Then by Definition 2.1.1, there exists $z \in S^1$, such that

$$\Sigma_{\mathrm{MO}}^{\omega}(M) \setminus \{p\} \subset \{x \in \mathbb{C} \mid \langle x, z \rangle < \langle p, z \rangle\}. \tag{2.6.15}$$

By Proposition 2.6.19 and Lemma 2.6.10, the set $\langle \Sigma_{\mathrm{MO}}^{\omega}(M), z \rangle$ is a compact interval, and by (2.6.15), it holds that $\langle p, z \rangle$ is its upper endpoint. By Proposition 2.6.12, there is a point $(s^*, x^*) \in S^1 \times M$ such that

$$\lim_{t \to \infty} \langle \rho_{\omega}^t(s^*, x^*), z \rangle = \langle p, z \rangle. \tag{2.6.16}$$

So (2.6.14) holds, since otherwise there must be a sequence $t_k \to \infty$ such that $\lim_{k \to \infty} \rho_{\omega}^{t_k}(s^*, x^*) =: \tilde{p} \neq p$. But this means $\langle \tilde{p}, z \rangle \in \langle \Sigma_{\mathrm{MO}}^{\omega}(M), z \rangle$, and hence $\langle \tilde{p}, z \rangle < \langle p, z \rangle$. This contradicts (2.6.16). $\qquad\square$

In our setting, the Morse spectrum has a very simple structure. In fact, it is a closed disc around the origin.

Theorem 2.6.25. *Assume that ρ_{ω} is continuous. Let $M \subset X$ be a compact and invariant set such that $S^1 \times M$ is chain transitive for $\Psi|_{S^1 \times M}$. Then for every $\omega \in \mathbb{R}$, the Morse spectrum $\Sigma_{\mathrm{MO}}^{\omega}(M)$ is the disc D_r around the origin with radius $r := \max_{x \in M} |f_{\omega}^*(x)|$, where the maximum is taken only over those points, where $f_{\omega}^*(x)$ exists.*

Proof. By [Ste09, Theorem 2.2.7], $\Sigma_{\mathrm{MO}}^{\omega}(M)$ is compact. By using Lemma 2.6.20, Lemma 2.6.21, and Proposition 2.6.22, convexity of $\Sigma_{\mathrm{MO}}^{\omega}(M)$ can be shown with the same arguments as in [CFJ07, Theorem 2.7 (i)].

By Lemma 2.6.23, the exposed points of $\Sigma_{\mathrm{MO}}^{\omega}(M)$ are limits of ρ_{ω}. Lemma 2.6.23 implies $\|p\| = \|f_{\omega}^*(b)\| \leq r$ for every exposed point $p \in \Sigma_{\mathrm{MO}}^{\omega}(M)$. So all exposed points of $\Sigma_{\mathrm{MO}}^{\omega}(M)$ are contained in D_r. As $\Sigma_{\mathrm{MO}}^{\omega}(M) \in \mathbb{C}$ is compact and convex, it is the convex hull of its exposed points, compare [Str35, Section 2]. Because D_r is convex, this implies $\Sigma_{\mathrm{MO}}^{\omega}(M) \subset D_r$. On the other hand by definition, for every $(z, b) \in S^1 \times M$, where the harmonic limit exists, it holds that $z f_{\omega}^*(b) \in \Sigma_{\mathrm{MO}}^{\omega}(M)$. In particular, let $b_0 \in M$ be such that $r = |f_{\omega}^*(b_0)|$. Then $z f_{\omega}^*(b_0) \in \Sigma_{\mathrm{MO}}^{\omega}(M)$ for every $z \in S^1$. Because D_r is the convex hull of $\{z f_{\omega}^*(b_0) \mid z \in S^1\}$, and $\Sigma_{\mathrm{MO}}^{\omega}(M)$ is convex, it follows that $\Sigma_{\mathrm{MO}}^{\omega}(M) \supset D_r$. $\qquad\square$

It might be surprising, that the Morse spectrum always is a disc. The following example illustrates, how every point in the disc can be obtained in the case of a rotation on S^1.

Example 2.6.26. Consider $X := S^1$, $\Phi_t x := \mathrm{e}^{-\mathrm{i}t\omega}x$ for $\omega \in \mathbb{R} \setminus \{0\}$, and $f(x) := 2\Re x = x + \overline{x}$. For $(z, x) \in S^1 \times S^1$ and $T > 0$, it holds that

$$
\begin{aligned}
z \int_0^T \mathrm{e}^{\mathrm{i}t\omega} f(\Phi_t x)\mathrm{d}t &= z \int_0^T \mathrm{e}^{\mathrm{i}t\omega}(\mathrm{e}^{-\mathrm{i}t\omega}x + \mathrm{e}^{\mathrm{i}t\omega}\overline{x})\mathrm{d}t \\
&= z \int_0^T (x + \mathrm{e}^{2\mathrm{i}t\omega}\overline{x})\mathrm{d}t \\
&= z(Tx + \frac{1}{2\mathrm{i}\omega}(\mathrm{e}^{\mathrm{i}T\omega} - 1)\overline{x}) \\
&= Tzx + \mathrm{i}\frac{1 - \mathrm{e}^{\mathrm{i}T\omega}}{2\omega} z\overline{x}.
\end{aligned}
$$

Now consider $\lambda \in \mathbb{C}$, $\varepsilon^k \to 0$, $T^k \to \infty$, and (ε^k, T^k)-chains ζ^k in $S^1 \times S^1$ with $\rho_\omega(\zeta^k) \to \lambda$. Then one can show that also $\tilde{\rho}_\omega(\zeta^k) \to \lambda$, where

$$
\tilde{\rho}_\omega(\zeta^k) := \frac{1}{\sum_{i=0}^{n-1} T_i} \sum_{i=0}^{n-1} T_i z_i x_i.
$$

To see this, note that

$$
\begin{aligned}
|\rho_\omega(\zeta^k) - \tilde{\rho}_\omega(\zeta^k)| &= \left| \frac{1}{\sum_{j=0}^{n-1} T_j} \sum_{j=0}^{n-1} \mathrm{i}\frac{1 - \mathrm{e}^{\mathrm{i}T_j\omega}}{2\omega} z\overline{x}_j \right| \\
&\leq \left| \frac{1}{nT^k} \cdot \frac{n}{\omega} \right| \\
&= \frac{1}{\omega T^k},
\end{aligned}
$$

which tends to 0 as $k \to \infty$.

Next, consider the specific (ε^k, T^k)-chains ζ^k of length $n^k := k$ given by

$$
\begin{aligned}
\varepsilon^k &:= 2\pi/k, \\
T_i &:= T^k, \\
z_{i+1} &:= \mathrm{e}^{\mathrm{i}\omega T} z_i \text{ and} \\
x_{i+1} &:= \mathrm{e}^{\mathrm{i}(\delta^k - \omega T)} z_i
\end{aligned}
$$

for $\delta^k := \alpha\varepsilon^k$, and arbitrary $\alpha \in [0, 1]$, $i = 0, \ldots, n^k - 1$. For these chains, $z_i = \mathrm{e}^{\mathrm{i}\omega iT} z_0$ and $x_i = \mathrm{e}^{\mathrm{i}(i\delta^k - \omega iT_i)} z_0$. Thus $T_i z_i x_i = T\mathrm{e}^{\mathrm{i}i\delta^k} z_0 x_0$. Hence

$$
\tilde{\rho}_\omega(\zeta^k) = \frac{1}{n^k T^k} \sum_{i=0}^{n^k-1} T^k \mathrm{e}^{\mathrm{i}i\delta^k} z_0 x_0 = \frac{z_0 x_0}{k} \sum_{i=0}^{k-1} \mathrm{e}^{\mathrm{i}i\delta^k}.
$$

For $\alpha \neq 0$, this implies

$$\tilde{\rho}_\omega(\zeta^k) = \frac{z_0 x_0}{k} \cdot \frac{e^{ik\delta^k} - 1}{e^{i\delta^k} - 1} = \frac{z_0 x_0}{k} \cdot \frac{e^{i\alpha 2\pi} - 1}{e^{i\alpha 2\pi/k} - 1} \rightarrow z_0 x_0 i \frac{1 - e^{2i\alpha\pi}}{2\alpha\pi}.$$

For $\alpha = 0$, one has $\delta^k = 0$, and thus $\tilde{\rho}_\omega(\zeta^k) = z_0 x_0$. Let $\gamma : [0, 1] \rightarrow D_1$ be given by

$$\gamma(\alpha) := \begin{cases} 1 & \text{if } \alpha = 0, \\ i\frac{1 - e^{2i\alpha\pi}}{2\alpha\pi} & \text{otherwise.} \end{cases}$$

The curve γ is continuous and it holds that $\lambda(\alpha) = z_0 x_0 \cdot \gamma(\alpha)$. As $\gamma(0) = 1$ and $\gamma(1) = 0$, it holds that $\bigcup_{(z_0, x_0) \in S^1 \times S^1, \alpha \in [0,1]} \lambda(\alpha) = D_1$. ⌟

2.6.4 Morse spectrum and rotational factor maps

Now we turn wo the question, what connection there is between the Morse spectrum and the existence of rotational factor maps.

If we identify $\mathbb{C}$ with $\mathbb{R}^2$, then for every $\omega \in \mathbb{R}$, by Lemma 2.6.3, the map ρ_ω is a growth rate in the sense of [Ste09, Definition 2.1.1] with the exception that it is in general not continuous in x. But we will additionally assume this continuity, where needed (e. g., in Theorem 2.6.29). So we spend some thoughts on continuity of ρ.

Lemma 2.6.27. *If $f : X \rightarrow \mathbb{R}$ and $\Phi : \mathbb{R}^+ \times X \rightarrow X$ are continuous, then the map $\rho : \mathbb{R} \times [0, \infty) \times S^1 \times X \rightarrow \mathbb{C}$, $(\omega, t, z, x) \mapsto \rho_\omega^t(z, x)$ is continuous.*

Proof. By Proposition 2.3.1, $f_\omega^t(x)$ is continuous in $(\omega, t, x) \in \mathbb{R} \times [0, \infty) \times X$. So $\rho_\omega^t(z, x) = z f_\omega^t(x)$ is continuous in z and in (ω, t, x). Clearly, the continuity in (ω, t, x) is uniform with respect to z, because $|z f_{\omega_1}^{t_1}(x_1) - z f_{\omega_1}^{t_1}(x_2)| = |z| \cdot |f_{\omega_1}^{t_1}(x_1) - f_{\omega_1}^{t_1}(x_2)| = |f_{\omega_1}^{t_1}(x_1) - f_{\omega_1}^{t_1}(x_2)|$. So by [Ama95, Lemma 8.1], the growth rate $\rho_\omega^t(z, x)$ is continuous. □

As f is bounded, also $\rho_\omega(\zeta)$ is bounded.

Lemma 2.6.28. *For every $\omega \in \mathbb{R}$, all $\varepsilon, T > 0$, and all (ε, T)-chains ζ in $S^1 \times X$, it holds that $|\rho_\omega(\zeta)| \leq \sup_{x \in X} |f(x)| < \infty$.*

Proof. Let $\omega \in \mathbb{R}$, $\varepsilon, T > 0$. Let ζ be an (ε, T)-chain in $S^1 \times X$ given by the points $(z_0, x_0), \ldots, (z_n, x_n) \in S^1 \times X$, and times $T_0, \ldots, T_{n-1} > T$. Let $F := \sup_{x \in X} |f(x)| < \infty$. Then

$$|\rho_\omega(\zeta)| = \left| \frac{1}{\sum_{i=0}^{n-1} T_i} \sum_{i=0}^{n-1} z_i \int_0^{T_i} e^{it\omega} f(\Phi_t x_i) dt \right|$$

$$\leq \frac{1}{\sum_{i=0}^{n-1} T_i} \sum_{i=0}^{n-1} |z_i| \int_0^{T_i} |e^{it\omega}| |f(\Phi_t x_i)| dt$$

$$= \frac{1}{\sum_{i=0}^{n-1} T_i} \sum_{i=0}^{n-1} \int_0^{T_i} |f(\Phi_t x_i)| \mathrm{d}t$$

$$\leq \frac{1}{\sum_{i=0}^{n-1} T_i} \sum_{i=0}^{n-1} T_i F$$

$$= F. \qquad \square$$

The following theorem gives some information on the connection between harmonic limits and the Morse spectrum over limit sets.

Theorem 2.6.29. *Assume that f and Φ are continuous. Let $M \subset X$ be compact and invariant, $x \in M$, $\omega \in \mathbb{R}$, and $t_k \subset \mathbb{R}$ with $t_k \to \infty$ such that $\lim_{k\to\infty} f_\omega^{t_k}(x)$ exists. Then $\lim_{k\to\infty} f_\omega^{t_k}(x) \in \Sigma_{\mathrm{MO}}^\omega(\omega(x))$. In particular, $f_\omega^*(x) \in \Sigma_{\mathrm{MO}}^\omega(\omega(x))$, if $f_\omega^*(x)$ exists.*

Note that $\omega(x)$ is compact and invariant, as M is compact, see [Ama95, Theorem 17.2]. Further note, that Φ restricted to $\omega(x)$ is chain transitive by [CK00, Proposition B.2.28]. So $\Sigma_{\mathrm{MO}}^\omega(\omega(x))$ is defined. This theorem is a special case of [Ste09, Theorem 2.3.6]. As Stender works on more general spaces than we do, his proof is much more complex than we need it. So we provide a simpler version of the proof.

Proof of Theorem 2.6.29. Fix $\omega \in \mathbb{R}$, $\delta > 0$, $\varepsilon > 0$, and $T > 0$. Let $x \in S$, and $(s_k)_{k\in\mathbb{N}} \subset \mathbb{R}^+$ be a sequence with $s_k \to \infty$ for $k \to \infty$ such that $\lambda := \lim_{k\to\infty} f_\omega^{s_k}(x) = \lim_{k\to\infty} \rho_\omega^{s_k}(1, x)$ exists. It suffices to show that there is an (ε, T)-chain ξ in $S^1 \times \omega(x)$ such that

$$|\rho_\omega(\xi) - \lambda| \leq \delta. \qquad (2.6.17)$$

To be precise, (2.6.17) implies that for all sequences $\varepsilon_k \to 0$, $T_k \to 0$ and $\delta_k \to 0$ there are (ε_k, T_k)-chains ξ^k in $S^1 \times \omega(x)$ with $|\rho_\omega(\xi^k) - \lambda| \leq \delta_k$ for all k. As $\rho_\omega(\xi^k)$ is bounded (compare Lemma 2.6.28), we may assume that it converges. As $\delta_k \to 0$, we get $|\rho_\omega(\xi^k) - \lambda| \to 0$, i.e., $\rho_\omega(\xi^k) \to \lambda$.

As $\rho_\omega : [T, 2T] \times S^1 \times X \to \mathbb{C}$ is continuous (compare Lemma 2.6.27), and $S^1 \times X$ is compact, there is $0 < \gamma \leq \varepsilon/2$ such that for all $(z_1, x_1), (z_2, x_2) \in S^1 \times X$ with $d\big((z_1, x_1), (z_2, x_2)\big) < \gamma$, it holds that

$$|\rho_\omega^t(z_1, x_1) - \rho_\omega^t(z_2, x_2)| \leq \frac{\delta}{2} \qquad (2.6.18)$$

for all $t \in [T, 2T]$.

As $\Phi : [T, 2T] \times X \to X$ is continuous, there is $0 < \beta \leq \gamma/2$ such that for all $t \in [T, 2T]$ and all $x_1, x_2 \in X$ with $d(x_1, x_2) < \beta$, it holds that

$$d(\Phi_t x_1, \Phi_t x_2) < \gamma. \qquad (2.6.19)$$

As $\Phi_t x \to \omega(x)$ for $t \to \infty$ (compare [Ama95, Theorem 17.2]), there is $S \geq 0$ such that for all $t \geq S$, it holds that

$$d\big(\Phi_t x, \omega(x)\big) < \beta. \tag{2.6.20}$$

Let $(z_0, x_0) := \Psi_S^\omega(1, x) = \big(e^{iS\omega}, \Phi_S x\big)$. Let $t_k := s_k - S$. We may assume that $t_k \geq 0$. Then $\lim_{k\to\infty} \rho_\omega^{t_k}(z_0, x_0) = \lambda$ by Lemma 2.6.5. Thus there is $K \in \mathbb{N}$ such that for all $k \geq K$, it holds that

$$|\rho_\omega^{t_k}(z_0, x_0) - \lambda| < \frac{\delta}{2}. \tag{2.6.21}$$

Choose k big enough such that $T_0 := t_k \geq \max\{t_K, 2T\}$. There is $n \in \mathbb{N}$ such that $T_0 = nT + r$ for some $r \in [T, 2T)$.

Define $\tau_0 := \cdots := \tau_{n-1} := T$ and $\tau_n := r$. Let $(z_{j+1}, x_{j+1}) := \Psi_{\tau_j}^\omega(z_j, x_j) = \big(e^{i\omega\tau_j} x_j, \Phi_{\tau_j} x_j\big)$, $j = 1, \ldots, n$. This constitutes a trivial (ε, T)-chain $\tilde{\xi}$ in $S^1 \times X$. By Lemma 2.6.17, triviality of the chain implies $\rho_\omega(\tilde{\xi}) = \rho_\omega^{T_0}(z_0, x_0)$. Hence, (2.6.21) implies

$$|\rho_\omega(\tilde{\xi}) - \lambda| < \frac{\delta}{2}. \tag{2.6.22}$$

In general, $\tilde{\xi}$ does not lie in $S^1 \times \omega(x)$. So we will construct a chain ξ in $S^1 \times \omega(x)$ from $\tilde{\xi}$ by, loosely speaking, "pushing" $\tilde{\xi}$ onto $S^1 \times \omega(x)$ (see Figure 2.3 for an illustration). By (2.6.20), for every $j = 0, \ldots, n$, there is $y_j \in \omega(x)$ with

$$d(x_j, y_j) < \beta < \frac{\gamma}{2}. \tag{2.6.23}$$

Let ξ denote the chain given by points (z_j, y_j) and times τ_j.

The chain ξ lies in $S^1 \times \omega(x)$, as $\omega(x)$ is invariant. Furthermore, ξ is an (ε, T)-chain. In order to see this, note that $\Psi_{\tau_j}^\omega(z_j, y_j) = (z_{j+1}, \Phi_{\tau_j} y_j)$. So

$$\begin{aligned}
d\big(\Psi_{\tau_j}^\omega(z_j, y_j), (z_{j+1}, y_{j+1})\big) &= d\big((z_{j+1}, \Phi_{\tau_j} y_j), (z_{j+1}, y_{j+1})\big) \\
&= d(\Phi_{\tau_j} y_j, y_{j+1}) \\
&\leq d(\Phi_{\tau_j} y_j, \Phi_{\tau_j} x_j) + d(\Phi_{\tau_j} x_j, y_{j+1}) \\
&= d(\Phi_{\tau_j} y_j, \Phi_{\tau_j} x_j) + d(x_{j+1}, y_{j+1}).
\end{aligned}$$

By (2.6.19) and (2.6.23), this implies $d\big(\Psi_{\tau_j}^\omega(z_j, y_j), (z_{j+1}, y_{j+1})\big) < \gamma + \beta \leq {}^{3}\!/\!_2 \cdot \gamma \leq {}^{3}\!/\!_4 \cdot \varepsilon < \varepsilon$. By definition, $\tau_j \geq T$ for all j. So ξ indeed is an (ε, T)-chain in $S^1 \times \omega(x)$.

It remains to show that (2.6.17) holds.

$$\begin{aligned}
|\rho_\omega(\xi) - \lambda| &\leq |\rho_\omega(\tilde{\xi}) - \lambda| + |\rho_\omega(\xi) - \rho_\omega(\tilde{\xi})| \\
&\overset{(2.6.22)}{\leq} \frac{\delta}{2} + |\rho_\omega(\xi) - \rho_\omega(\tilde{\xi})|
\end{aligned}$$

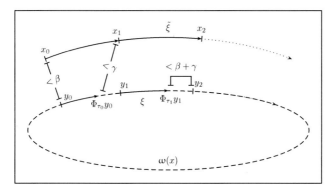

Figure 2.3: Constructing a chain $S^1 \times \xi$ in $\omega(x)$ from a trivial chain $\tilde{\xi}$ nearby
This figure only shows the components in X of $\tilde{\xi}$ and ξ.

$$= \frac{\delta}{2} + \left| \frac{1}{T_0} \sum_{j=0}^{n-1} \tau_j \rho_\omega^{\tau_j}(z_j, y_j) - \frac{1}{T_0} \sum_{j=0}^{n-1} \tau_j \rho_\omega^{\tau_j}(z_j, x_j) \right|$$

$$\leq \frac{\delta}{2} + \frac{1}{T_0} \sum_{j=0}^{n-1} \tau_j |\rho_\omega^{\tau_j}(z_j, y_j) - \rho_\omega^{\tau_j}(z_j, x_j)|$$

$$\overset{(2.6.18),\ (2.6.23)}{\leq} \frac{\delta}{2} + \frac{1}{T_0} \sum_{j=0}^{n-1} \tau_j \frac{\delta}{2}$$

$$= \delta \qquad\qquad\qquad \square$$

Let us have a look at the number

$$\sigma_\omega(x) := \max\{|\lambda| \mid \lambda \in \Sigma_{\mathrm{MO}}^\omega(\omega(x))\},$$

i.e., the radius of the Morse spectrum with respect to the limit set $\omega(x)$. Recall from Theorem 2.6.25, that $\Sigma_{\mathrm{MO}}^\omega(\omega(x))$ is the disc around the origin with radius $\sigma_\omega(x) = \max_{x \in \omega(x)} |f_\omega^\star(x)|$. So we get the following connection between the radius of the Morse spectrum and harmonic limits.

Theorem 2.6.30. *Assume that f and Φ are continuous. Let $M \subset X$ be compact and invariant, and $\omega \in \mathbb{R}$. Let $x_0 \in M$.*

1. *If $\sigma(x_0) = 0$, then $f_\omega^\star(x) = 0$ for all $x \in X$ with $\omega(x) \subset \omega(x_0)$.*

2. *If $\sigma(x_0) \neq 0$, then there is $x \in \omega(x_0)$ such that $f_\omega^\star(x) \neq 0$.*

Proof.

1. Assume that $\sigma_\omega(x_0) = 0$. Then $\Sigma_{\text{MO}}^\omega\big(\omega(x_0)\big) = \{0\}$. By Theorem 2.6.29, all limit points of $f_\omega^t(x)$ for $t \to \infty$ lie in $\Sigma_{\text{MO}}^\omega\big(\omega(x)\big)$. By boundedness of f, the harmonic average $f_\omega^t(x)$ is bounded, see Proposition 1.3.1. Hence $f_\omega^*(x)$ exists and equals zero.

2. Assume that $\sigma_\omega(x_0) \neq 0$. By Theorem 2.6.25, there is $x \in \omega(x_0)$ such that $|f_\omega^*(x)| = |\sigma_\omega(x_0)| \neq 0$.

$\square$

With this theorem, we can show the following result on the existence of rotational factor maps.

Corollary 2.6.31. *Assume that f and Φ are continuous. Let $M \subset X$ be compact and invariant, and $\omega \in \mathbb{R}$. If there is $x_0 \in M$ with $\sigma(x_0) \neq 0$, then $\overline{f_\omega^*}$ is a rotational factor map to frequency $\omega/2\pi$.*

Proof. By part 2. of Theorem 2.6.30, there is $x \in \omega(x_0)$ such that $f_\omega^*(x) \neq 0$. By Theorem 2.2.9, this implies that $\overline{f_\omega^*}$ is a rotational factor map by frequency $\omega/2\pi$. $\square$

Chapter 3

Control analysis

Consider the control system

$$\dot{x} = g(x, u) \tag{3.0.1}$$

on the set $X \subset \mathbb{R}^n$ for $g : \mathbb{X} \times U \to \mathbb{R}^n$ with controls $u : \mathbb{R} \to U \subset \mathbb{R}^m$ in some class $\mathcal{U}$ of so-called *admissible* controls. Assume that, for all $x \in X$, $u \in \mathcal{U}$, a unique solution of (3.0.1) exists, and denote this solution by $\Phi_t^u : \mathbb{R}^n \to \mathbb{R}^n$. Note that existence and uniqueness of the solution is guaranteed, if, e. g., all controls $u \in \mathcal{U}$ are bounded and piecewise continuous, g is continuous, and g is locally Lipschitz in its first argument, see Remark 3.4.1. We will also assume that $\mathcal{U}$ is shift invariant, i. e., that for all $u \in \mathcal{U}$ and all $t \in \mathbb{R}$, also $\theta_t u := u(\cdot + t) \in \mathcal{U}$. Hence we can additionally consider the map $\Psi_t : X \times \mathcal{U} \to X \times \mathcal{U}$ given by $(x, u) \mapsto (\Phi_t^u x, \theta_t u)$ for $t \geq 0$.

For fixed $u \in \mathcal{U}$, the map $\Phi^u : \mathbb{R}^+ \times X \to X$, $(t, x) \mapsto \Phi_t^u x$, is a map as considered in Chapter 2, which additionally depends on the parameter $u \in \mathcal{U}$. So we can analyze the rotational behaviour of such control systems via harmonic limits. To this end, we will apply the concepts and results from Chapter 2 to Φ_t^u in Section 3.2. In fact, we will consider, more generally, that for a fixed control $u \in \mathcal{U}$, a map $\Phi : \mathbb{R}^+ \times X \times \{\theta_t u \mid t \geq 0\} \to X$, $(t, x, u) \mapsto \Phi_t^u x$ is given, which need not stem from a differential equation, such that

$$\Phi_{s+t}^u = \Phi_s^{\theta_t u} \Phi_t^u \tag{3.0.2}$$

for all $s, t \geq 0$. Here, we can allow X to be any metric space instead of a subset of $\mathbb{R}^n$. Note that Φ_t^u is no semi-flow in general, because a semi-flow has to satisfy $\Phi_{s+t}^u = \Phi_s^u \Phi_t^u$ instead of (3.0.2), so some results from Chapter 2 cannot be applied here. In particular, the existence of a rotational factor map is not equivalent to the existence of a nonvanishing harmonic limit here. But in the case of periodic controls, we can modify those results such that they also hold in this setting. This is done in Subsection 3.2.2.

The map $\Psi : X \times \mathcal{U} \to X \times \mathcal{U}$ is a semi-flow. For some classes of control systems, Ψ even is a continuous flow, see [CK00, Chapter 4]. We will gather results for Ψ_t in Section 3.3, and analyze their rotational behaviour. In fact, we will consider all semiflows on $X \times \mathcal{U}$, not only those that are induced by a differential equation (3.0.1).

In Section 3.4, we analyze convergent systems, i. e., systems that, for every control $u \in \mathcal{U}$, have a bounded asymptotically stable solution, which is called a *steady-state solution*. For these systems, we can show that arbitrary frequencies can be produced by applying periodic controls, see Theorem 3.4.15 and Theorem 3.4.18. Furthermore, the control spectrum does not contain isolated points, see Theorem 3.4.28.

3.1 Preliminaries

Let us first collect some definitions and results, which we will use later in this chapter.

We will use again Lemma 1.1.2, Lemma 2.1.6, Lemma 2.1.7 and Lemma 2.1.8 from the previous preliminaries sections.

Antiderivatives   The first result is concerned with antiderivatives of $t \mapsto e^{at}p(t)$ for polynomials p.

Lemma 3.1.1. *Let $p \in \mathbb{C}[t]$ be a polynomial, and $a \in \mathbb{C} \setminus \{0\}$. Then there is a unique polynomial $q \in \mathbb{C}[t]$ such that $e^{at}q(t)$ is an antiderivative of $e^{at}p(t)$. It holds that*

$$q(t) = \sum_{j=0}^{\deg p} \frac{(-1)^j}{a^{j+1}} p^{(j)}(t).$$

Proof. We show the existence of such a polynomial q by induction over the degree of p. Assume that $\deg p = 0$, i. e., $p \equiv p(0)$. Clearly, $1/a \cdot e^{at}p(0)$ is an antiderivative of $e^{at}p(0)$, and $q(t) := 1/a \cdot p(0)$ is a polynomial.

Let $n \in \mathbb{N}$, and assume that we already know that $e^{at}p(t)$ has an antiderivative of the form $e^{at}q(t)$ for $q \in \mathbb{C}[t]$ if $\deg p = n-1$. Assume that $\deg p = n$. Then by partial integration, an antiderivative of $e^{at}p(t)$ can be given by $1/a \cdot e^{at}p(t) - \int 1/a \cdot e^{at}p'(t)dt$. As $1/a \cdot p'$ is a polynomial of degree $n - 1$, there is a polynomial $\tilde{q} \in \mathbb{C}[t]$ such that $e^{at}\tilde{q}(t)$ is an antiderivative of $1/a \cdot e^{at}p'(t)$. Thus $1/a \cdot e^{at}p(t) - e^{at}\tilde{q}(t) = e^{at}\left(1/a \cdot p(t) - \tilde{q}(t)\right)$ is an antiderivative of $e^{at}p(t)$. As $q(t) := 1/a \cdot p(t) - \tilde{q}(t)$ is a polynomial, this completes the induction.

Next show the uniqueness of q. Assume that there are $q_1, q_2 \in \mathbb{C}[t]$ such that $e^{at}q_j(t)$ is an antiderivative of $e^{at}p(t)$ for $j = 1, 2$. As antiderivatives are unique up to addition of a constant, there is $c \in \mathbb{C}$ such that $e^{at}q_1(t) - e^{at}q_2(t) = c$ for all $t \in \mathbb{R}$. This implies $e^{-at}c = q_1(t) - q_2(t) \in \mathbb{C}[t]$, which can only be true if $c = 0$ (remember that $a \neq 0$). So $e^{at}q_1(t) = e^{at}q_2(t)$ for all t, and thus $q_1 = q_2$.

Finally, show that

$$q(t) = \sum_{j=0}^{\deg p} \frac{(-1)^j}{a^{j+1}} p^{(j)}(t).$$

Let $F : \mathbb{C}[t] \to \mathbb{C}[t]$ be given by

$$p \mapsto \sum_{j=0}^{\deg p} \frac{(-1)^j}{a^{j+1}} p^{(j)}.$$

So we have to show that $e^{at}F(p)(t)$ is an antiderivative of $e^{at}p(t)$ for every $p \in \mathbb{C}[t]$. By the induction above, it suffices to show that this holds for polymials p of degree 0, and that $F(p) = 1/a \cdot p - F\left(1/a \cdot p'\right)$ holds for polynomials p with $\deg p > 0$.

Assume that $\deg p = 0$, i.e., $p \equiv p(0)$. Then $F(p) = 1/a \cdot p(0)$. Clearly, $1/a \cdot e^{at}p(0)$ is an antiderivative of $e^{at}p(t) = e^{at}p(0)$. Now assume that $\deg p > 0$. Then

$$
\begin{aligned}
F(p) - \frac{1}{a}p + F\left(\frac{1}{a}p'\right) &= \sum_{j=0}^{\deg p} \frac{(-1)^j}{a^{j+1}} p^{(j)} - \frac{1}{a}p + \sum_{j=0}^{\deg(1/a \cdot p')} \frac{(-1)^j}{a^{j+1}} \left(\frac{1}{a}p'\right)^{(j)} \\
&= \sum_{j=1}^{\deg p} \frac{(-1)^j}{a^{j+1}} p^{(j)} + \sum_{j=0}^{\deg p'} \frac{(-1)^j}{a^{j+2}} p^{(j+1)} \\
&= \sum_{j=1}^{\deg p} \frac{(-1)^j}{a^{j+1}} p^{(j)} - \sum_{j=1}^{\deg p} \frac{(-1)^j}{a^{j+1}} p^{(j)} \\
&= 0. \qquad \qquad \square
\end{aligned}
$$

Topology We will consider products of metric spaces endowed with the product topology later. The following lemma shows that this topology can be induced by certain metrics.

Lemma 3.1.2. *Let X_1, X_2 be metric spaces with metrics d_1 and d_2, respectively. Let $d_{\mathbb{R}^2}$ be a metric on $\mathbb{R}^2$. Then $d(x_1, x_2) := d_{\mathbb{R}^2}\left(d_1(x_1, x_1), d_2(x_1, x_1)\right)$ for $x_1 \in X_1$, $x_2 \in X_2$ defines a metric on the product space $X := X_1 \times X_2$. This metric induces the product topology.*

Proof. This follows from [Sch79, Theorem 2.2.2] and the equivalence of norms on $\mathbb{R}^2$. $\qquad \square$

Next, define the weak* topology, which we will use in connection with the set $\mathcal{U}$ of control functions.

Definition 3.1.3. Let E be a normed real (or complex) vector space, and denote by E^* its dual. The weakest topology on E^*, such that, for every $x \in E$, the map $\hat{x} : E^* \to \mathbb{R}$ (or $E^* \to \mathbb{C}$ respectively) given by $f \mapsto \hat{x}(f) := f(x)$ is continuous, is called *weak* topology.* $\lrcorner$

Remark 3.1.4. More generally, one can define the weak* topology for locally compact spaces instead of normed spaces, compare [Wer05, Definition VIII.3.2]. $\lrcorner$

Chapter 3 Control analysis

Example 3.1.5. Let $\mathbb{K} \in \{\mathbb{R}, \mathbb{C}\}$. Let μ be a σ-finite measure on $\mathbb{K}^m$. Consider $E := L^1(\mu)$. Then $E^* = L^\infty(\mu)$, see [Els05, Satz VII.3.2]. Here the weak* topology on $L^\infty(\mu)$ is the weakest topology, such that the map $L^\infty(\mu) \to \mathbb{K}$ given by $g \mapsto \int f g \mathrm{d}\mu$ is continuous for all $f \in L^1(\mu)$.

In this case, weak* convergence can be formulated as follows: A sequence $(g_n) \subset L^\infty(\mu)$ converges in the weak* topology to some $g \in L^\infty(\mu)$ if and only if it holds that $\lim_{n\to\infty} \int f g_n \mathrm{d}\mu = \int f g \mathrm{d}\mu$ for all $f \in L^1(\mu)$. ⌐

The weak* topology can be induced by the following kind of metrics.

Definition 3.1.6. Consider the set $\mathcal{U} := \{u : \mathbb{R} \to U \mid u \text{ measurable}\} \subset L^\infty(\mathbb{R}, \mathbb{R}^m)$ of admissible control functions for some compact set $U \subset \mathbb{R}^m$. A metric d on $\mathcal{U}$ of the form

$$d(u, v) = \sum_{n=1}^\infty \frac{1}{2^n} \frac{\left| \int \langle u(t) - v(t), x_n(t) \rangle \mathrm{d}t \right|}{1 + \left| \int \langle u(t) - v(t), x_n(t) \rangle \mathrm{d}t \right|},$$

where $\{x_n : \mathbb{R} \to \mathbb{R}^m \mid n \in \mathbb{N}\}$ is a countable dense subset of $L^1(\mathbb{R}, \mathbb{R}^m)$, and $\langle \cdot, \cdot \rangle$ denotes the Euclidean inner product in $\mathbb{R}^m$, is called a weak* metric—compare [CK00, Lemma 4.2.1] and the proof of [DS58, Theorem V.5.1]. ⌐

We will need the following lemma, when we analyze almost-periodic controls.

Lemma 3.1.7. *Let* $\mathcal{U}$ *be a shift-invariant set of almost periodic functions with values in some compact convex set* $U \subset \mathbb{R}^m$. *Endow* $\mathcal{U}$ *with a weak* metric (see Definition 3.1.6). It holds that, for all* $\tau \in \mathbb{R}$ *and all* $\delta > 0$, *there is* $\varepsilon > 0$ *such that, if* $d\big(u(t+\tau), u(t)\big) \le \varepsilon$ *for all* $t \in \mathbb{R}$, *then also*

$$d(\theta_{t+\tau} u, \theta_t u) \le \delta \tag{3.1.1}$$

for all $t \in \mathbb{R}$.

Remark 3.1.8. Lemma 3.1.7 does not hold for arbitrary metrics. Consider, e. g., the discrete metric

$$d(u, v) := \begin{cases} 1 & \text{if } u \ne v, \\ 0 & \text{if } u = v. \end{cases}$$

Let $\tau > 0$ and $1 > \delta > 0$. Then (3.1.1) can only be satisfied if $\theta_{t+\tau} u = \theta_t u$ for all $t \in \mathbb{R}$, i.e., if u is τ-periodic, but not for general almost periodic functions u. ⌐

Proof. Let $\tau \in \mathbb{R}$ and $\delta > 0$. Let $N \in \mathbb{N}$ be such that $\sum_{n=N+1}^\infty 2^{-n} < \delta/2$. Let x_n, $n \in \mathbb{N}$, be those functions, by which the metric on $\mathcal{U}$ is defined. Let $\xi := \max_{n=1}^N \int_\mathbb{R} \|x_n(s)\| \mathrm{d}s$. Let $\varepsilon > 0$ be such that

$$\frac{\varepsilon \xi}{1 + \varepsilon \xi} < \frac{\delta}{2}.$$

Assume that $d\big(u(t+\tau), u(t)\big) \leq \varepsilon$ for all $t \in \mathbb{R}$. Then

$$
\begin{aligned}
d(\theta_{t+\tau}u, \theta_t u) &= \sum_{n=1}^{\infty} \frac{1}{2^n} \frac{\left|\int_{\mathbb{R}} \langle \theta_{t+\tau}u(s) - \theta_t u(s), x_n(s)\rangle \mathrm{d}s\right|}{1 + \left|\int_{\mathbb{R}} \langle \theta_{t+\tau}u(s) - \theta_t u(s), x_n(s)\rangle \mathrm{d}s\right|} \\
&\leq \sum_{n=1}^{\infty} \frac{1}{2^n} \frac{\int_{\mathbb{R}} \|\theta_{t+\tau}u(s) - \theta_t u(s)\| \|x_n(s)\| \mathrm{d}s}{1 + \int_{\mathbb{R}} \|\theta_{t+\tau}u(s) - \theta_t u(s)\| \|x_n(s)\| \mathrm{d}s} \\
&= \sum_{n=1}^{\infty} \frac{1}{2^n} \frac{\int_{\mathbb{R}} \|u(s+t+\tau) - u(s+t)\| \|x_n(s)\| \mathrm{d}s}{1 + \int_{\mathbb{R}} \|u(s+t+\tau) - u(s+t)\| \|x_n(s)\| \mathrm{d}s} \\
&\leq \sum_{n=1}^{\infty} \frac{1}{2^n} \frac{\int_{\mathbb{R}} \varepsilon \|x_n(s)\| \mathrm{d}s}{1 + \int_{\mathbb{R}} \varepsilon \|x_n(s)\| \mathrm{d}s} \\
&\leq \sum_{n=1}^{N} \frac{1}{2^n} \frac{\varepsilon \xi}{1 + \varepsilon \xi} + \sum_{n=N+1}^{\infty} \frac{1}{2^n} \cdot 1 \\
&< \sum_{n=1}^{N} \frac{1}{2^n} \frac{\delta}{2} + \frac{\delta}{2} \\
&< \delta. \qquad\qquad \square
\end{aligned}
$$

In this chapter, we need the following variant of the Birkhoff Ergodic Theorem. Recall [CFS82, Theorem 1.2.1]. The difference between these two theorems is, that we additionally assume boundedness here, but do not need the full semi-flow property.

Theorem 3.1.9 (Birkhoff Ergodic Theorem for continuous time). *Consider a metric space X, a map $S : \mathbb{R}^+ \times X \to X$ given by $(t, x) \mapsto S_t x$, a finite S_1-invariant measure μ, and a function $f : X \to \mathbb{C}$ that is p-integrable with respect to μ, $1 \leq p < \infty$. Further assume that $S_{s+t}x = S_s S_t x$ for all $x \in X$, all $s \in \mathbb{R}^+$, and all $t \in \mathbb{N}_0$, and assume that $f\big(S_t(x)\big)$ for μ-almost all $x \in X$ is bounded and measurable in (t, x). Then the limit*

$$
f^*(x) := \lim_{T \to \infty} \frac{1}{T} \int_0^T f(S_t x) \mathrm{d}t
$$

exists for μ-almost all $x \in X$. Furthermore, f^* is p-integrable, and it holds that

$$
\int_X f^* \mathrm{d}\mu = \int_X f \mathrm{d}\mu.
$$

Proof. Note that for every $T \geq 1$, it holds that

$$
\frac{1}{T} \int_0^T f\big(S_t(x)\big) \mathrm{d}t = \frac{\lfloor T \rfloor}{T} \cdot \frac{1}{\lfloor T \rfloor} \int_0^{\lfloor T \rfloor} f\big(S_t(x)\big) \mathrm{d}t + \frac{1}{T} \int_{\lfloor T \rfloor}^T f\big(S_t(x)\big) \mathrm{d}t. \tag{3.1.2}
$$

Further note that

$$\frac{\lfloor T \rfloor}{T} \to 1 \tag{3.1.3}$$

and

$$\frac{1}{T} \int_{\lfloor T \rfloor}^{T} f\big(S_t(x)\big) \mathrm{d}t \to 0 \tag{3.1.4}$$

for almost all $x \in X$ and $T \to \infty$. Equation (3.1.4) holds, because $f\big(S_t(x)\big)$ is bounded in t for almost all $x \in X$.

So from equations (3.1.2), (3.1.3) and (3.1.4), it follows that

$$\lim_{T \to \infty} \frac{1}{T} \int_0^T f\big(S_t(x)\big) \mathrm{d}t = \lim_{T \to \infty} \frac{1}{\lfloor T \rfloor} \int_0^{\lfloor T \rfloor} f\big(S_t(x)\big) \mathrm{d}t = \lim_{n \to \infty} \frac{1}{n} \int_0^n f\big(S_t(x)\big) \mathrm{d}t, \tag{3.1.5}$$

if one of the limits exists. Thus it suffices to show that

$$\lim_{n \to \infty} \frac{1}{n} \int_0^n f\big(S_t(x)\big) \mathrm{d}t \tag{3.1.6}$$

exists for almost all x. This can be shown using the discrete-time version of the Birkhoff Ergodic Theorem [KH06, Theorem 4.1.2].

Let

$$g(x) := \int_0^1 f\big(S_t(x)\big) \mathrm{d}t.$$

Then it holds that

$$
\begin{aligned}
\frac{1}{n} \int_0^n f\big(S_t(x)\big) \mathrm{d}t &= \frac{1}{n} \sum_{k=0}^{n-1} \int_k^{k+1} f\big(S_t(x)\big) \mathrm{d}t \\
&= \frac{1}{n} \sum_{k=0}^{n-1} \int_k^{k+1} f\Big(S_{t-k}\big(S_k(x)\big)\Big) \mathrm{d}t \\
&= \frac{1}{n} \sum_{k=0}^{n-1} \int_0^1 f\Big(S_t\big(S_k(x)\big)\Big) \mathrm{d}t \\
&= \frac{1}{n} \sum_{k=0}^{n-1} g\big(S_k(x)\big) \\
&= \frac{1}{n} \sum_{k=0}^{n-1} g\big(S_1^k(x)\big).
\end{aligned}
\tag{3.1.7}
$$

By the discrete-time Birkhoff Ergodic Theorem [KH06, Theorem 4.1.2], the limit

$$\lim_{n \to \infty} \frac{1}{n} \sum_{k=0}^{n-1} g\big(S_1^k(x)\big) \tag{3.1.8}$$

exists for almost every $x \in X$. Note that all preconditions of that theorem are satisfied: S_1 is measurable and preserves μ by assumption. The map g is integrable, because

$$
\begin{aligned}
\int_X g \mathrm{d}\mu &= \int_X \int_0^1 f\big(S_t(x)\big) \mathrm{d}t \mathrm{d}\mu(x) \\
&\overset{(*)}{=} \int_0^1 \int_X f\big(S_t(x)\big) \mathrm{d}\mu(x) \mathrm{d}t \\
&\overset{(**)}{=} \int_0^1 \int_X f(x) \mathrm{d}\mu(x) \mathrm{d}t \\
&= \int_X f(x) \mathrm{d}\mu(x).
\end{aligned}
\tag{3.1.9}
$$

Here, the equation marked with $(*)$ holds due to Fubini's Theorem [Bau92, Korollar 23.7], because $(t, x) \mapsto f\big(S_t(x)\big)$ is bounded and measurable, and hence integrable on $[0, 1] \times X$; the equation marked with $(**)$ holds due to invariance of μ. So by (3.1.7), also the limit (3.1.6) exists, which completes the first part of the proof.

Note that equations (3.1.5) and (3.1.7) imply for almost all $x \in X$, that

$$
f^*(x) = \lim_{T \to \infty} \frac{1}{T} \int_0^T f\big(S_t(x)\big) \mathrm{d}t = \lim_{n \to \infty} \frac{1}{n} \sum_{k=0}^{n-1} g\big(S_1^k(x)\big).
$$

Further note that

$$
\left| \frac{1}{n} \sum_{k=0}^{n-1} g\big(S_1^k(x)\big) \right| \leq \frac{1}{n} \sum_{k=0}^{n-1} \sup_{x \in X} |g(x)| = \sup_{\xi \in X} |g(\xi)| \leq \sup_{\xi \in X} |f(\xi)|
$$

for all $n \in \mathbb{N}$. Clearly, $1/n \cdot \sum_{k=0}^{n-1} g\big(S_1^k(x)\big)$ is integrable in x for every $n \in \mathbb{N}$, because g is integrable and μ is invariant. Also $\sup_{\xi \in X} |f(\xi)|$ is integrable in x, because it is constant, and μ is finite. So we can apply Lebesgue's Dominated Convergence Theorem [Bau92, Satz 15.6], and get that the map f^* is p-integrable, and that

$$
\int f^*(x) \mathrm{d}\mu(x) = \int \lim_{n \to \infty} \frac{1}{n} \sum_{k=0}^{n-1} g\big(S_1^k(x)\big) \mathrm{d}\mu(x) = \lim_{n \to \infty} \int \frac{1}{n} \sum_{k=0}^{n-1} g\big(S_1^k(x)\big) \mathrm{d}\mu(x).
$$

Hence

$$
\begin{aligned}
\int f^*(x) \mathrm{d}\mu(x) &= \lim_{n \to \infty} \frac{1}{n} \sum_{k=0}^{n-1} \int g\big(S_1^k(x)\big) \mathrm{d}\mu(x) \\
&\overset{(*)}{=} \lim_{n \to \infty} \frac{1}{n} \sum_{k=0}^{n-1} \int g(x) \mathrm{d}\mu(x)
\end{aligned}
$$

145

$$\overset{(3.1.9)}{=} \lim_{n \to \infty} \frac{1}{n} \sum_{k=0}^{n-1} \int f(x) \mathrm{d}\mu(x)$$

$$= \int f(x) \mathrm{d}\mu(x),$$

where the equation marked with $(*)$ holds due to invariance of μ. $\qquad \square$

3.2 Pointwise analysis

First, we perform a pointwise analysis, i.e., we fix a control $u \in \mathcal{U}$, and assume that, for a metric space X, a map $\Phi : \mathbb{R}^+ \times X \times \{\theta_t u \mid t \geq 0\} \to X$, $(t, x, u) \mapsto \Phi_t^u x$, is given that satisfies (3.0.2). This is a map as considered in Chapter 2, which additionally depends on $u \in \mathcal{U}$. So we can apply those concepts and results from Chapter 2 to this new setting that do not require Φ^u to be a semi-flow. Note that, as in Chapter 2, we do not assume continuity of Φ^u in general. But we will assume that $f(\Phi_t^u x)$ is locally integrable, whenever we consider this term for some $f : X \to \mathbb{C}$ and $x \in X$.

We start by defining harmonic averages and limits.

Definition 3.2.1 (Harmonic average and limit). For any function $f : X \to \mathbb{C}$, such that $f(\Phi_t^u x)$ is locally integrable with respect to t, all $\omega \in \mathbb{R}$, and $x \in X$, define the *harmonic average*

$$f_\omega^T(x, u) := \frac{1}{T} \int_0^T \mathrm{e}^{\mathrm{i}\omega t} f(\Phi_t^u x) \mathrm{d}t;$$

furthermore, define the *harmonic limit*

$$f_\omega^*(x, u) := \lim_{T \to \infty} f_\omega^T(x, u),$$

if the limit exists. $\qquad \lrcorner$

Example 3.2.2. Consider the control system

$$\dot{x} = u$$

in $\mathbb{R}$ for controls $u : \mathbb{R} \to \mathbb{R}$. The solution is given by $\Phi_t^u x = \int_0^t u(\tau) \mathrm{d}\tau + x$. Fix $u := \cos$ and $f := \mathrm{id}$. Then $\Phi_t^{\cos} x = \sin t + x$, and hence

$$f_\omega^*(x, \cos) = \begin{cases} x & \text{if } \omega = 0, \\ \pm \frac{1}{2}\mathrm{i} & \text{if } \omega = \pm 1, \\ 0 & \text{otherwise,} \end{cases}$$

as we will show now.

It holds that

$$f_0^T(x, \cos) = \frac{1}{T} \int_0^T (\sin t + x)\mathrm{d}t$$

$$= \frac{1}{T}[-\cos t + xt]_{t=0}^T$$

$$= \frac{1}{T}(-\cos T + xT + 1)$$

$$= \frac{1}{T}(1 - \cos T) + x$$

$$\to x \text{ as } T \to \infty.$$

This proves the case $\omega = 0$.

For the case $\omega = \pm 1$, note that

$$f_{\pm 1}^*(x, \cos) = \lim_{T \to \infty} \frac{1}{T} \int_0^T \mathrm{e}^{\pm it}(\sin t + x)\mathrm{d}t$$

$$= \lim_{T \to \infty} \frac{1}{T} \int_0^T \mathrm{e}^{\pm it} \sin t \mathrm{d}t + \lim_{T \to \infty} \frac{1}{T} \int_0^T \mathrm{e}^{\pm it} \mathrm{d}t \cdot x$$

$$= \lim_{T \to \infty} \frac{1}{T} \int_0^T \mathrm{e}^{\pm it} \sin t \mathrm{d}t + \lim_{T \to \infty} \frac{1}{\pm iT}(\mathrm{e}^{\pm iT} - 1) \cdot x$$

$$= \lim_{T \to \infty} \frac{1}{T} \int_0^T \mathrm{e}^{\pm it} \sin t \mathrm{d}t,$$

and apply Lemma 2.1.8. This yields $f_{\pm 1}^*(x, \cos) = \pm 1/2 \cdot i$.

Finally, for $\omega \notin \{0, \pm 1\}$,

$$f_\omega^*(x, \cos) = \lim_{T \to \infty} \frac{1}{T} \int_0^T \mathrm{e}^{i\omega T}(\sin t + x)\mathrm{d}t$$

$$= \lim_{T \to \infty} \frac{1}{T} \int_0^T \mathrm{e}^{i\omega T} \sin t \mathrm{d}t + \lim_{T \to \infty} \frac{1}{T} \int_0^T \mathrm{e}^{i\omega T} \mathrm{d}t \cdot x$$

$$= \lim_{T \to \infty} \frac{1}{T} \int_0^T \mathrm{e}^{i\omega T} \sin t \mathrm{d}t,$$

and hence, by Lemma 2.1.7, $f_\omega^*(x, \cos) = 0.$ ⌟

3.2.1 General controls

For the following, we fix a control $u \in \mathcal{U}$ with no additional assumptions.

Continuity and boundedness First, we look at two simple properties: continuity and boundedness.

Proposition 3.2.3. *If $f : X \to \mathbb{C}$ and $\Phi^u : \mathbb{R}^+ \times X \to X$ are continuous, and if we set $f_\omega^0(x, u) := f(x)$, then $f_\omega^T(x, u)$ is continuous in $(\omega, T, x) \in \mathbb{R} \times [0, \infty) \times X$.*

Proof. This follows from Proposition 2.3.1. $\square$

Proposition 3.2.4. *For all $x \in X$ and $\omega \in \mathbb{R}$, it holds that*

$$|f_\omega^T(x, u)| \leq \sup_{y \in X} |f(y)|.$$

Furthermore, $|f_\omega^*(x, u)| \leq \sup_{y \in X} |f(y)|$, if $f_\omega^*(x)$ exists.

Proof. This follows directly from Definition 3.2.1. $\square$

Example 3.2.5. Recall the system $\dot{x} = u$ in $\mathbb{R}$ from Example 3.2.2. Let $u := \cos$ and $f := \mathrm{id}$. Then with Lemma 1.1.4 and Lemma 2.1.6, one can compute that

$$f_\omega^T(x, \cos) = \begin{cases} \frac{1}{T}(1 - \cos T) + x & \text{if } \omega = 0, \\ \pm \frac{1}{T}\left(\mathrm{i}x(1 - \mathrm{e}^{\pm \mathrm{i}T}) - \frac{\mathrm{i}}{2}\mathrm{e}^{\pm \mathrm{i}T}\sin T\right) \pm \frac{1}{2}\mathrm{i} & \text{if } \omega = \pm 1, \\ \frac{1}{T}\left(\mathrm{i}x\frac{1 - \mathrm{e}^{\mathrm{i}\omega T}}{\omega} + \frac{\mathrm{e}^{\mathrm{i}\omega T}[\cos T - \mathrm{i}\omega \sin T] - 1}{\omega^2 - 1}\right) & \text{otherwise.} \end{cases} \tag{3.2.1}$$

As $\Phi_t^{\cos} x = \sin t + x$ and $f = \mathrm{id}$ are continuous, $f_\omega^T(x, \cos)$ is continuous in $(\omega, T, x) \in \mathbb{R} \times [0, \infty) \times \mathbb{R}$ by Proposition 3.2.3. In fact, e. g.,

$$\lim_{T \to 0} f_\omega^T(x, \cos) = \lim_{T \to 0} \begin{cases} \frac{1}{T}(1 - \cos T) + x \\ \pm \frac{1}{T}\left(\mathrm{i}x(1 - \mathrm{e}^{\pm \mathrm{i}T}) - \frac{\mathrm{i}}{2}\mathrm{e}^{\pm \mathrm{i}T}\sin T\right) \pm \frac{1}{2}\mathrm{i} \\ \frac{1}{T}\left(\frac{\mathrm{i}x(1 - \mathrm{e}^{\mathrm{i}\omega T})}{\omega} + \frac{\mathrm{e}^{\mathrm{i}\omega T}[\cos T - \mathrm{i}\omega \sin T] - 1}{\omega^2 - 1}\right) \end{cases}$$

$$= \lim_{T \to 0} \begin{cases} \sin T + x \\ x\mathrm{e}^{\pm \mathrm{i}T} + \frac{1}{2}\mathrm{e}^{\pm \mathrm{i}T}\sin T \mp \frac{1}{2}\mathrm{e}^{\pm \mathrm{i}T}\cos T \pm \frac{1}{2}\mathrm{i} \\ x\mathrm{e}^{\mathrm{i}\omega T} + \frac{\mathrm{i}\omega \mathrm{e}^{\mathrm{i}\omega T}[\cos T - \mathrm{i}\omega \sin T] + \mathrm{e}^{\mathrm{i}\omega T}[-\sin T - \mathrm{i}\omega \cos T]}{\omega^2 - 1} \end{cases}$$

$$= \begin{cases} x \\ x \mp \frac{\mathrm{i}}{2} \pm \frac{1}{2}\mathrm{i} \\ x + \frac{\mathrm{i}\omega - \mathrm{i}\omega}{\omega^2 - 1} \end{cases}$$

$$= x = f(x) = f_\omega^0(x, \cos)$$

by l'Hôpital's Rule (compare [Kön04, p. 150]), where the subdomains are as in (3.2.1). Similarly,

$$
\begin{aligned}
\lim_{\omega \to 0} f_\omega^T(x, \cos) &= \lim_{\omega \to 0} \frac{1}{T} \left(\frac{\mathrm{i}x(1 - \mathrm{e}^{\mathrm{i}\omega T})}{\omega} + \frac{\mathrm{e}^{\mathrm{i}\omega T}[\cos T - \mathrm{i}\omega \sin T] - 1}{\omega^2 - 1} \right) \\
&= \lim_{\omega \to 0} \frac{1}{T} \left(Tx\mathrm{e}^{\mathrm{i}\omega T} + \frac{\mathrm{e}^{\mathrm{i}\omega T}[\cos T - \mathrm{i}\omega \sin T] - 1}{\omega^2 - 1} \right) \\
&= \frac{1}{T} \left(Tx + \frac{\cos T - 1}{-1} \right) \\
&= \frac{1}{T}(1 - \cos T) + x = f_0^T(x, \cos).
\end{aligned}
$$

Note that the boundedness result in Proposition 3.2.4 is not of much use here, because $\sup_{y \in \mathbb{R}} f(y) = \infty$. See Example 3.2.9 for a more meaningful example for boundedness. ⌟

Behaviour along trajectories and existence Corollary 2.3.5 showed how the harmonic limit behaves along trajectories. Unfortunately, this result cannot be adapted to the pointwise control setting as easily. We will give a similar result in the case of periodic controls in Proposition 3.2.17, though. In general, it does not hold that $f_\omega^*(\Phi_t^u x, u) = \mathrm{e}^{-\mathrm{i}t\omega} f_\omega^*(x, u)$ for all $t > 0$. Compare the following example.

Example 3.2.6. Recall the system $\dot{x} = u$ from Example 3.2.2. Let $u := \cos$ and $f := \mathrm{id}$. Then $\Phi_t^{\cos} x = x + \sin t$ and $\mathrm{id}_0^*(x, \cos) = x$, which implies $\mathrm{id}_0^*(\Phi_t^{\cos} x, \cos) = \Phi_t^{\cos} x = x + \sin t$. Note that $\mathrm{e}^{-\mathrm{i}t0} \mathrm{id}_0^*(x, \cos) = \mathrm{id}_0^*(x, \cos) = x$. So $\mathrm{id}_0^*(\Phi_t^{\cos} x, \cos) = \mathrm{e}^{-\mathrm{i}t0} \mathrm{id}_0^*(x, \cos)$ holds if and only if $x + \sin t = x$, i.e., for $t = k\pi$, $k \in \mathbb{Z}$.

Similarly, $\mathrm{id}_1^*(x, \cos) = \frac{1}{2} \cdot \mathrm{i} = \mathrm{id}_1^*(\Phi_t^{\cos} x, \cos)$. So $\mathrm{e}^{-\mathrm{i}t} \mathrm{id}_1^*(x, \cos) = \mathrm{id}_1^*(\Phi_t^{\cos} x, \cos)$ holds if and only if $x = 0$ or $\mathrm{e}^{-\mathrm{i}t} = 1$, i.e., if $x = 0$ or $t = 2k\pi$, $k \in \mathbb{Z}$. ⌟

Likewise, we cannot give an existence result like in Theorem 2.3.14 in the pointwise setting. For this, we either have to assume periodicity of the control (see Theorem 3.2.19) or pass to the global point of view (see Theorem 3.3.8). But the following result on the number of possible frequencies still holds.

Proposition 3.2.7. *If $x \in X$ and $f : X \to \mathbb{C}$ are such that the limit $\lim_{T \to \infty} \frac{1}{T} \cdot \int_0^T |f(\Phi_t^u x)|^2 \mathrm{d}t$ exists, then there is a set $\Omega \subset \mathbb{R}$ with at most countably many elements, such that for all $\omega \in \mathbb{R} \setminus \Omega$, where $f_\omega^*(x, u)$ exists, it holds that $f_\omega^*(x, u) = 0$, i.e., there are at most countably many $\omega \in \mathbb{R}$ for which $f_\omega^*(x, u)$ exists and is nonzero.*

Proof. This follows from Remark 2.3.8 and Theorem 2.3.7 applied to the map $g(t) := f(\Phi_t^u x)$. □

Asymptotics If the asymptotic behaviour of the system is known, one can make use of the following proposition.

Proposition 3.2.8. *Let $x \in X$ and $T_0 \geq 0$. Let $g : [T_0, \infty) \to \mathbb{C}$ be locally integrable. Let $f : X \to \mathbb{C}$. If $g(t) - f(\Phi_t^u x) \to 0$ for $t \to \infty$. Then it holds for every $\omega \in \mathbb{R}$, that $f_\omega^*(x, u) = \lim_{T \to \infty} 1/T \cdot \int_{T_0}^T e^{i\omega t} g(t) \mathrm{d}t$, provided that one of the limits exists.*

Proof. See Proposition 2.3.26. □

Let the following example illustrate this proposition.

Example 3.2.9. Recall the system $\dot{x} = u$ from Example 3.2.2. Let $u := 1/(t^2+1)$ and $f := \sin$. Then $\Phi_t^u x = x + \arctan t$. Note that $\lim_{t \to \infty} \Phi_t^u x = x + \pi/2$, and hence by Proposition 3.2.8,

$$\sin_\omega^*(x, u) = \lim_{T \to \infty} \frac{1}{T} \int_0^T e^{i\omega t} \sin\left(x + \frac{\pi}{2}\right) \mathrm{d}t = \begin{cases} \sin\left(x + \frac{\pi}{2}\right) & \text{if } \omega = 0, \\ 0 & \text{otherwise.} \end{cases}$$

Furthermore, $|\sin_\omega^*(x, u)| \leq 1 = \sup_{y \in \mathbb{R}} |\sin(y)|$, as shown in Proposition 3.2.4. ⌐

Periodicity Even if the control u is not periodic, the map $t \mapsto f(\Phi_t^u x)$ might be periodic, quasi-periodic, or almost periodic. In that case, we get the following results.

Proposition 3.2.10. *Assume that $t \mapsto f(\Phi_t^u x)$ is almost periodic for some $x \in X$ and $f : X \to \mathbb{C}$. Then $f_\omega^*(x, u)$ exists for every $\omega \in \mathbb{R}$.*

Proof. See Proposition 2.3.29. □

For almost periodic integrands, one can show how fast the limit converges, compare Proposition 2.3.30.

Proposition 3.2.11. *Assume that $x \in X$ and $f : X \to \mathbb{C}$ are such that $t \mapsto f(\Phi_t^u x)$ is quasi-periodic with a locally Lebesgue integrable generating function (see Definition 1.1.3) and periods τ_j, $j = 1, \ldots, n$. If $\omega \in \mathbb{R}$ is such that the numbers $\omega, 2\pi/\tau_1, \ldots, 2\pi/\tau_n$ are rationally independent, then $f_\omega^*(x, u)$ equals zero.*

Proof. This follows from Proposition 2.3.32. □

Proposition 3.2.12. *Assume that $x \in X$ and $f : X \to \mathbb{C}$ are such that $t \mapsto f(\Phi_t^u x)$ is τ-periodic, $\tau > 0$. If $\omega \in \mathbb{R}$ is such that $\omega\tau/2\pi \notin \mathbb{Z}$, then $f_\omega^*(x, u)$ equals zero.*

Proof. This follows from Proposition 2.3.33. □

Proposition 3.2.13. *Assume that $t \mapsto f(\Phi_t^u x)$ is almost periodic for some $x \in X$ and $f : X \to \mathbb{C}$. Then it holds that $\lim_{T \to \infty} 1/T \cdot \int_0^T |f(\Phi_t^u x)|^2 \mathrm{d}t = \sum_{\omega \in \mathbb{R}} |f_\omega^*(x, u)|^2$.*

Proof. This follows from Lemma 2.3.36. □

Example 3.2.14. Recall the system $\dot{x} = u$ from Example 3.2.2. Let $u := \cos$ and $f := \mathrm{id}$. Then $\Phi_t^{\cos} x = x + \sin t$. So $f(\Phi_t^{\cos} x)$ is 2π-periodic, and Proposition 3.2.12 implies that, for $\omega \notin \mathbb{Z}$, it holds that $f_\omega^*(x, \cos) = 0$. In fact, we have computed in Example 3.2.2 that $f_\omega^*(x, \cos) = 0$ for $\omega \notin \{0, \pm 1\}$. ⌟

Rotational factor maps   We can also introduce the concept of rotational factor maps for this setting.

Definition 3.2.15. Let $\mathcal{F}$ be a class of functions $X \to \mathbb{C}$. Suppose that there is a map $F \in \mathcal{F}$, such that $F \not\equiv 0$ and

$$F \circ \Phi_t^u = \mathrm{e}^{\mathrm{i} t \omega} \cdot F \tag{3.2.2}$$

holds for some $\omega \in \mathbb{R}$ and all t. Then we say that Φ_t^u admits the *rotational factor map* $F \in \mathcal{F}$ *with frequency* $\omega/2\pi$. ⌟

There is no such strong connection between rotational factor maps and harmonic limits in this pointwise control setting as we had in Theorem 2.2.9, because there is no analogon to Corollary 2.3.5, see Example 3.2.6. Only the following holds.

Proposition 3.2.16. *Assume that Φ_t^u admits the rotational factor map F with frequency $\omega/2\pi$. Then there is $f : X \to \mathbb{C}$ such that $f_{-\omega}^*(\cdot, u) = F(\cdot)$. In particular, $f := F$ satisfies this equation.*

Proof. See part 2. of Theorem 2.2.9. □

This means, in the pointwise control setting for general controls, all rotational factor maps can be detected with a proper function f, but not every nonzero harmonic limit corresponds to a rotational factor map.

3.2.2 Periodic controls

Now we have a closer look at the special case, when the control u is τ-periodic for some $\tau > 0$. This also includes the case of constant controls. Of course, the results for general controls in Subsection 3.2.1 still hold for periodic controls. Additionally, the following results can be shown.

A result similar to Corollary 2.3.5 on the behaviour of the harmonic limit along trajectories holds in the periodic case. The difference to Corollary 2.3.5 is, that it only holds at times that are multiples of the period of the control u.

Proposition 3.2.17. *Assume that the control $u \in \mathcal{U}$ is τ-periodic, $\tau > 0$. If a function $f : X \to \mathbb{C}$, $\omega \in \mathbb{R}$, and $x \in X$ are such that $f_\omega^*(x, u)$ exists, then $f_\omega^*(\Phi_{k\tau}^u x, u) = \mathrm{e}^{-\mathrm{i} k \tau \omega} f_\omega^*(x, u)$ for all $k \in \mathbb{N}_0$. Particularly, $f_\omega^*(\Phi_{k\tau}^u x, u)$ exists for all $k \in \mathbb{N}_0$, and $|f_\omega^*(\Phi_{k\tau}^u x, u)|$ is independent of k.*

Proof. The proof is analogous to the proof of Corollary 2.3.5. Nevertheless, we provide a full proof here.

Let $f : X \to \mathbb{C}$, $\omega \in \mathbb{R}$ and $x \in X$ be such that $f_\omega^*(x, u)$ exists. Then for any $k \in \mathbb{N}_0$, it holds that

$$
\mathrm{e}^{-\mathrm{i}k\tau\omega} f_\omega^*(x, u) = \mathrm{e}^{-\mathrm{i}k\tau\omega} \lim_{T\to\infty} f_\omega^T(x, u) = \lim_{T\to\infty} \mathrm{e}^{-\mathrm{i}k\tau\omega} f_\omega^T(x, u)
$$

$$
= \lim_{T\to\infty} \frac{1}{T} \int_0^T \mathrm{e}^{\mathrm{i}\omega(s-k\tau)} f(\Phi_s^u x)\mathrm{d}s \overset{(*)}{=} \lim_{T\to\infty} \frac{1}{T} \int_{k\tau}^{T+k\tau} \mathrm{e}^{\mathrm{i}\omega(s-k\tau)} f(\Phi_s^u x)\mathrm{d}s
$$

$$
= \lim_{T\to\infty} \frac{1}{T} \int_0^T \mathrm{e}^{\mathrm{i}\omega s} f(\Phi_{s+k\tau}^u x)\mathrm{d}s \overset{(3.0.2)}{=} \lim_{T\to\infty} \frac{1}{T} \int_0^T \mathrm{e}^{\mathrm{i}\omega s} f\big(\Phi_s^{\theta_{k\tau}u}(\Phi_{k\tau}^u x)\big)\mathrm{d}s
$$

$$
\overset{(**)}{=} \lim_{T\to\infty} \frac{1}{T} \int_0^T \mathrm{e}^{\mathrm{i}\omega s} f\big(\Phi_s^u(\Phi_{k\tau}^u x)\big)\mathrm{d}s = f_\omega^*(\Phi_{k\tau}^u x, u),
$$

where the equation marked with $(**)$ holds due to periodicity of u. Thus the limit $f_\omega^*(\Phi_{k\tau}^u x, u)$ exists. Furthermore, it follows from this calculation, that

$$
|f_\omega^*(\Phi_{k\tau}^u x, u)| = |f_\omega^*(x, u)|
$$

for every $k \in \mathbb{N}_0$.

To see that the equation marked with $(*)$ holds, first note that

$$
\int_0^{k\tau} \mathrm{e}^{\mathrm{i}\omega(s-k\tau)} f(\Phi_s^u x)\mathrm{d}s
$$

is finite due to local integrability, which implies $\lim_{T\to\infty} 1/T \cdot \int_0^{k\tau} \mathrm{e}^{\mathrm{i}\omega(s-k\tau)} f(\Phi_s^u x)\mathrm{d}s = 0$. Furthermore,

$$
\frac{1}{T} \int_T^{T+k\tau} \mathrm{e}^{\mathrm{i}\omega(s-k\tau)} f(\Phi_s^u x)\mathrm{d}s = \frac{1}{T} \mathrm{e}^{-\mathrm{i}k\tau\omega}\big((T+k\tau)f_\omega^{T+k\tau}(x, u) - T f_\omega^T(x, u)\big)
$$

$$
= \mathrm{e}^{-\mathrm{i}k\tau\omega}\left(f_\omega^{T+k\tau}(x, u) + \frac{k\tau}{T} f_\omega^{T+k\tau}(x, u) - f_\omega^T(x, u)\right),
$$

which tends to 0 as $T \to \infty$, because

$$
\lim_{T\to\infty} f_\omega^T(x, u) = \lim_{T\to\infty} f_\omega^{T+k\tau}(x, u) = f_\omega^*(x, u) < \infty. \qquad \square
$$

For the next theorem, we will look at the time-τ map of the system, i.e., at the map $\Phi_\tau^u : X \to X$. This map defines a dynamical system in discrete time, which satisfies $\Phi_\tau^u \Phi_{k\tau}^u x = \Phi_\tau^{\theta_{k\tau}u} \Phi_{k\tau}^u x = \Phi_{(k+1)\tau}^u x$ for every $k \in \mathbb{N}_0$. So we could also analyze the discrete-time harmonic limit $\lim_{n\to\infty} 1/n \cdot \sum_{j=0}^{n-1} \mathrm{e}^{\mathrm{i}j\tau\omega} f\big((\Phi_\tau^u)^j x\big)$ instead of $f_\omega^*(x, u)$. Note that those limits do not coincide. It can even occur that one limit is different from zero and the other limit vanishes, because $\Phi_t^u x$ is only evaluated at $t = \tau j$, $j \in \mathbb{N}_0$, in the discrete-time harmonic limit, which might not capture the whole rotational behaviour of the system. Compare the following example.

Example 3.2.18. Consider the control system given by the differential equation $\dot{x} = -x + u$ in $\mathbb{R}$. For $u_0(t) := \sin t + \cos t$, the solution is given by $\Phi_t^{u_0} x = e^{-t} x + \sin t$. Let $f(x) := x + 1$. Then by Proposition 2.3.26 and Lemma 2.1.8,

$$
\begin{aligned}
f_1^*(x, u_0) &= \lim_{T \to \infty} \frac{1}{T} \int_0^T e^{it} (e^{-t} x + \sin t + 1) dt \\
&= \frac{1}{2\pi} \int_0^{2\pi} e^{it} (\sin t + 1) dt \\
&= \frac{1}{2\pi} \int_0^{2\pi} e^{it} \sin t\, dt + \frac{1}{2\pi} \int_0^{2\pi} e^{it} dt \\
&= \frac{1}{2} i.
\end{aligned}
$$

But $(\Phi_{2\pi}^{u_0})^j x = \Phi_{2j\pi}^{u_0} x = e^{-2j\pi} x + \sin(2j\pi) = e^{-2j\pi} x \to 0$ for $j \to \infty$. So

$$
\lim_{n \to \infty} \frac{1}{n} \sum_{j=0}^{n-1} e^{2\pi i j} f\big((\Phi_{2\pi}^{u_0})^j x\big) = \lim_{n \to \infty} \frac{1}{n} \sum_{j=0}^{n-1} f(0) = 1
$$

by Proposition 1.3.15.

If we consider $f(x) := x$ instead, we get $f_1^*(x) = 1/2 \cdot i$ and

$$
\lim_{n \to \infty} \frac{1}{n} \sum_{j=0}^{n-1} e^{2\pi i j} f\big((\Phi_{2\pi}^{u_0})^j x\big) = f(0) = 0
$$

by an analogous computation.

If we consider $f(t) := e^{i \arcsin t}$, we get

$$
\begin{aligned}
f_1^*(x, u_0) &= \lim_{T \to \infty} \frac{1}{T} \int_0^T e^{it} f\big(e^{-t} x + \sin t\big) dt \\
&\overset{(*)}{=} \lim_{T \to \infty} \frac{1}{T} \int_0^T e^{it} f(\sin t) dt \\
&= \lim_{T \to \infty} \frac{1}{T} \int_0^T e^{it} e^{it} dt \\
&= \lim_{T \to \infty} \frac{1}{2iT} (e^{2iT} - 1) \\
&= 0,
\end{aligned}
$$

where the equation marked with $(*)$ holds due to Proposition 3.2.8, because f is continuous, and hence $f\big(e^{-t} x + \sin t\big) - f(\sin t) \to 0$ as $t \to \infty$. But similarly, by Proposition 1.3.15,

$$
\lim_{n \to \infty} \frac{1}{n} \sum_{j=0}^{n-1} e^{2\pi i j} f\big((\Phi_{2\pi}^{u_0})^j\big) = \lim_{n \to \infty} \frac{1}{n} \sum_{j=0}^{n-1} f(0) = 1.
$$

Theorem 3.2.19. *Assume that the control $u \in \mathcal{U}$ is τ-periodic, $\tau > 0$. Assume that there is a finite measure μ on X, such that the map $f : X \to \mathbb{C}$ is of class $L^1(\mu)$, and such that Φ_τ^u preserves μ. Further assume that $t \mapsto f(\Phi_t^u x)$ is bounded for μ-almost every $x \in X$. Then for every $\omega \in \mathbb{R}$, there is a null set $\Xi_\omega \subset X$ such that the harmonic limit $f_\omega^*(x, u)$ exists for all $x \in X \setminus \Xi_\omega$, and the map $x \mapsto f_\omega^*(x, u)$ is of class $L^1(\mu, \mathbb{C})$.*

Proof. This follows from the Birkhoff Ergodic Theorem 3.1.9 applied to the map $S : \mathbb{R}^+ \times (S^1 \times X) \to S^1 \times X$, given by $S_t(z, x) := (e^{it\tau\omega} z, \Phi_{t\tau}^u x)$ for $\omega \in \mathbb{R}$. More precisely, let $f : X \to \mathbb{C}$ be of class $L^1(\mu)$, and $\omega \in \mathbb{R}$. Define $g : S^1 \times X \to \mathbb{C}$ by $(z, x) \mapsto z f(x)$. Consider the finite S_1-invariant measure $\nu := \lambda \times \mu$ on $S^1 \times X$, where λ is the Lebesgue measure on S^1. Then g is of class $L^1(\nu)$. Note that $S_t S_k(z, x) = S_{k+t}(z, x)$ for all $k \in \mathbb{N}_0$, $t \in \mathbb{R}^+$, $z \in S^1$, $x \in X$, because $S_t S_k(z, x) = S_t(e^{ik\tau\omega} z, \Phi_{k\tau}^u x) = (e^{it\tau\omega} e^{ik\tau\omega} z, \Phi_{t\tau}^u \Phi_{k\tau}^u x) = (e^{it\tau\omega} e^{ik\tau\omega} z, \Phi_{t\tau}^{\theta_{k\tau} u} \Phi_{k\tau}^u x) = (e^{i(k+t)\tau\omega} z, \Phi_{(k+t)\tau}^u x) = S_{k+t}(z, x)$. So the Birkhoff Ergodic Theorem is applicable, and it follows that

$$\lim_{T \to \infty} \frac{1}{T} \int_0^T g\big(S_t(z, x)\big) \mathrm{d}t = \lim_{T \to \infty} \frac{1}{T} \int_0^T e^{it\tau\omega} z f(\Phi_{t\tau}^u x) \mathrm{d}t = z f_\omega^*(x, u)$$

exists for ν-almost all $(z, x) \in S^1 \times X$.

If $z_0 f_\omega^*(x_0, u)$ exists for some $(z_0, x_0) \in S^1 \times X$, then clearly also $z f_\omega^*(x_0, u)$ exists for all $z \in S^1$. Thus $f_\omega^*(x, u)$ exists for μ-almost all $x \in X$.

By the Birkhoff Ergodic Theorem, the map $S^1 \times X \to \mathbb{C}$, $(z, x) \to z f_\omega^*(x, u)$ is ν-integrable. By Fubini's Theorem [Bau92, Korollar 23.7], this implies μ-integrability of $x \mapsto f_\omega^*(x, u)$. □

Existence of a finite invariant measure, which is needed in this theorem, can be shown for a fairly large class of systems by the Theorem of Krylov and Bogolubov, see [KH06, Theorem 4.1.1], i.e., for continuous Φ_τ^u and compact X. Unfortunately, even for those systems, the null set $\Xi_\omega \subset X$ of points where f_ω^* does not exist can depend on ω.

Note that, in Theorem 3.2.19, the set $\Xi_\omega \subset X$ of points, where f_ω^* does not exist, depends on ω. We will use the Wiener-Wintner Ergodic Theorem (Theorem 2.3.18) in the following theorem to show that, for ergodic systems, Ξ_ω does not depend on ω.

Theorem 3.2.20. *Assume that the control $u \in \mathcal{U}$ is τ-periodic, $\tau > 0$. Further assume that there is a finite measure μ on X such that Φ_τ^u preserves μ and is ergodic. Let $f : X \to \mathbb{C}$ be integrable such that $t \mapsto f(\Phi_t^u x)$ is bounded for almost all $x \in X$. Then there is a null set $\Xi \subset X$ such that, for all $x \in X \setminus \Xi$, the harmonic limit $f_\omega^*(x, u)$ exists for all $\omega \in \mathbb{R}$.*

Proof. This follows from Theorem 3.1.9 together with Theorem 1.3.14. □

We now give a second definition of rotational factor maps, this time tailored to the time-τ map Φ_τ^u. Note that we speak of rotational factor maps *by angle* ω here, in contrast to the rotational factor maps *with frequency* $\omega/2\pi$ in Definition 3.2.15.

Definition 3.2.21. Assume that the control $u \in \mathcal{U}$ is τ-periodic, $\tau > 0$. Let $\mathcal{F}$ be a class of functions mapping $X \to \mathbb{C}$. Suppose that there is a map $F \in \mathcal{F}$, such that $F \not\equiv 0$ and

$$F \circ \Phi_\tau^u = e^{i\tau\omega} \cdot F \tag{3.2.3}$$

holds for some $\omega \in \mathbb{R}$. Then we say that Φ_τ^u *admits the rotational factor map* $F \in \mathcal{F}$ *by angle* ω. ⌟

Corollary 3.2.22. *Assume that the control* $u \in \mathcal{U}$ *is* τ*-periodic,* $\tau > 0$. *Let* $f : X \to \mathbb{C}$. *If* $f_\omega^*(x, u)$ *exists for all* x *in some nonvoid set* $M \subset X$, *and does not vanish everywhere in* M, *then* Φ_t^u *admits the rotational factor map* $F := \mathbf{1}_M \cdot \overline{f_\omega^*(\cdot, u)}$ *by angle* ω *in the class of bounded functions mapping* $X \to \mathbb{C}$.

Proof. This follows from Proposition 3.2.17. ☐

3.2.3 Harmonic spectrum

Let us define the harmonic spectrum as the set of all $\omega \in \mathbb{R}$ for which the harmonic limit f_ω^* to Φ_t^u does not vanish everywhere. Remember that we still perform a pointwise analysis, so let a control $u \in \mathcal{U}$ be fixed. We also fix a map $f : X \to \mathbb{C}$.

Definition 3.2.23 (Harmonic spectrum). For every (positively) Φ^u-invariant set $\Xi \subset X$, the *harmonic spectrum* is defined by

$$\Sigma_H^u(\Xi) := \{\omega \in \mathbb{R} \mid \text{there is } x \in \Xi \text{ such that } f_\omega^*(x, u) \neq 0\}.$$

If $\Xi = X$, we omit the argument. ⌟

The set $1/(2\pi) \cdot \Sigma_H^u(\Xi)$ can not be interpreted as the set of frequencies that can be observed in Ξ under f, because a nonvanishing harmonic limit does not imply the existence of a rotational factor map in the general pointwise control setting. Nevertheless, one can expect to get useful information on the dynamical properties from this set.

Example 3.2.24. Recall Example 3.2.2. There $\Sigma_H^{cos} = \{0, \pm 1\}$. ⌟

3.3 Global analysis

Now we turn to a global approach, i. e., we consider a semi-flow $\Psi : \mathbb{R}^+ \times (X \times \mathcal{U}) \to X \times \mathcal{U}$, where $X \subset \mathbb{R}^n$, and $\mathcal{U}$ is a set of control functions mapping $\mathbb{R} \to U \subset \mathbb{R}^m$. Consider the standard topology on X, and assume that $\mathcal{U}$ also is endowed with some topology. Then we will consider the product topology on $X \times \mathcal{U}$. Note that, if we speak of invariance of a set $A \subset X \times \mathcal{U}$ under Ψ, we always mean *positive* invariance in this section, i. e., that $\Psi_t(x, u) \in A$ for all points $(x, u) \in A$, and times $t \geq 0$.

Remark 3.3.1. If $\mathcal{U} \subset L_\infty$, i. e., if the controls are Borel measurable and essentially bounded, then we can endow $\mathcal{U}$ with the weak* topology, see Definition 3.1.3. Otherwise, we can endow $\mathcal{U}$ with, e. g., the trivial or the discrete topology. ⌟

As a special case, we will consider $\Psi_t(x, u) = (\Phi_t^u x, \theta_t u)$ for a map $\Phi : \mathbb{R}^+ \times X \times \mathcal{U} \to X$ that satisfies $\Phi_{s+t}^u = \Phi_s^{\theta_t u} \Phi_t^u$ with a shift-invariant set $\mathcal{U}$. Solutions of control systems (3.0.1) are of this type. In the following, we apply the results from Chapter 2 to the general semi-flow Ψ. If stronger results can be derived in the special case $(\Phi_t^u x, \theta_t u)$, we will have a closer look at this case.

First, we transfer the concept of harmonic averages and limit to the control setting.

Definition 3.3.2 (Harmonic average and limit). For all $\omega \in \mathbb{R}$, $x \in X$ and $u \in \mathcal{U}$, and for any function $f : X \times \mathcal{U} \to \mathbb{C}$, such that $f(\Psi_t(x, u))$ is locally integrable with respect to t, define the *harmonic average* by

$$f_\omega^T(x, u) := \frac{1}{T} \int_0^T \mathrm{e}^{\mathrm{i}\omega t} f(\Psi_t(x, u)) \mathrm{d}t;$$

furthermore, define the *harmonic limit* by

$$f_\omega^*(x, u) := \lim_{T \to \infty} f_\omega^T(x, u),$$

if the limit exists. ⌟

For any map $g : \mathbb{R} \to \mathbb{C}$, and $\omega \in \mathbb{R}$, it holds that $\mathrm{e}^{\mathrm{i}\omega t} g(t)$ is locally integrable if and only if $g(t)$ is locally integrable, compare [Bau92, Satz 12.2]. Thus local integrability of the map $f(\Psi_t(x, u))$ implies that the integral $\int_0^T \mathrm{e}^{\mathrm{i}\omega t} f(\Psi_t(x, u)) \mathrm{d}t$ is finite for all $T \geq 0$. The map $t \to f(\Psi_t(x, u))$ is locally integrable if, e. g., f is measurable and bounded, and $t \mapsto \Psi_t(x, u)$ is Borel measurable. This particularly is the case for continuous f, if $X \times \mathcal{U}$ is compact, and $t \mapsto \Psi_t(x, u)$ is Borel measurable.

Remark 3.3.3. Consider the special case $\Psi_t(x, u) = (\Phi_t^u x, \theta_t u)$. For some topologies on $\mathcal{U}$—including the trivial topology and the weak* topology—the time shift θ is continuous. So for these topologies, it suffices to assume, that f is Borel measurable and bounded, and that $t \mapsto \Phi_t^u x$ is Borel measurable, in order to guarantee local integrability of $t \to f(\Psi_t(x, u))$. ⌟

The first property we transfer is continuity of the harmonic average.

Proposition 3.3.4. *If $f : X \times \mathcal{U} \to \mathbb{C}$ and $\Psi : \mathbb{R}^+ \times (X \times \mathcal{U}) \to X \times \mathcal{U}$ are continuous, and if we set $f_\omega^0(x, u) := f(x, u)$, then $f_\omega^T(x, u)$ is continuous in $(\omega, T, x, u) \in \mathbb{R} \times [0, \infty) \times X \times \mathcal{U}$.*

Proof. See Proposition 2.3.1. $\qquad\square$

Corollary 3.3.5. *Consider the special case* $\Psi_t(x, u) = (\Phi_t^u x, \theta_t u)$. *If* $f : X \times \mathcal{U} \to \mathbb{C}$, $\Phi : \mathbb{R}^+ \times X \times \mathcal{U} \to X$, *and* θ *are continuous, and if we set* $f_\omega^0(x, u) := f(x, u)$, *then* $f_\omega^T(x, u)$ *is continuous in* $(\omega, T, x, u) \in \mathbb{R} \times [0, \infty) \times X \times \mathcal{U}$.

Proof. The flow Ψ is continuous as Φ and θ are continuous. So continuity follows from Proposition 3.3.4. $\qquad\square$

The next property we transfer is boundedness. The harmonic average is bounded above by $\sup_{(y,v)\in X\times\mathcal{U}}|f(y, v)|$. The same holds for the harmonic limit, if it exists.

Proposition 3.3.6. *For all* $x \in X$, $u \in \mathcal{U}$, *and* $\omega \in \mathbb{R}$, *it holds that*

$$|f_\omega^T(x, u)| \leq \sup_{(y,v)\in X\times\mathcal{U}} |f(y, v)|.$$

Furthermore,

$$|f_\omega^*(x, u)| \leq \sup_{(y,v)\in X\times\mathcal{U}} |f(y, v)|,$$

if $f_\omega^*(x, u)$ *exists.*

Proof. See Proposition 2.3.4. $\qquad\square$

The following result shows how the harmonic limit behaves along trajectories. We get that, if $f_\omega^*(x, u)$ exists at some point $(x, u) \in X \times \mathcal{U}$ for $\omega \in \mathbb{R}$, then the harmonic limit exists at every point of the trajectory starting in (x, u), and, beyond that, the modulus of the harmonic limit is constant along the trajectory.

Proposition 3.3.7. *If a function* $f : X \times \mathcal{U} \to \mathbb{C}$, $\omega \in \mathbb{R}$, *and* $(x, u) \in X \times \mathcal{U}$ *are such that* $f_\omega^*(x, u)$ *exists, then* $f_\omega^*\big(\Psi_t(x, u)\big) = \mathrm{e}^{-\mathrm{i}t\omega} f_\omega^*(x, u)$ *for all* $t \geq 0$. *Particularly,* $f_\omega^*\big(\Psi_t(x, u)\big)$ *exists for all* $t \geq 0$ *and* $|f_\omega^*\big(\Psi_t(x, u)\big)|$ *is independent of* t.

Proof. See Corollary 2.3.5. $\qquad\square$

If we have a finite invariant measure μ on $X \times \mathcal{U}$, then for every μ-integrable function f and all $\omega \in \mathbb{R}$, the harmonic limit exists almost everywhere.

Theorem 3.3.8. *Assume that there is a finite invariant measure* μ *on* $X \times \mathcal{U}$, *such that the map* $f : X \times \mathcal{U} \to \mathbb{C}$ *is of class* $L^1(\mu)$. *Then for every* $\omega \in \mathbb{R}$, *the harmonic limit* $f_\omega^*(x, u)$ *exists for almost all* $(x, u) \in X \times \mathcal{U}$, *and the map* $(x, u) \mapsto f_\omega^*(x, u)$ *is of class* $L^1(\mu, \mathbb{C})$.

Proof. See Theorem 2.3.14. $\qquad\square$

Existence of a finite invariant measure, which is needed in Theorem 3.3.8, can be shown for a fairly large class of systems by the Theorem of Krylov and Bogolubov, see [NS89, Theorem VI.9.05], i. e., for continuous Φ and compact X. Unfortunately, even for those systems, the null set $\Xi_\omega \subset X$ of points, where f_ω^* does not exist, can depend on ω. In the special case $\Psi_t(x, u) = (\Phi_t^u x, \theta_t u)$, continuity of Ψ is guaranteed, e. g., if Φ is continuous and if $\mathcal{U}$ is endowed with the weak* topology.

Note that, in Theorem 3.3.8, the set $\Xi_\omega \subset X$ of points, where f_ω^* does not exist, depends on ω. We will use the Wiener-Wintner Ergodic Theorem (Theorem 2.3.18) in the following theorem to show that, for ergodic systems, Ξ_ω does not depend on ω.

Theorem 3.3.9. *Assume that there is an ergodic invariant measure μ on $X \times \mathcal{U}$. Let $f : X \times \mathcal{U} \to \mathbb{C}$ be integrable. Then there is a null set $\Xi \subset X \times \mathcal{U}$ such that, for all $(x, u) \in X \times \mathcal{U} \setminus \Xi$, the harmonic limit $f_\omega^*(x, u)$ exists for all $\omega \in \mathbb{R}$.*

Proof. See Theorem 2.3.17. ☐

The following result shows that under a weak condition, there can only be countably many $\omega \in \mathbb{R}$ for which the harmonic limit is different from zero.

Proposition 3.3.10. *Let $f : X \times \mathcal{U} \to \mathbb{C}$ and $(x, u) \in X \times \mathcal{U}$ be such that*

$$\lim_{T \to \infty} \frac{1}{T} \int_0^T |f(\Psi_t(x, u))|^2 \mathrm{d}t$$

exists. Then there is a set $\Omega \subset \mathbb{R}$ with at most countably many elements, such that for all $\omega \in \mathbb{R} \setminus \Omega$, where $f_\omega^*(x, u)$ exists, it holds that $f_\omega^*(x, u) = 0$.

Proof. See Theorem 2.3.7. ☐

If we know the asymptotic behaviour of the system, then we can use the following results.

Proposition 3.3.11. *Let $X \times \mathcal{U}$ be a metric space, $(x, u) \in X \times \mathcal{U}$ and $T_0 \geq 0$. Let $g : [T_0, \infty) \to \mathbb{C}$ be locally integrable. If $g(t) - f(\Psi_t(x, u)) \to 0$ for $t \to \infty$, then it holds for every $\omega \in \mathbb{R}$, that*

$$f_\omega^*(x, u) = \lim_{T \to \infty} \frac{1}{T} \int_{T_0}^T \mathrm{e}^{\mathrm{i}\omega t} g(t) \mathrm{d}t,$$

provided that one of the limits exists.

Proof. See Proposition 2.3.26. ☐

Corollary 3.3.12. *Let $X \times \mathcal{U}$ be a metric space, and $f : X \times \mathcal{U} \to \mathbb{C}$. Assume that f is continuous, and let $(x, u) \in X \times \mathcal{U}$. If there is $(y, v) \in X \times \mathcal{U}$ such that $d(\Psi_t(x, u), \Psi_t(y, v)) \to 0$ for $t \to \infty$, then $f_\omega^*(x, u) = f_\omega^*(y, v)$ for every $\omega \in \mathbb{R}$, provided that one of the limits exists.*

Proof. See Corollary 2.3.27. ☐

Periodicity   Along almost periodic trajectories, the harmonic limit always exists.

Lemma 3.3.13. *If* $(x, u) \in X \times \mathcal{U}$ *and* $f : X \times \mathcal{U} \to \mathbb{C}$ *are such that* $t \mapsto f\big(\Psi_t(x, u)\big)$ *is almost periodic, then* $f_\omega^*(x, u)$ *exists for all* $\omega \in \mathbb{R}$.

Proof. See Proposition 2.3.29.   □

Proposition 3.3.14. *Consider the special case* $\Psi_t(x, u) = (\Phi_t^u x, \theta_t u)$. *Let* $\mathcal{U}$ *be endowed with a weak* metric (see Definition 3.1.6). Assume that* $u \in \mathcal{U}$ *is almost periodic. If* $x \in X$ *is such that* $\Phi_t^u x$ *is almost periodic, then, for continuous maps* f, *the harmonic limit* $f_\omega^*(x, u)$ *exists for all* $\omega \in \mathbb{R}$.

Proof. By Lemma 3.3.13, it suffices to show that $t \mapsto f\big(\Psi_t(x, u)\big)$ is almost periodic. Assume that u and $\Phi_t^u x$ are almost periodic. Let $L_u(\varepsilon)$ denote for $\varepsilon > 0$ the interval length for u.

Then for all $\varepsilon > 0$ and all intervals $I \subseteq \mathbb{R}$ of length $L_u(\varepsilon)$, there is a translation number $\tau \in I$ such that for all $t \in \mathbb{R}$, it holds that $d\big(u(t + \tau), u(t)\big) \leq \varepsilon$. By Lemma 3.1.7, it follows that for every $\delta > 0$ and any interval $I \subset \mathbb{R}$ of length $L_u\big(\varepsilon(\delta)\big)$, there is a translation number $\tau \in I$ such that $d(\theta_{t+\tau} u, \theta_t u) \leq \delta$ holds for all $t \in \mathbb{R}$. This means, that $t \mapsto \theta_t u$ is almost periodic.

Endow $X \times \mathcal{U}$ with the maximum metric $d_{\max}$ on $X \times \mathcal{U}$, i. e.,

$$d_{\max}\big((a, b), (c, d)\big) := \max\{d(a, c), d(b, d)\}.$$

Note that $d_{\max}$ induces the product topology, see Lemma 3.1.2.

For two almost periodic functions, one can find common translation numbers, see Bohr's proof of Theorem III in [Boh47, Section 48]. So for every $\varepsilon > 0$, there is an interval length $L(\varepsilon)$ such that, in every interval $I \subset \mathbb{R}$ of length $L(\varepsilon)$, there is a translation number τ such that $d(\theta_{t+\tau} u, \theta_t u) \leq \varepsilon$ and $d(\Phi_{t+\tau}^u x, \Phi_t^u x) \leq \varepsilon$ for all $t \in \mathbb{R}$. Hence

$$
\begin{aligned}
d_{\max}\big(\Psi_{t+\tau}(x, u)\big) &= d_{\max}\big((\Phi_{t+\tau}^u x, \tau_{t+\tau} u), (\Phi_t^u x, \tau_t u)\big) \\
&= \max\{d(\Phi_{t+\tau}^u x, \Phi_t^u x), d(\theta_{t+\tau} u, \theta_t u)\} \leq \varepsilon
\end{aligned}
$$

for all $t \in \mathbb{R}$, which means almost periodicity of $\Psi_t(x, u)$.

As f is continuous, also $f\big(\Psi_t(x, u)\big)$ is almost periodic by Lemma 1.1.2.   □

For almost periodic integrands, one can show how fast the limit converges, compare Proposition 2.3.30.

Now we turn to the question, which frequencies can occur at periodic and quasi-periodic orbits. With *(quasi-)periodic orbit* we mean an orbit such that $f\big(\Psi_t(x, u)\big)$ is (quasi-)periodic (see Definition 1.1.3). In the case, where $f\big(\Psi_t(x, u)\big)$ is quasi-periodic with periods $\tau_1, \ldots, \tau_n$, the following proposition shows that $f_\omega^*(x, u) = 0$ if $\omega, {}^{2\pi}/_{\tau_1}, \ldots, {}^{2\pi}/_{\tau_n}$ are rationally independent, i. e., if there is no nonzero tuple

$(c_0, c_1, \ldots, c_n) \in \mathbb{Q}^n$ such that $c_0 \omega + c_1 {}^{2\pi}/_{\tau_1} + \cdots + c_n {}^{2\pi}/_{\tau_n} = 0$. To put it the other way round, $f^*_\omega(x, u) \neq 0$ can only hold if the numbers $\omega, {}^{2\pi}/_{\tau_1}, \ldots, {}^{2\pi}/_{\tau_n}$ are rationally dependent, i. e., if ${}^{2\pi}/_{\tau_1}, \ldots, {}^{2\pi}/_{\tau_n}$ are rationally dependent or $\omega = c_1 {}^{2\pi}/_{\tau_1} + \cdots + c_n {}^{2\pi}/_{\tau_n}$ for some $c_1, \ldots, c_n \in \mathbb{Q}$.

Proposition 3.3.15. *Assume that $(x, u) \in X \times \mathcal{U}$ and $f : X \times \mathcal{U} \to \mathbb{C}$ are such that $t \mapsto f\big(\Psi_t(x, u)\big)$ is quasi-periodic with a locally Lebesgue integrable generating function (see Definition 1.1.3), and periods τ_j, $j = 1, \ldots, n$. If $\omega \in \mathbb{R}$ is such that the numbers $\omega, {}^{2\pi}/_{\tau_1}, \ldots, {}^{2\pi}/_{\tau_n}$ are rationally independent, then $f^*_\omega(x, u) = 0$.*

Proof. See Proposition 2.3.32 □

In the periodic case, we get a stronger result, i. e., the harmonic limit can only be different from zero if the period is a multiple of ${}^{2\pi}/_\omega$. If, e. g., $(x, u) \in X \times \mathcal{U}$ and $f : X \times \mathcal{U} \to \mathbb{C}$ are such that $f\big(\Psi_t(x, u)\big)$ is 2π-periodic, then $f^*_\omega(x, u) \neq 0$ implies $2\pi = k \cdot {}^{2\pi}/_\omega$ for some $k \in \mathbb{Z}$, i. e., $\omega \in \mathbb{Z}$.

Proposition 3.3.16. *Assume that $(x, u) \in X \times \mathcal{U}$ and $f : X \times \mathcal{U} \to \mathbb{C}$ are such that $t \mapsto f\big(\Psi_t(x, u)\big)$ is τ-periodic, $\tau > 0$. If $\omega \in \mathbb{R}$ is such that ${}^{\omega\tau}/_{2\pi} \notin \mathbb{Z}$, then $f^*_\omega(x, u) = 0$.*

Proof. See Proposition 2.3.33. □

Remark 3.3.17. If $\Psi_t(x, u) = (\Phi^u_t x, \theta_t u)$, then τ-periodicity of $\Psi_t(x, u)$ is equivalent to τ-periodicity of $\Phi^u_t x$ and $\theta_t u$. ⌐

Proposition 3.3.15 and Proposition 3.3.16 only gave necessary criteria for the harmonic limit $f^*_\omega(x, u)$ to be different from zero. The following theorem will show the other direction for periodic points. Under an additional condition, at a τ-periodic point $x \in X$, for every frequency where a nonvanishing harmonic limit can possibly exist by Proposition 3.3.16, i. e., for every ω that is a multiple of ${}^{2\pi}/_\tau$, one can find a continuous function $f : X \to \mathbb{C}$ such that $f^*_\omega(x, u) \neq 0$.

Proposition 3.3.18. *Let $\omega > 0$. Let $X \times \mathcal{U}$ be a metric space. Assume that $(x, u) \in X \times \mathcal{U}$ is such that $t \mapsto \Psi_t(x, u)$ is ${}^{2\pi}/_\omega$-periodic. If there is a nontrivial open interval $J \subset \big[0, {}^{2\pi}/_\omega\big)$ such that $t \mapsto \Psi_t(x, u)$ is injective on J, then for every $k \in \mathbb{Z}$, there is $f : X \times \mathcal{U} \to \mathbb{C}$ continuous, such that $f^*_{k\omega}(x, u) \neq 0$.*

Proof. See Theorem 2.3.38. □

Corollary 3.3.19. *Consider the special case $\Psi_t(x, u) = (\Phi^u_t x, \theta_t u)$. Let $\omega > 0$. Let $X \times \mathcal{U}$ be a metric space. Assume that $(x, u) \in X \times \mathcal{U}$ is such that $t \mapsto \Phi^u_t x$ is ${}^{2\pi}/_\omega$-periodic. If ${}^{2\pi}/_\omega$ is the prime period of u, then for every $k \in \mathbb{Z}$, there is $f : X \times \mathcal{U} \to \mathbb{C}$ continuous, such that $f^*_{k\omega}(x, u) \neq 0$.*

Proof. As both $t \mapsto \Phi_t^u x$ and u are $2\pi/\omega$-periodic, also $\Psi_t(x, u)$ is $2\pi/\omega$-periodic, compare Remark 3.3.17. So by Proposition 3.3.18, it suffices to show that there is a nontrivial open interval $J \subset [0, 2\pi/\omega)$ such that $p(t) := \Psi_t(x, u)$ is injective on J.

As $2\pi/\omega$ is the prime period of u, it holds that $t \mapsto \theta_t u$ is injective on $J := (0, 2\pi/\omega)$. Hence also $p(t)$ is injective on J. □

In Chapter 2, harmonic limits were introduced in order to analyze the existence of rotational factor maps. We transfer this concept to the global control setting, too.

Definition 3.3.20. Let $\mathcal{F}$ be a class of functions mapping $X \times \mathcal{U} \to \mathbb{C}$. Suppose that there is a map $F \in \mathcal{F}$, such that $F \not\equiv 0$ and

$$F \circ \Psi_t = \mathrm{e}^{\mathrm{i}t\omega} \cdot F \tag{3.3.1}$$

holds for some $\omega \in \mathbb{R}$ and all t. Then we say that Ψ_t *admits the rotational factor map* $F \in \mathcal{F}$ *with frequency* $\omega/2\pi$. ⌟

Proposition 3.3.21. *Let* $f : X \times \mathcal{U} \to \mathbb{C}$. *If* f_ω^* *exists on some nonvoid set* $M \subset X \times \mathcal{U}$ *and does not vanish everywhere in* M, *then* Ψ_t *admits the rotational factor map* $F := \mathbf{1}_M \cdot \overline{f_\omega^*}$ *with frequency* $\omega/2\pi$ *in the class of bounded functions mapping* $X \times \mathcal{U} \to \mathbb{C}$.

Proof. See Theorem 2.2.9. □

Proposition 3.3.22. *Assume that* Ψ_t *admits the rotational factor map* F *with frequency* $\omega/2\pi$. *Then there is* $f : X \times \mathcal{U} \to \mathbb{C}$ *such that* $f_{-\omega}^* = F$. *In particular,* $f := F$ *satisfies this equation.*

Proof. See part 2. of Theorem 2.2.9. □

3.3.1 Control spectrum

In Subsection 3.2.3, we introduced the harmonic spectrum as the set of all $\omega \in \mathbb{R}$ for which the harmonic limit f_ω^* does not vanish everywhere. This was done for a fixed control $u \in \mathcal{U}$ and a fixed function $f : X \to \mathbb{C}$. Now we let u and f be arbitrary, i. e., we consider the union of all harmonic spectra. This set is called *control spectrum*.

Definition 3.3.23 (Control spectrum). For every Ψ-invariant set $\Xi \subset X \times \mathcal{U}$ and all classes $\mathcal{F}$ of functions mapping $\Xi \to \mathbb{C}$, the *control spectrum* is defined by

$$\Sigma_{\mathrm{C}}(\Xi, \mathcal{F}) := \{\omega \in \mathbb{R} \mid \text{there are } (x, u) \in \Xi \text{ and } f \in \mathcal{F} \text{ such that } f_\omega^*(x, u) \neq 0\}.$$

If $\Xi = X \times \mathcal{U}$, we omit the first argument. If it is clear from the context, which class $\mathcal{F}$ is used, or if the choice does not matter, we omit the second argument. ⌟

We will usually deal with the special case of $\Sigma_C(\Xi)$ for $\Xi = N \times \mathcal{V}$, where $\mathcal{V} \subset \mathcal{U}$ is shift-invariant, and $N \subset X$ is (positively) Φ^u-invariant for every $u \in \mathcal{V}$. Moreover, we usually will have some set $\mathcal{W} \subset \mathcal{U}$, which generates the invariant set $\mathcal{V}$, i.e., $\mathcal{V} = \mathcal{W}^*$, where

$$\mathcal{W}^* := \bigcup_{t \geq 0} \theta_t \mathcal{W}.$$

In this case, we will write $\Sigma_C(N, \mathcal{W}) := \Sigma_C(N \times \mathcal{W}^*)$, or $\Sigma_C(N, \mathcal{W}, \mathcal{F})$ if the choice of $\mathcal{F}$ matters. Furthermore, we will frequently deal with classes $\mathcal{F}$ of functions that do not depend on the control functions, i.e., functions of the form $f(x, u) = \tilde{f}(x)$. To simplify our notation, for a class $\tilde{\mathcal{F}}$ of functions $N \to \mathbb{C}$, we will simply write $\Sigma_C(N, \mathcal{W}, \tilde{\mathcal{F}}) := \Sigma_C\big(N, \mathcal{W}, \{f : \Xi \to \mathbb{C} \mid f(x, u) = \tilde{f}(x), f \in \tilde{\mathcal{F}}\}\big)$.

For all $u \in \mathcal{W}$, the spectrum $\Sigma_H^u(N)$ defined in Subsection 3.2.3 is contained in the control spectrum $\Sigma_C(N \times \mathcal{W}^*)$, and the union of all harmonic spectra equals the control spectrum.

Proposition 3.3.24. *Let $\mathcal{W} \subset \mathcal{U}$, and let $N \subset X$ be (positively) Φ^u-invariant for every $u \in \mathcal{W}$. Then $\bigcup_{u \in \mathcal{W}^*} \Sigma_H^u(N) = \Sigma_C(N \times \mathcal{W}^*)$. If N additionally is negatively Φ^u-invariant for every $u \in \mathcal{W}$, then $\bigcup_{u \in \mathcal{W}} \Sigma_H^u(N) = \Sigma_C(N \times \mathcal{W}^*)$.*

Proof. Let $\omega \in \bigcup_{u \in \mathcal{W}^*} \Sigma_H^u(N)$. Then there is $u \in \mathcal{W}^*$ such that $\omega \in \Sigma_H^u(N)$, i.e., there are $x \in N$ and $u \in \mathcal{W}^*$ such that $f_\omega^*(x, u) \neq 0$. This implies $\omega \in \Sigma_C(N \times \mathcal{W}^*)$. As $\mathcal{W} \subset \bigcup_{t \in \mathbb{R}} \theta_t \mathcal{W} = \mathcal{W}^*$, this particularly means that $\bigcup_{u \in \mathcal{W}} \Sigma_H^u(N) \subset \Sigma_C(N \times \mathcal{W}^*)$.

On the other hand, let $\omega \in \Sigma_C(N \times \mathcal{W}^*)$. Then there are $x_0 \in N$, $u_0 \in \mathcal{W}$ and $\tau \geq 0$ such that $0 \neq f_\omega^*(x_0, \theta_\tau u_0)$. As $\theta_\tau u_u \in \mathcal{W}^*$, this implies $\omega \in \bigcup_{u \in \mathcal{W}^*} \Sigma_H^u(N)$. If N is negatively invariant, there is $x_1 \in N$ such that $\Phi_\tau^{u_0} x_1 = x_0$. Then

$$0 \neq e^{i\omega\tau} f_\omega^*(x_0, \theta_\tau u_0) = e^{i\omega\tau} f_\omega^*(\Phi_\tau^{u_0} x_1, \theta_\tau u_0)$$
$$= e^{i\omega\tau} f_\omega^*\big(\Psi_\tau(x_1, u_0)\big) \overset{\text{Proposition 3.3.7}}{=} f_\omega^*(x_1, u_0).$$

This implies $\omega \in \Sigma_H^{u_0}(N) \subset \bigcup_{u \in \mathcal{W}} \Sigma_H^u(N)$. □

The following monotonicity result holds.

Proposition 3.3.25. *Let $\Xi_1 \subset \Xi_2 \subset X \times \mathcal{U}$ and $\mathcal{F}_1 \subset \mathcal{F}_2$. Then $\Sigma_C(\Xi_1, \mathcal{F}_1) \subset \Sigma_C(\Xi_2, \mathcal{F}_2)$.*

Proof. This is clear by Definition 3.3.23. □

Look at the following example for an illustration of this concept.

Example 3.3.26. Consider the control system

$$\dot{x} = y + u$$
$$\dot{y} = -x + u$$

in $\mathbb{R}^2$ for controls $u \in \{-\sin^2\}^*$, i.e., controls $u(t) = -\sin^2(t+s)$ for $s \in [0, 2\pi)$. The solutions of this system to initial values $x(0) = x_0$ and $y(0) = y_0$ with controls $u_s(t) := -\sin^2(t+s)$ are

$$x(t) = a\sin t + b\cos t + \frac{1}{3}\sin(2t + 2s) - \frac{1}{6}\cos(2t + 2s) - \frac{1}{2}$$

and

$$y(t) = a\cos t - b\sin t + \frac{1}{3}\sin(2t + 2s) + \frac{1}{6}\cos(2t + 2s) + \frac{1}{2},$$

where $a, b \in \mathbb{R}$ are constants depending on s and the initial values, which are given by

$$a := y_0 - \frac{1}{2} - \frac{1}{6}\cos 2s - \frac{1}{3}\sin 2s$$

and

$$b := x_0 + \frac{1}{2} + \frac{1}{6}\cos 2s - \frac{1}{3}\sin 2s.$$

So for $f(x, y) = x$ and $\omega = \pm 1$, we get by Lemma 2.1.8 and Lemma 2.1.7, that

$$f_{\pm 1}^*(x_0, y_0; u_s) = \frac{1}{2}(\pm ia + b),$$

and for $\omega = 2$, we get by the same lemmas, that

$$f_{\pm 2}^*(x_0, y_0; u_s) = \frac{1}{12}e^{-2is}(\pm 2i - 1).$$

Lemma 2.1.7 finally implies that $f_\omega^*(x_0, y_0; u_s) = 0$ for $\omega \in \mathbb{R} \setminus \{\pm 1, \pm 2\}$.

As one can find initial values x_0 and y_0 and a parameter s such that $\pm ia + b \neq 0$, it follows that $\Sigma_C(\mathbb{R}^2, \{-\sin^2\}, \{f\}) = \{\pm 1, \pm 2\}$. ⌟

3.3.2 Dependence of control spectra on control range

Consider a family $\mathcal{U}^\rho$ of sets of admissible control functions, $\rho \geq 0$, and a set $N \subset X$ that is (positively) Φ^u-invariant for all $\rho \geq 0$ and every $u \in \mathcal{U}^\rho$. Denote the control spectrum in dependence of the parameter ρ by $\Sigma_C^\rho(N) := \Sigma_C(N, \mathcal{U}^\rho)$, or $\Sigma_C^\rho(N, \mathcal{F}) := \Sigma_C(N, \mathcal{U}^\rho, \mathcal{F})$ if we want to emphasize the choice of $\mathcal{F}$. Of particular interest is the case where there is a set $U \subset \mathbb{R}^m$ such that, for all $\rho \geq 0$, every control $u \in \mathcal{U}^\rho$ has values in ρU.

We will look at an example first.

Example 3.3.27. Consider the scalar linear control system on $[-1, 1]$ given by

$$\dot{x} = -ax + u(t), \tag{3.3.2}$$

where $a > 0$ and $u(t) \in [-\rho, \rho]$ for all t and some $\rho > 0$ with $a \geq \rho$. This last inequality ensures that the system stays in $[-1, 1]$. The solution of (3.3.2) is given by $\phi(t, x, u) = e^{-at}x + \int_0^t e^{-a(t-s)}u(s)\mathrm{d}s$.

Let $f(x, u) := x$ and $\omega \in \mathbb{R}$, and compute the harmonic limit. For every $T > 0$ and $x \in [-1, 1]$, it holds that

$$
\begin{aligned}
f_\omega^T(x, u) &= \frac{1}{T}\int_0^T e^{\mathrm{i}\omega t}\phi(t, x, u)\mathrm{d}t \\
&= \frac{1}{T}\int_0^T e^{\mathrm{i}\omega t}e^{-at}x\mathrm{d}t + \frac{1}{T}\int_0^T e^{\mathrm{i}\omega t}\int_0^t e^{-a(t-s)}u(s)\mathrm{d}s\mathrm{d}t \\
&= \frac{1}{T}\int_0^T e^{(\mathrm{i}\omega - a)t}x\mathrm{d}t + \frac{1}{T}\int_0^T e^{(\mathrm{i}\omega - a)t}\int_0^t e^{as}u(s)\mathrm{d}s\mathrm{d}t \\
&= \frac{1}{T(\mathrm{i}\omega - a)}e^{(\mathrm{i}\omega - a)t}x\bigg|_{t=0}^{T} + \frac{1}{T(\mathrm{i}\omega - a)}e^{(\mathrm{i}\omega - a)t}\int_0^t e^{as}u(s)\mathrm{d}s\bigg|_{t=0}^{T} \\
&\quad - \frac{1}{T(\mathrm{i}\omega - a)}\int_0^T e^{(\mathrm{i}\omega - a)t}e^{at}u(t)\mathrm{d}t \\
&= \frac{1}{T(\mathrm{i}\omega - a)}e^{(\mathrm{i}\omega - a)T}x - \frac{1}{T(\mathrm{i}\omega - a)} + \frac{1}{T(\mathrm{i}\omega - a)}e^{(\mathrm{i}\omega - a)T}\int_0^T e^{as}u(s)\mathrm{d}s - 0 \\
&\quad - \frac{1}{T(\mathrm{i}\omega - a)}\int_0^T e^{\mathrm{i}\omega t}u(t)\mathrm{d}t \\
&= \frac{1}{T(\mathrm{i}\omega - a)}\left[e^{(\mathrm{i}\omega - a)T}x - 1 + e^{(\mathrm{i}\omega - a)T}\int_0^T e^{as}u(s)\mathrm{d}s\right] \\
&\quad - \frac{1}{T(\mathrm{i}\omega - a)}\int_0^T e^{\mathrm{i}\omega t}u(t)\mathrm{d}t.
\end{aligned}
$$

The term in square brackets is bounded, as

$$
\begin{aligned}
\left|e^{(\mathrm{i}\omega - a)T}x - 1 + e^{(\mathrm{i}\omega - a)T}\int_0^T e^{as}u(s)\mathrm{d}s\right| &\leq e^{-aT} + 1 + e^{-aT}\rho\int_0^T e^{as}\mathrm{d}s \\
= e^{-aT} + 1 + e^{-aT}\cdot\rho\,\frac{1}{a}e^{as}\bigg|_{s=0}^{T} &= e^{-aT} + 1 + e^{-aT}\cdot\rho\left[\frac{1}{a}e^{aT} - \frac{1}{a}\right] \\
= e^{-aT} + 1 + \frac{\rho}{a} - \frac{\rho}{a}e^{-aT} &\leq 1\cdot 1 + 1 + \frac{\rho}{a} \leq 1 + 1 + \frac{a}{a} = 3.
\end{aligned}
$$

This means, the harmonic limit $f_\omega^*(x, u)$ exists, if and only if

$$
-\lim_{T \to \infty}\frac{1}{T(\mathrm{i}\omega - a)}\int_0^T e^{\mathrm{i}\omega t}u(t)\mathrm{d}t \tag{3.3.3}
$$

exists, and in that case, both limits are equal. Note that (3.3.3) does not depend on x, so f_ω^* either exists for all $x \in [-1, 1]$ or does not exist anywhere.

If we choose $u(t) = \rho \sin(\omega t)$ as the control function for some $0 < \rho \leq a$, we get for $\omega \neq 0$, that

$$
\begin{aligned}
-\frac{1}{T(\mathrm{i}\omega - a)} \int_0^T \mathrm{e}^{\mathrm{i}\omega t} u(t) \mathrm{d}t &= -\frac{1}{T(\mathrm{i}\omega - a)} \int_0^T \mathrm{e}^{\mathrm{i}\omega t} \rho \sin(\omega t) \mathrm{d}t \\
&= \frac{\mathrm{i}\rho \cos(\omega T) \sin(\omega T) - \rho \sin^2(\omega T)}{2T(\mathrm{i}\omega - a)\omega} - \frac{\mathrm{i}\rho}{2(\mathrm{i}\omega - a)},
\end{aligned}
$$

which tends to $-\mathrm{i}\rho/2(\mathrm{i}\omega - a)$ for $T \to \infty$. So we get $f_\omega^*\big(x, \rho \sin(\omega \cdot)\big) = -\mathrm{i}\rho/2(\mathrm{i}\omega - a) \neq 0$.

If we choose $u \equiv \rho$ as the control function for some $0 < \rho \leq a$, we get for $\omega = 0$, that $f_\omega^*(x) = \rho/a \neq 0$.

So for every $\omega \in \mathbb{R}$, we find an arbitrarily small control function such that $f_\omega^* \neq 0$. If we let $\mathcal{U}^\rho := \{\rho \sin \mid 0 < \rho \leq a\} \cup \{u \equiv \rho \mid 0 < \rho \leq a\}$, then $\Sigma_{\mathrm{C}}^\rho([-1,1], \{f\}) = \mathbb{R}$ for all $\rho > 0$. ⌟

In this example, $\mathrm{id}_\omega^*(x, u)$ does not depend on x. This might seem contradictory to Corollary 2.3.5, where it is shown that $\mathrm{id}_\omega^*(\Phi_t x) = \mathrm{e}^{-\mathrm{i}\omega t} \mathrm{id}_\omega^*(x)$. But note that we have to look at the flow Ψ_t on $X \times U$ here. So Proposition 3.3.7 implies $\mathrm{id}_\omega^*\big(\Psi_t(x, u)\big) = \mathrm{e}^{-\mathrm{i}\omega t} \mathrm{id}_\omega^*(x, u)$ in this example, which can be true even if id_ω^* does not depend on its first argument.

The preceding example shows that, for some systems, is is possible to create arbitrary frequencies by applying arbitrarily small controls, i. e., that $\Sigma_{\mathrm{C}}^\rho = \mathbb{R}$ for all ρ. We will have a look at some more examples.

Example 3.3.28. Consider the system $\dot{x} = \big(a + u(t)\big)x$ on $\mathbb{R}$ for $u(t) \in [-\rho, \rho]$ with $a \in \mathbb{R}$ and $\rho > 0$. Solutions are given by

$$
\Phi(t, x, u) = x \mathrm{e}^{at + \int_0^t u(s)\mathrm{d}s}.
$$

Denote the set of all control functions with values in $[-\rho, \rho]$ by $\mathcal{W}$.

Case 1: First assume that $\rho < -a$, and choose $f : \mathbb{R} \to \mathbb{C}$ continuous. Then

$$
0 < \mathrm{e}^{at + \int_0^t u(s)\mathrm{d}s} \leq \mathrm{e}^{at + \int_0^t |u(s)|\mathrm{d}s} \leq \mathrm{e}^{at + \rho t} \to 0
$$

for $t \to \infty$. So $\Phi(t, x, u) \to 0$ for all $x \in \mathbb{R}$ and any u. Hence $f_\omega^*(x, u) = 0$ for all $\omega \neq 0$. Furthermore, $f_0^*(x, u) = f(0)$. That means $\Sigma_{\mathrm{C}}^\rho\big(\mathbb{R}, \mathcal{W}, C(\mathbb{R}, \mathbb{C})\big) = \{0\}$.

Note that, as $\rho > 0$, this case can only occur if $a < 0$. So we can interpret this result as follows: Without control, the system exponentially tends to zero with rate $|a|$. If the control radius ρ is lower than the decay rate $|a|$, then it is not possible to create any rotation because, loosely speaking, the controls are not strong enough to counteract the convergence to zero.

Case 2: Now assume that $\rho > |a|$. Let $b := \rho - |a| > 0$. For every $\omega \neq 0$, define the control

$$u_\omega(t) := \begin{cases} b - a & \text{if } t \in \left[\frac{2k\pi}{\omega}, \frac{(2k+1)\pi}{\omega}\right) \text{ for } k \in \mathbb{Z}, \\ -b - a & \text{if } t \in \left[\frac{(2k+1)\pi}{\omega}, \frac{(2k+2)\pi}{\omega}\right) \text{ for } k \in \mathbb{Z}, \end{cases} \tag{3.3.4}$$

which has values in $[-\rho, \rho]$. Note that we will not repeat the subdomains in piecewise defined functions in the following, as they will always be the same as in (3.3.4). Then

$$\int_0^t u_\omega(s)\mathrm{d}s := -at + \begin{cases} bt - \frac{2kb\pi}{\omega}, \\ -bt + \frac{(2k+2)b\pi}{\omega}, \end{cases}$$

and hence

$$\Phi(t, x, u_\omega) = \begin{cases} xe^{bt - 2kb\pi/\omega}, \\ xe^{-bt + (2k+2)b\pi/\omega}. \end{cases}$$

Let $f(x, u) := x$, and compute the harmonic limit. Note that $\Phi(t, x, u_\omega)$ is $2\pi/\omega$-periodic.

$$\begin{aligned}
f_\omega^*(x, u_\omega) &= \frac{\omega}{2\pi} \int_0^{2\pi/\omega} e^{i\omega t} \Phi(t, x, u_\omega) \mathrm{d}t \\
&= \frac{\omega}{2\pi} \int_0^{\pi/\omega} e^{i\omega t} xe^{bt} \mathrm{d}t + \frac{\omega}{2\pi} \int_{\pi/\omega}^{2\pi/\omega} e^{i\omega t} xe^{-bt + 2b\pi/\omega} \mathrm{d}t \\
&= \frac{x\omega}{2\pi} \left[\int_0^{\pi/\omega} e^{(i\omega + b)t} \mathrm{d}t + e^{2b\pi/\omega} \int_{\pi/\omega}^{2\pi/\omega} e^{(i\omega - b)t} \mathrm{d}t \right] \\
&= \frac{x\omega}{2\pi} \left[\frac{1}{i\omega + b} \left(e^{(i\omega + b)\pi/\omega} - 1 \right) + e^{2b\pi/\omega} \frac{1}{i\omega - b} \left(e^{(i\omega - b)2\pi/\omega} - e^{(i\omega - b)\pi/\omega} \right) \right] \\
&= \frac{x\omega}{2\pi} \left[\frac{1}{i\omega + b} \left(e^{i\pi + b\pi/\omega} - 1 \right) + e^{2b\pi/\omega} \frac{1}{i\omega - b} \left(e^{i2\pi - 2b\pi/\omega} - e^{i\pi - b\pi/\omega} \right) \right] \\
&= \frac{x\omega}{2\pi} \left[\frac{1}{i\omega + b} \left(-e^{b\pi/\omega} - 1 \right) + e^{2b\pi/\omega} \frac{1}{i\omega - b} \left(e^{-2b\pi/\omega} + e^{-b\pi/\omega} \right) \right] \\
&= \frac{x\omega}{2\pi} \left[\frac{1}{i\omega + b} \left(-e^{b\pi/\omega} - 1 \right) + \frac{1}{i\omega - b} \left(1 + e^{b\pi/\omega} \right) \right] \\
&= \frac{x\omega}{2\pi} \left[-\frac{1}{i\omega + b} + \frac{1}{i\omega - b} \right] \left(e^{b\pi/\omega} + 1 \right) \\
&= \frac{x\omega}{2\pi} \cdot \frac{-2b}{\omega^2 + b^2} \cdot \left(e^{b\pi/\omega} + 1 \right)
\end{aligned}$$

So $f_\omega^*(x, u_\omega) \neq 0$ if $x \neq 0$. Hence $\Sigma_C^\rho(\mathbb{R}, \mathcal{W}, \{f\}) = \mathbb{R} \setminus \{0\}$.

This can be interpreted as follows: If the control radius ρ exceeds the absulute value of the growth rate a, then it is possible to create rotation by any frequency by applying appropriate controls.

Case 3: It remains to examine the case $0 < \rho \le a$. For every $\omega \neq 0$, define the control

$$u_\omega(t) := \begin{cases} \rho & \text{if } t \in \left[\frac{2k\pi}{\omega}, \frac{(2k+1)\pi}{\omega}\right) \text{ for } k \in \mathbb{Z}, \\ \rho & \text{if } t \in \left[\frac{(2k+1)\pi}{\omega}, \frac{(2k+2)\pi}{\omega}\right) \text{ for } k \in \mathbb{Z}, \end{cases}$$

which has values in $[-\rho, \rho]$. As in Case 2, we will not repeat the subdomains in piecewise defined function from now on. Then we have, similarly to Case 2,

$$\int_0^t u_\omega(s)\mathrm{d}s := \begin{cases} \rho t - \frac{2k\rho\pi}{\omega} \\ -\rho t + \frac{(2k+2)\rho\pi}{\omega}, \end{cases}$$

and hence

$$\Phi(t, x, u_\omega) = \begin{cases} x\mathrm{e}^{(a+\rho)t - 2k\rho\pi/\omega} \\ x\mathrm{e}^{(a-\rho)t + (2k+2)\rho\pi/\omega}. \end{cases}$$

Let $f : \mathbb{R} \to \mathbb{C}$ be given by $x \mapsto \mathrm{e}^{\mathrm{i}\ln|x|}$, and compute the harmonic limit. Note that

$$f_\omega^*(x, u_\omega) = \lim_{K\to\infty} f_\omega^{(2K+2)\pi/\omega}(x, u_\omega)$$

$$= \lim_{K\to\infty} \frac{\omega}{(2K+2)\pi} \sum_{k=0}^K \left[\int_{2k\pi/\omega}^{(2k+1)\pi/\omega} \mathrm{e}^{\mathrm{i}\omega t} f\big(\Phi(t, x, u_\omega)\big)\mathrm{d}t \right. $$

$$\left. + \int_{(2k+1)\pi/\omega}^{(2k+2)\pi/\omega} \mathrm{e}^{\mathrm{i}\omega t} f\big(\Phi(t, x, u_\omega)\big)\mathrm{d}t \right],$$

and that

$$\mathrm{e}^{\mathrm{i}\omega t} f\big(\Phi(t, x, u_\omega)\big) = \mathrm{e}^{\mathrm{i}\omega t} \begin{cases} \mathrm{e}^{\mathrm{i}\ln|x\mathrm{e}^{(a+\rho)t - 2k\rho\pi/\omega}|} \\ \mathrm{e}^{\mathrm{i}\ln|x\mathrm{e}^{(a-\rho)t + (2k+2)\rho\pi/\omega}|} \end{cases}$$

$$= \mathrm{e}^{\mathrm{i}\omega t} \begin{cases} \mathrm{e}^{\mathrm{i}\left(\ln|x| + (a+\rho)t - 2k\rho\pi/\omega\right)} \\ \mathrm{e}^{\mathrm{i}\left(\ln|x| + (a-\rho)t + (2k+2)\rho\pi/\omega\right)} \end{cases}$$

$$= \begin{cases} \mathrm{e}^{\mathrm{i}\left(\ln|x| + (\omega+a+\rho)t - 2k\rho\pi/\omega\right)} \\ \mathrm{e}^{\mathrm{i}\left(\ln|x| + (\omega+a-\rho)t + (2k+2)\rho\pi/\omega\right)}. \end{cases}$$

With this, one can compute that $f_\omega^*(x, u_\omega) = 0$ if and only if one of the following cases holds:

- if $a/\omega \notin \mathbb{Z}$,

- if $a/\omega, \rho/\omega \in \mathbb{Z}$ with same parity,

- $|x| = 1$.

If none of these cases holds, i. e., if

- $a/\omega \in \mathbb{Z}$, and

- $\rho/\omega \notin \mathbb{Z}$ or $\rho/\omega \in \mathbb{Z}$ with a parity different from that of a/ω, and

- $|x| \neq 1$,

then one can compute that

$$f_\omega^*(x, u_\omega) = \frac{[(-1)^L e^{i\pi\rho/\omega} - 1] \cdot i\omega\rho \ln|x|}{(2\omega + L\omega + \rho)(2\omega + L\omega - \rho) \cdot \pi} \neq 0,$$

where $L := a/\omega$. So $\Sigma_{\mathbb{C}}(\mathbb{R}, \mathcal{W}, \{f\}) = \{a/k \mid k \in \mathbb{Z} \setminus \{0\}\}$.

Example 3.3.29. Consider $\dot{x} = Ax + Bu$ with $A \in \mathbb{R}^{n \times n}$, $B \in \mathbb{R}^{n \times m}$ and $u(t) \in \rho U$ for some $\rho > 0$ and a set $U \subset \mathbb{R}^m$. Denote this set of control functions by $\mathcal{U}$. Assume that A is Hurwitz.

The solution of this system is given by $\Phi_t(x, u) = e^{At}x + \int_0^t e^{A(t-s)}Bu(s)\mathrm{d}s$. Let $\lambda \in \operatorname{spec} A$ and let $\Pi : \mathbb{R}^n \to \mathbb{C}^n$ be the projection onto the generalized eigenspace of λ along the other eigenspaces. Note that $\Pi(e^{At}x) = e^{\lambda t}p_x(t)$ for some polynomial $p_x \in \mathbb{C}[t]^n$ (compare Lemma 2.5.5). In particular,

$$p_x(t) = \sum_{j=0}^n \frac{1}{j!}(A - \lambda I)^j t^j \Pi x. \tag{3.3.5}$$

Then for any $\omega \in \mathbb{R}$,

$$\begin{aligned}
\Pi_\omega^T(x, u) &= \frac{1}{T}\int_0^T e^{i\omega t}\Pi\left[e^{At}x + \int_0^t e^{A(t-s)}Bu(s)\mathrm{d}s\right]\mathrm{d}t \\
&= \frac{1}{T}\int_0^T e^{i\omega t}\Pi e^{At}x\mathrm{d}t + \frac{1}{T}\int_0^T e^{i\omega t}\int_0^t \Pi e^{A(t-s)}Bu(s)\mathrm{d}s\mathrm{d}t \\
&= \frac{1}{T}\int_0^T e^{i\omega t}e^{\lambda t}p_x(t)\mathrm{d}t + \frac{1}{T}\int_0^T e^{i\omega t}\int_0^t e^{\lambda(t-s)}p_{Bu(s)}(t-s)\mathrm{d}s\mathrm{d}t \\
&= \frac{1}{T}\int_0^T e^{(\lambda+i\omega)t}p_x(t)\mathrm{d}t + \frac{1}{T}\int_0^T e^{(\lambda+i\omega)t}\int_0^t e^{-\lambda s}p_{Bu(s)}(t-s)\mathrm{d}s\mathrm{d}t.
\end{aligned}$$

By Lemma 3.1.1, for every $x \in \mathbb{R}^n$m there is a polynomial $\tilde{p}_x \in \mathbb{C}[t]^n$ such that $e^{(\lambda+i\omega)t}\tilde{p}_x(t)$ is an antiderivative of $e^{(\lambda+i\omega)t}p_x(t)$. So

$$\frac{1}{T}\int_0^T e^{(\lambda+i\omega)t}p_x(t)\mathrm{d}t \cdot f(x) = \frac{1}{T}e^{(\lambda+i\omega)T}\tilde{p}_x(T) - \frac{1}{T}\tilde{p}_x(0) \to 0$$

for $T \to \infty$, as $\Re\lambda < 0$. Thus

$$f_\omega^*(x, u) = \lim_{T \to \infty} \frac{1}{T} \int_0^T e^{(\lambda + i\omega)t} \int_0^t e^{-\lambda s} p_{Bu(s)}(t - s) \mathrm{d}s \mathrm{d}tm \qquad (3.3.6)$$

if the limit exists.

Note that if $B \cdot \mathbb{R}^m \subset E_\lambda^\perp$, then $p_{Bu(s)}(t) = 0$. So in this case, $f_\omega^*(x, u) = 0$. Otherwise, let $y \in \mathbb{R}^m$ with $\Pi By \neq 0$. For $\rho > 0$, choose $u_\rho(s) := y \cdot \rho \cos \omega s = y \cdot \frac{\rho}{2}\big(e^{i\omega s} + e^{-i\omega s}\big)$ as the control function. Note that p_x is linear in x, compare (3.3.5). So

$$\frac{1}{T} \int_0^T e^{(\lambda + i\omega)t} \int_0^t e^{-\lambda s} p_{Bu_\rho(s)}(t - s) \mathrm{d}s \mathrm{d}t$$

$$= \frac{\rho}{2T} \int_0^T e^{(\lambda + i\omega)t} \int_0^t e^{-\lambda s} p_{By}(t - s)\big(e^{i\omega s} + e^{-i\omega s}\big) \mathrm{d}s \mathrm{d}t$$

$$= \frac{\rho}{2T} \int_0^T e^{(\lambda + i\omega)t} \int_0^t e^{-\lambda s} p_{By}(t - s) e^{i\omega s} \mathrm{d}s \mathrm{d}t$$

$$+ \frac{\rho}{2T} \int_0^T e^{(\lambda + i\omega)t} \int_0^t e^{-\lambda s} p_{By}(t - s) e^{-i\omega s} \mathrm{d}s \mathrm{d}t$$

$$= \frac{\rho}{2T} \int_0^T e^{(\lambda + i\omega)t} \int_0^t e^{-(\lambda - i\omega)s} p_{By}(t - s) \mathrm{d}s \mathrm{d}t$$

$$+ \frac{\rho}{2T} \int_0^T e^{(\lambda + i\omega)t} \int_0^t e^{-(\lambda + i\omega)s} p_{By}(t - s) \mathrm{d}s \mathrm{d}t$$

$$= \frac{\rho}{2T} \int_0^T e^{2i\omega t} \int_0^t e^{(\lambda - i\omega)(t - s)} p_{By}(t - s) \mathrm{d}s \mathrm{d}t + \frac{\rho}{2T} \int_0^T \int_0^t e^{(\lambda + i\omega)(t - s)} p_{By}(t - s) \mathrm{d}s \mathrm{d}t.$$

Substituting $r := t - s$ in the inner integrals yields

$$\frac{1}{T} \int_0^T e^{(\lambda + i\omega)t} \int_0^t e^{-\lambda s} p_{Bu_\rho(s)}(t - s) \mathrm{d}s \mathrm{d}t$$

$$= \frac{\rho}{2T} \int_0^T e^{2i\omega t} \int_0^t e^{(\lambda - i\omega)r} p_{By}(r) \mathrm{d}r \mathrm{d}t + \frac{\rho}{2T} \int_0^T \int_0^t e^{(\lambda + i\omega)r} p_{By}(r) \mathrm{d}r \mathrm{d}t.$$

Recall the existence of a polynomial $\tilde{p}_{By} \in \mathbb{C}[t]^n$ such that $e^{(\lambda + i\omega)t}\tilde{p}_{By}(t)$ is an antiderivative of $e^{(\lambda + i\omega)t} p_{By}(t)$. Similarly, by Lemma 3.1.1, there is a polynomial $\hat{p}_{By} \in \mathbb{C}[t]$ such that $e^{(\lambda - i\omega)t}\hat{p}_{By}(t)$ is an antiderivative of $e^{(\lambda - i\omega)t} p_{By}(t)$. So

$$\frac{1}{T} \int_0^T e^{(\lambda + i\omega)t} \int_0^t e^{-\lambda s} p_{Bu_\rho(s)}(t - s) \mathrm{d}s \mathrm{d}t$$

$$= \frac{\rho}{2T} \int_0^T e^{2i\omega t}\big(e^{(\lambda - i\omega)t}\hat{p}_{By}(t) - \hat{p}_{By}(0)\big) \mathrm{d}t \qquad (3.3.7)$$

$$+ \frac{\rho}{2T} \int_0^T \big(e^{(\lambda + i\omega)t}\tilde{p}_{By}(t) - \tilde{p}_{By}(0)\big) \mathrm{d}t$$

$$= \frac{\rho}{2T} \int_0^T \left(e^{(\lambda+i\omega)t} \hat{p}_{By}(t) - e^{2i\omega t} \hat{p}_{By}(0) \right) dt$$

$$+ \frac{\rho}{2T} \int_0^T \left(e^{(\lambda+i\omega)t} \tilde{p}_{By}(t) - \tilde{p}_{By}0) \right) dt$$

$$= \frac{\rho}{2T} \int_0^T e^{(\lambda+i\omega)t} \left(\hat{p}_{By}(t) + \tilde{p}_{By}(t) \right) dt$$

$$- \frac{\rho}{2T} \int_0^T e^{2i\omega t} \hat{p}_{By}(0) dt - \frac{\rho}{2T} \int_0^T \tilde{p}_{By}(0) dt$$

$$= \frac{\rho}{2T} \int_0^T e^{(\lambda+i\omega)t} \left(\hat{p}_{By}(t) + \tilde{p}_{By}(t) \right) dt$$

$$- \frac{\rho}{4i\omega T} \left(e^{2i\omega T} - 1 \right) \hat{p}_{By}(0) - \frac{\rho}{2} \tilde{p}_{By}(0).$$

Again, there is a polynomial $q \in \mathbb{C}[t]^n$ such that $e^{(\lambda+i\omega)t} q(t)$ is an antiderivative of $e^{(\lambda+i\omega)t} \left(\hat{p}_{By}(t) + \tilde{p}_{By}(t) \right)$. So

$$\frac{\rho}{2T} \int_0^T e^{(\lambda+i\omega)t} \left(\hat{p}_{By}(t) + \tilde{p}_{By}(t) \right) dt = \frac{\rho}{2T} \left(e^{(\lambda+i\omega)T} q(T) - q(0) \right).$$

This tends to zero as $T \to \infty$, because $\Re\lambda < 0$. Similarly,

$$\lim_{T \to \infty} \rho/(4i\omega T) \cdot \left(e^{2i\omega T} - 1 \right) \hat{p}_{By}(0) = 0.$$

This, together with (3.3.6) and (3.3.7), implies that

$$f_\omega^*(x, u_\rho) = -\frac{\rho}{2} \tilde{p}_{By}(0).$$

Now we want to show that $\tilde{p}_{By}(0) \neq 0$. To see this, recall from (3.3.5), that $p_{By} = \sum_{j=0}^n 1/j! \cdot (A - \lambda I)^j t^j \Pi By$. The polynomial $\tilde{p}_{By}$ was defined such that $e^{(\lambda+i\omega)t} \tilde{p}_{By}(t)$ is an antiderivative of $e^{(\lambda+i\omega)t} p_{By}(t)$. As $\lambda + i\omega \neq 0$ (because $\Re\lambda < 0$), Lemma 3.1.1 implies $\tilde{p}_{By}(t) = \sum_{j=0}^{\deg p} (-1)^j (\lambda + i\omega)^{-j-1} p_{By}^{(j)}(t)$. Note that $p_{By}^{(j)}(0) = (A - \lambda I)^j \Pi By$. So

$$\tilde{p}_{By}(0) = \sum_{j=0}^{\deg p} \frac{(-1)^j}{(\lambda + i\omega)^{j+1}} (A - \lambda I)^j \Pi By.$$

Let $S \in \mathbb{C}^{n \times n}$ be such that $J := SAS^{-1}$ is in Jordan normal form. Then

$$\tilde{p}_{By}(0) = S \cdot \sum_{j=0}^{\deg p} \frac{(-1)^j}{(\lambda + i\omega)^{j+1}} (J - \lambda I)^j \cdot S^{-1} \Pi By.$$

As Π is the projection onto the generalized eigenspace to λ, it holds that $(J - \lambda I)^j \cdot S^{-1}\Pi$ does not depend on the Jordan blocks that belong to the other eigenvalues.

So without changing the result, we can assume that $J + i\omega I$ is invertible, which implies that $A + i\omega I$ is invertible. It holds that

$$
\begin{aligned}
\left[-\frac{1}{\lambda + i\omega}(A - \lambda I) - I \right] &= \frac{1}{\lambda + i\omega} \cdot \left[-A + \lambda I - (\lambda + i\omega)I \right] \\
&= \frac{1}{\lambda + i\omega} \cdot \left[-A - i\omega I \right] \\
&= -\frac{1}{\lambda + i\omega} \cdot \left[A + i\omega I \right].
\end{aligned}
\tag{3.3.8}
$$

So also $-\frac{1}{\lambda + i\omega}(A - \lambda I) - I$ is invertible.

Thus we can compute $\tilde{p}_{By}$ with the formula for partial sums of the geometric series.

$$
\begin{aligned}
\tilde{p}_{By}(0) &= \sum_{j=0}^{\deg p} \frac{(-1)^j}{(\lambda + i\omega)^{j+1}} (A - \lambda I)^j \Pi B y \\
&= \frac{1}{\lambda + i\omega} \cdot \sum_{j=0}^{\deg p} \left[-\frac{1}{\lambda + i\omega} \cdot (A - \lambda I) \right]^j \cdot \Pi B y \\
&= \frac{1}{\lambda + i\omega} \cdot \left[-\frac{1}{\lambda + i\omega}(A - \lambda I) - I \right]^{-1} \\
&\quad \cdot \left(\left[-\frac{1}{\lambda + i\omega}(A - \lambda I) \right]^{\deg p + 1} - I \right) \cdot \Pi B y \\
&= -\left[A + i\omega I \right]^{-1} \cdot \left(\left[-\frac{1}{\lambda + i\omega}(A - \lambda I) \right]^{\deg p + 1} - I \right) \cdot \Pi B y \\
&= -\left[A + i\omega I \right]^{-1} \left[-\frac{1}{\lambda + i\omega}(A - \lambda I) \right]^{\deg p + 1} \cdot \Pi B y + \left[A + i\omega I \right]^{-1} \cdot \Pi B y \\
&= -\left[A + i\omega I \right]^{-1} \frac{1}{\left(-(\lambda + i\omega) \right)^{\deg p + 1}} (A - \lambda I)^{\deg p + 1} \cdot \Pi B y \\
&\quad + \left[A + i\omega I \right]^{-1} \cdot \Pi B y.
\end{aligned}
$$

It holds that $(A - \lambda I)^{\deg p + 1} \cdot \Pi = 0$. So $\tilde{p}_{By}(0) = [A + i\omega I]^{-1} \cdot \Pi B y \neq 0$.

So it is possible to produce arbitrary frequencies by applying arbitrarily small controls, i. e., $\Sigma_C^\rho(\mathbb{R}^n, \mathcal{U}, \{\mathrm{id}_{\mathbb{R}^n}\}) = \mathbb{R}$ for all $\rho > 0$. ⌟

As a last example, consider the normal form of the Hopf bifurcation in $\mathbb{R}^2$. This example is treated only numerically here.

Example 3.3.30 (Hopf bifurcation). Consider the normal form of the Hopf bifurcation in $\mathbb{R}^2$

$$
\dot{x} = Ax + (\beta - \|x\|^2)x,
\tag{3.3.9}
$$

where $A := \begin{pmatrix} 0 & -1 \\ 1 & 0 \end{pmatrix}$ and $\beta > 0$. This system has a stable circular limit cycle for $\beta > 0$, which is the circle around the origin with radius $\sqrt{\beta}$. The 2π-periodic solution on the limit cycle is given by $y_1(t) = \sqrt{\beta}\cos t$, $y_2(t) = \sqrt{\beta}\sin t$. One can verify stability with [Ama95, Satz 23.9] by computing the divergence of the right-hand side of (3.3.9) along the periodic solution, i. e.,

$$\operatorname{div}\big(Ax + (\beta - \|x\|^2)x\big) = \beta - 3x_1^2 - x_2^2 + \beta - 3x_2^2 - x_1^2 = 2\beta - 4x_1^2 - 4x_2^2$$

for $x_1(t) = \sqrt{\beta}\cos t$, $x_2(t) = \sqrt{\beta}\sin t$, which yields

$$2\beta - 4\beta\cos^2(t) - 4\beta\sin^2(t) = -2\beta < 0.$$

Apply an additive control on (3.3.9), i. e., consider

$$\dot{x} = Ax + (\beta - \|x\|^2)x + Bu, \tag{3.3.10}$$

where $B \in \mathbb{R}^{n \times m}$ and $u(t) \in \rho U$ for $U \subset \mathbb{R}^m$ compact and convex, and $\rho > 0$. Consider the map $f : \mathbb{R}^2 \to \mathbb{C}$ given by $f(x_1, x_2) := x_1$. Numerical computations indicate that $f_\omega^*\big((1,0), \rho\sin\omega t\big) \neq 0$ for all $\rho > 0$ and $\omega \in \mathbb{R}$. This particularly means that $\Sigma_{\mathrm{C}}^\rho = \mathbb{R}$ for $\mathcal{U}^\rho := \{\rho\sin\alpha t \mid \alpha \in \mathbb{R}\}$ and all $\rho > 0$. So it is possible to produce arbitrary frequencies by applying arbitrarily small controls.

In Figure 3.1, one can see an approximation of $f_\omega^*\big((1,0), \rho\sin\omega t\big)$ for $\rho = 0.1$ and $\omega \in [0.5, 1.5]$. Around $\omega = 1$, i. e., around the frequency $1/2\pi$, there is a region, where the harmonic limit is big compared to frequencies that deviate more from $1/2\pi$. Note that this is exactly the frequency of the stable periodic solution. This means that, if the system is perturbed by a control with period close to that of the original stable periodic solution, it still exhibits rotational behaviour with a frequency close to that of the original stable periodic solution. Numerical computations indicate that the

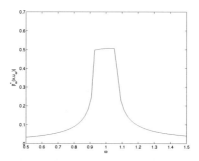

Figure 3.1: Approximation of $f_\omega^*(x, u_\omega)$ for system (3.3.10)
$f(x_1, x_2) := x_1$, $x = (1,0)$, $u_\omega(t) = 0.1\sin\omega t$

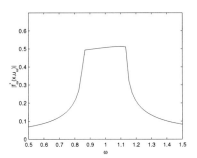

Figure 3.2: Approximation of $f_\omega^*(x, u_\omega)$ for system (3.3.10)
$f(x_1, x_2) := x_1$, $x = (1, 0)$, $u_\omega(t) = 0.2 \sin \omega t$

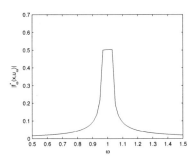

Figure 3.3: Approximation of $f_\omega^*(x, u_\omega)$ for system (3.3.10)
$f(x_1, x_2) := x_1$, $x = (1, 0)$, $u_\omega(t) = 0.05 \sin \omega t$

width of this region depends on the control radius ρ, see Figure 3.2 and Figure 3.3. However, it is not completely understood yet, why this kind of plateau appears in the first place, particularly why it has such sharp edges. ⌟

3.4 Convergent systems

In this section, we focus on so-called *convergent systems*. Those are nonlinear control systems with the property that, for every bounded control, there is a bounded asymptotically stable solution, which is called a *steady-state solution*. Compare [PvN05; PvN06; PWN07].

Consider the system

$$\dot{x} = g(x, u) \tag{3.4.1}$$

on $\mathbb{R}^n$ with a continuous function $g : \mathbb{R}^{n+m} \to \mathbb{R}^n$, which is locally Lipschitz with respect to x, for controls $u : \mathbb{R} \to \mathbb{R}^m$ taken from some class $\mathcal{U}$ of bounded piecewise continuous control functions.

Remark 3.4.1. Note that, by these assumptions on Lipschitz and piecewise continuity, solutions of (3.4.1) always exist and are unique. This follows from [BS98, Theorem A]. ⌟

We will first use the pointwise approach as discussed in Section 3.2, i.e., for each $u \in \mathcal{U}$, we will look at the solution of (3.4.1), which we denote by $\phi(t, x, u) \in \mathbb{R}^n$. If we do not want to specify the initial value x, we simply write $\phi(t, u)$ for a solution of (3.4.1). Later, in Subsection 3.4.3, we will have a look at the global approach, i.e., at the extended solutions $\Psi_t(x, u) = \big(\phi(t, x, u), \theta_t u\big)$.

3.4.1 Stability

Before we give a definition of convergent systems, we need to specify the concept of stability we will be using. Compare [PvN06, Section 2.1.1], particularly Definitions 2.1, 2.2 and 2.3.

Definition 3.4.2. Let $u \in \mathcal{U}$. A solution $\bar{\phi}(t, u)$ of (3.4.1) that is defined for all times $t > t_*$ for some t_*, is called *asymptotically stable in a set* $\mathcal{X} \subset \mathbb{R}^n$, if

1. it is asymptotically stable, i.e.,

   a) for every $t_0 > t_*$ and $\varepsilon > 0$, there is $\delta = \delta(\varepsilon, t_0) > 0$ such that all solutions $\phi(t, u)$ of (3.4.1) with $\|\phi(t_0, u) - \bar{\phi}(t_0, u)\| < \delta$ satisfy $\|\phi(t, u) - \bar{\phi}(t, u)\| < \varepsilon$ for all $t \geq t_0$, and

   b) for every $t_0 > t_*$, there is $\delta = \delta(t_0) > 0$ such that all solutions $\phi(t, u)$ of (3.4.1) with $\|\phi(t_0, u) - \bar{\phi}(t_0, u)\| < \delta$ satisfy $\phi(t, u) - \bar{\phi}(t, u) \to 0$ for $t \to \infty$;

2. for every solution $\phi(t, u)$ of (3.4.1) that starts in $\mathcal{X}$ at some time $t_0 > t_*$, it holds that $\phi(t, u) - \bar{\phi}(t, u) \to 0$ for $t \to \infty$.

If $\mathcal{X} = \mathbb{R}^n$, then $\bar{\phi}(t, u)$ is called *globally asymptotically stable*. ⌟

Remark 3.4.3. Property 1. a) in Definition 3.4.2 is called (Lyapunov) stability. See Figure 3.4 for an illustration. There, the solution $\bar{\phi}(t, u)$ is stable, if every solution starting between the dotted curves stays between the dashed curves for positive time. ⌟

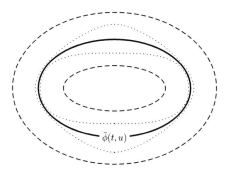

Figure 3.4: Stability of a solution.
See Remark 3.4.3 for an explanation.

Definition 3.4.4. Let $u \in \mathcal{U}$. A solution $\bar{\phi}(t, u)$ of (3.4.1) that is defined for all times $t > t_*$ for some t_* is called *uniformly asymptotically stable in a set* $\mathcal{X} \subset \mathbb{R}^n$, if

1. it is uniformly asymptotically stable, i. e.,

   a) for all $\varepsilon > 0$, there is $\delta = \delta(\varepsilon) > 0$ such that for every $t_0 > t_*$ and all solutions $\phi(t, u)$ of (3.4.1) with $\|\phi(t_0, u) - \bar{\phi}(t_0, u)\| < \delta$, it holds that $\|\phi(t, u) - \bar{\phi}(t, u)\| < \varepsilon$ for all $t \geq t_0$, and

   b) there is $\delta > 0$ such that, for all $\varepsilon > 0$, there is $T(\varepsilon) > 0$ such that all solutions $\phi(t, u)$ of (3.4.1) with $\|\phi(t_0, u) - \bar{\phi}(t_0, u)\| < \delta$ for some $t_0 \geq t_*$ satisfy $\phi(t, u) - \bar{\phi}(t, u) < \varepsilon$ for all $t \geq t_0 + T(\varepsilon)$; and

2. for every compact set $K \subset \mathcal{X}$ and all $\varepsilon > 0$, there is a time $T(\varepsilon, K) > 0$ such that for all solutions $\phi(t, u)$ of (3.4.1) with $\phi(t_0, u) \in K$ for some $t_0 > t_*$ and all $t \geq t_0 + T(\varepsilon, K)$, it holds that $\|\phi(t, u) - \bar{\phi}(t, u)\| < \varepsilon$.

If $\mathcal{X} = \mathbb{R}^n$, then $\bar{\phi}(t, u)$ is called *globally uniformly asymptotically stable*. ⌟

Remark 3.4.5. Property 1. a) in Definition 3.4.4 is called uniform stability. See Figure 3.5 for an illustration. There, the solution $\bar{\phi}(t, u)$ is uniformly stable, if every solution starting between the dotted curves stays between the dashed curves for positive time. ⌟

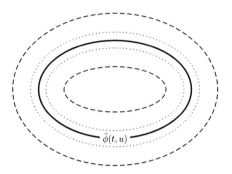

Figure 3.5: Uniform stability of a solution.
See Remark 3.4.5 for an explanation.

Definition 3.4.6. Let $u \in \mathcal{U}$. A solution $\bar{\phi}(t, u)$ of (3.4.1) that is defined for all times $t > t_*$ for some t_* is called *exponentially stable in some set* $\mathcal{X} \subset \mathbb{R}^n$, if

1. it is exponentially stable, i.e., there are constants $\delta > 0$, $C > 0$, and $\beta > 0$ such that for every $t_0 > t_*$ and all solutions $\phi(t, u)$ of (3.4.1) with $\|\phi(t_0, u) - \bar{\phi}(t_0, u)\| < \delta$, it holds that $\|\phi(t, u) - \bar{\phi}(t, u)\| < Ce^{-\beta(t-t_0)}$ for all $t \geq t_0$, and

2. there are constants $C > 0$ and $\beta > 0$, such that every solution $\phi(t, u)$ that starts in $\mathcal{X}$ at some time $t_0 > t_*$ satisfies

$$\|\phi(t, u) - \bar{\phi}(t, u)\| \leq Ce^{-\beta(t-t_0)}\|\phi(t_0, u) - \bar{\phi}(t_0, u)\|$$

for all $t \geq t_0$.

If $\mathcal{X} = \mathbb{R}^n$, then $\bar{\phi}(t, u)$ is called *globally exponentially stable*, and property 1. can be dropped, as it follows from property 2. ⌟

Remark 3.4.7. Note that property 2. in all of these three stability concepts implies that all solutions starting in $\mathcal{X}$ at some time $t_0 > t_*$ exist for all times $t \geq t_0$. ⌟

3.4.2 Definition and properties of convergent systems

Now we define convergent systems, and discuss some of their properties.

Definition 3.4.8. System (3.4.1) is called *(uniformly/exponentially) convergent in some set* $\mathcal{X} \subset \mathbb{R}^n$ *for a class* $\mathcal{U}$ *of piecewise continuous controls*, if, for all controls $u \in \mathcal{U}$, the following properties are satisfied:

1. There is a bounded solution $\bar{\phi}(t, u)$ that is defined on $\mathbb{R}$; and

2. the solution $\bar{\phi}(t, u)$ is (uniformly/exponentially) asymptotically stable in $\mathcal{X}$.

The solution $\bar{\phi}$ is called a *steady-state solution*, and is not unique in general (for *uniformly/exponentially* convergent systems it is unique, compare Proposition 3.4.11). ⌟

Remark 3.4.9. Note that the asymptotic stability of the steady-state solution implies that all solutions starting in $\mathcal{X}$ at some time $t_0 > t_*$ exist for all times $t \geq t_0$. Compare Remark 3.4.7. ⌟

Remark 3.4.10. Note that it is possible to define convergence not only for differential equations like (3.4.1), but also for differential inclusions $\dot{x} \in G(x, u)$, where $G : \mathbb{R}^{n+m} \to \mathfrak{P}(\mathbb{R}^n)$. Compare [PWN07]. ⌟

For some sufficient conditions for convergence, see [PvN06, Section 2.2.4].

Proposition 3.4.11. *If system* (3.4.1) *is uniformly convergent in* $\mathcal{X}$ *for* $\mathcal{U}$*, then for every* $u \in \mathcal{U}$*, the steady-state solution* $\bar{\phi}(t, u)$ *is uniquely determined.*

Proof. Compare [PvN06, Property 2.15]. □

Proposition 3.4.12. *Exponential convergence in* $\mathcal{X}$ *implies asymptotic convergence in* $\mathcal{X}$*.*

Proof. It suffices to show that exponential stability in $\mathcal{X}$ of a solution implies asymptotic stability in $\mathcal{X}$. Let $u \in \mathcal{U}$, and assume that $\bar{\phi}(t, u)$ is an exponentially stable solution of (3.4.1) defined on (t_*, ∞) with constants $C > 0$ and $\beta > 0$.

For every $\varepsilon > 0$, define $\delta(\varepsilon) := \varepsilon/C$. Let $t_0 > t_*$, and let $\phi(t, u)$ be a solution of (3.4.1) starting in $\mathcal{X}$ at time t_0. Then by exponential stability, it holds that

$$\|\phi(t, u) - \bar{\phi}(t, u)\| \leq C\mathrm{e}^{-\beta(t - t_0)} \|\phi(t_0, u) - \bar{\phi}(t_0, u)\| \tag{3.4.2}$$

for all $t \geq t_0$. So if $\|\phi(t_0, u) - \bar{\phi}(t_0, u)\| < \delta$, then $\|\phi(t, u) - \bar{\phi}(t, u)\| < C\mathrm{e}^{-\beta(t - t_0)}\delta(\varepsilon) = \varepsilon\mathrm{e}^{-\beta(t - t_0)} \leq \varepsilon$ for all $t \geq t_0$.

For every $\varepsilon > 0$, define $T(\varepsilon) := 1/\beta \cdot \ln(C/\varepsilon)$. Let $t_0 > t_*$, and let $\phi(t, u)$ be a solution of (3.4.1) with $|\phi(t_0, u) - \bar{\phi}(t_0, u)\| < 1$. Then by exponential stability (3.4.2), it holds for $t \geq t_0 + T(\varepsilon)$, that

$$\|\phi(t, u) - \bar{\phi}(t, u)\| < C\mathrm{e}^{-\beta(t - t_0)} \leq C\mathrm{e}^{-\beta(t_0 + T(\varepsilon) - t_0)}$$
$$= C\mathrm{e}^{-\beta T(\varepsilon)} = C\mathrm{e}^{-\ln(C/\varepsilon)}$$
$$= C\frac{\varepsilon}{C} = \varepsilon.$$

Let $K \subset \mathbb{R}^n$ be compact. For every $\varepsilon > 0$, define $T(\varepsilon, K) := 1/\beta \cdot \ln\left(C/\varepsilon \cdot \operatorname{diam} K\right)$. Let $t_0 > t_*$ and $\phi(t, u)$ be a solution of (3.4.1) such that $\phi(t_0, u) \in K$. Then by exponential stability (3.4.2), it holds for $t \geq t_0 + T(\varepsilon)$, that

$$\|\phi(t, u) - \bar{\phi}(t, u)\| \leq C \mathrm{e}^{-\beta(t - t_0)} \operatorname{diam} K \leq C \mathrm{e}^{-\beta(t_0 + T(\varepsilon) - t_0)} \operatorname{diam} K$$
$$= C \mathrm{e}^{-\beta T(\varepsilon)} \operatorname{diam} K = C \mathrm{e}^{-\ln(C/\varepsilon \cdot \operatorname{diam} K)} \operatorname{diam} K$$
$$= C \frac{\varepsilon}{C \operatorname{diam} K} \operatorname{diam} K = \varepsilon. \qquad \square$$

Convergence has some interesting consequences on the harmonic limit.

Proposition 3.4.13. *Assume that System* (3.4.1) *is convergent in $\mathcal{X}$ for $\mathcal{U}$. For all initial points $x \in \mathcal{X}$ and all controls $u \in \mathcal{U}$, denote by $\bar{\phi}(t, x, u)$ a steady-state solution, to which the solution to u starting in x converges. Then for any continuous $f : X \to \mathbb{C}$ and all $\omega \in \mathbb{R}$, $x \in \mathcal{X}$, and $u \in \mathcal{U}$, it holds that*

$$f_\omega^*(x, u) = \lim_{T \to \infty} \frac{1}{T} \int_0^T \mathrm{e}^{\mathrm{i}\omega t} f \circ \bar{\phi}(t, x, u) \mathrm{d}t, \qquad (3.4.3)$$

provided that the limits exist.

Proof. Let $u \in \mathcal{U}$ and $x \in \mathcal{X}$, and denote by $\phi(t, x, u)$ the solution of (3.4.1) starting in x. By convergence, the steady-state solution $\bar{\phi}(t, x, u)$ is asymptotically stable, i.e., it holds that $\phi(t, x, u) - \bar{\phi}(t, x, u) \to 0$ for $t \to \infty$. So continuity of f and Lemma 2.3.25 imply (3.4.3), provided that either $f_\omega^*(x, u)$ or $\lim_{T \to \infty} 1/T \cdot \int_0^T \mathrm{e}^{\mathrm{i}\omega t} f \circ \bar{\phi}(t, u) \mathrm{d}t$ exists. $\square$

The following proposition is cited from [PvN06].

Proposition 3.4.14. *Assume that System* (3.4.1) *is uniformly convergent in $\mathcal{X}$ for $\mathcal{U}$. If $u \in \mathcal{U}$ is constant, then the steady-state solution $\bar{\phi}(t, u)$ is constant. If $u \in \mathcal{U}$ is τ-periodic for some $\tau > 0$, then the steady-state solution $\bar{\phi}(t, u)$ is τ-periodic.*

Proof. See [PvN06, Property 2.23]. $\square$

This implies, that in globally uniformly convergent systems, it is possible to create arbitrary frequencies by applying periodic controls.

Theorem 3.4.15. *Let $\mathcal{U}$ be a set of $2\pi/\omega$-periodic controls. Assume that the System* (3.4.1) *is uniformly convergent in $\mathcal{X}$ for $\mathcal{U}$ to the steady-state solution $\bar{\phi}(t, u)$. Then for every $k \in \mathbb{N}$, there is a continuous map $f : \mathbb{R}^n \to \mathbb{C}$ such that $f_{k\omega}^*(x, u) \neq 0$ for all $x \in \mathcal{X}$.*

Proof. Let $u \in \mathcal{U}$ and $x \in \mathcal{X}$. Recall from Proposition 3.4.11, that the steady-state solution is uniquely determined, because the system is *uniformly* convergent. By Proposition 3.4.14, the steady-state solution $\bar{\phi}(t, u)$ is $2\pi/\omega$-periodic. So Proposition 3.4.13 implies that

$$f_{k\omega}^*(x, u) = \frac{\omega}{2\pi} \int_0^{2\pi/\omega} \mathrm{e}^{\mathrm{i}k\omega t} f\big(\bar{\phi}(t, u)\big) \mathrm{d}t$$

for every $k \in \mathbb{N}$ and all continuous maps f. So by Corollary 2.3.40, for every $k \in \mathbb{N}$, there is a continuous map $f : \mathbb{R}^n \to \mathbb{C}$ such that $f_{k\omega}^*(t, x) \neq 0$. $\qquad \square$

For exponentially convergent systems and Hölder continuous f, we get the following estimate on how fast the harmonic averages converge to the harmonic limit.

Proposition 3.4.16. *Let $\mathcal{U}$ be a set of $2\pi/\omega$-periodic controls. Assume that System (3.4.1) is exponentially convergent in $\mathcal{X}$ for $\mathcal{U}$. Then for Hölder continuous f, and all $x \in \mathcal{X}$, $u \in \mathcal{U}$, it holds that*

$$\left| f_\omega^*(x) - \frac{1}{T} \int_0^T \mathrm{e}^{\mathrm{i}\omega t} f\big(\phi(t, x, u)\big) \mathrm{d}t \right|$$
$$< \frac{1}{T} \left[\frac{2\pi F}{\omega} + \frac{LC^\alpha}{\alpha\beta} \left[1 - \mathrm{e}^{-\alpha\beta T} \right] \cdot \left\| \bar{\phi}(0, u) - \phi(0, x, u) \right\|^\alpha \right],$$

where

$$F := \max_{t \in [0, 2\pi/\omega)} |f\big(\bar{\phi}(t, u)\big)|,$$

and $C, \beta > 0$ are the constants of exponential convergence (see part 2. of Definition 3.4.6), and $L, \alpha > 0$ are the Hölder constants of f (see Definition 2.1.11). In particular, $\left| f_\omega^*(x) - 1/T \cdot \int_0^T \mathrm{e}^{\mathrm{i}\omega t} f\big(\phi(t, x, u)\big) \mathrm{d}t \right| \in \mathcal{O}(1/T)$ for $T \to \infty$.

Proof. As the system is convergent, it holds that

$$f_\omega^*(x) = \frac{\omega}{2\pi} \int_0^{2\pi/\omega} \mathrm{e}^{\mathrm{i}\omega t} f\big(\bar{\phi}(t, x, u)\big) \mathrm{d}t.$$

So for $T \geq \pi/\omega$, it holds that

$$f_\omega^*(x) - \frac{1}{T} \int_0^T \mathrm{e}^{\mathrm{i}\omega t} f\big(\phi(t, x, u)\big) \mathrm{d}t$$
$$= \frac{\omega}{2\pi} \int_0^{2\pi/\omega} \mathrm{e}^{\mathrm{i}\omega t} f\big(\bar{\phi}(t, u)\big) \mathrm{d}t - \frac{1}{T} \int_0^T \mathrm{e}^{\mathrm{i}\omega t} f\big(\phi(t, x, u)\big) \mathrm{d}t$$
$$= \frac{\omega}{n(T)2\pi} \int_0^{n(T)2\pi/\omega} \mathrm{e}^{\mathrm{i}\omega t} f\big(\bar{\phi}(t, u)\big) \mathrm{d}t - \frac{1}{T} \int_0^T \mathrm{e}^{\mathrm{i}\omega t} f\big(\phi(t, x, u)\big) \mathrm{d}t,$$

where $n(T) := \left\lfloor \frac{T\omega}{2\pi} + \frac{1}{2} \right\rfloor \in \mathbb{N}$. Thus

$$f_\omega^*(x) - \frac{1}{T} \int_0^T \mathrm{e}^{\mathrm{i}\omega t} f\big(\phi(t,x,u)\big) \mathrm{d}t$$

$$= \underbrace{\frac{1}{T} \int_0^T \mathrm{e}^{\mathrm{i}\omega t} f\big(\bar{\phi}(t,u)\big) \mathrm{d}t - \frac{1}{T} \int_0^T \mathrm{e}^{\mathrm{i}\omega t} f\big(\phi(t,x,u)\big) \mathrm{d}t}_{=:I_1(T)}$$

$$+ \underbrace{\frac{\omega}{n(T)2\pi} \int_0^{n(T)2\pi/\omega} \mathrm{e}^{\mathrm{i}\omega t} f\big(\bar{\phi}(t,u)\big) \mathrm{d}t - \frac{1}{T} \int_0^T \mathrm{e}^{\mathrm{i}\omega t} f\big(\bar{\phi}(t,u)\big) \mathrm{d}t}_{=:I_2(T)},$$

which implies

$$\left| f_\omega^*(x) - \frac{1}{T} \int_0^T \mathrm{e}^{\mathrm{i}\omega t} f\big(\phi(t,x,u)\big) \mathrm{d}t \right| \le |I_1(T)| + |I_2(T)|. \tag{3.4.4}$$

First, we look for an estimate of $I_2(T)$. It holds that

$$I_2(T) = \frac{\omega}{n(T)2\pi} \int_0^{n(T)2\pi/\omega} \mathrm{e}^{\mathrm{i}\omega t} f\big(\bar{\phi}(t,u)\big) \mathrm{d}t - \frac{1}{T} \int_0^{n(T)2\pi/\omega} \mathrm{e}^{\mathrm{i}\omega t} f\big(\bar{\phi}(t,u)\big) \mathrm{d}t$$

$$+ \frac{1}{T} \int_0^{n(T)2\pi/\omega} \mathrm{e}^{\mathrm{i}\omega t} f\big(\bar{\phi}(t,u)\big) \mathrm{d}t - \frac{1}{T} \int_0^T \mathrm{e}^{\mathrm{i}\omega t} f\big(\bar{\phi}(t,u)\big) \mathrm{d}t$$

$$= \left(\frac{\omega}{n(T)2\pi} - \frac{1}{T} \right) \int_0^{n(T)2\pi/\omega} \mathrm{e}^{\mathrm{i}\omega t} f\big(\bar{\phi}(t,u)\big) \mathrm{d}t$$

$$- \frac{1}{T} \int_{n(T)2\pi/\omega}^T \mathrm{e}^{\mathrm{i}\omega t} f\big(\bar{\phi}(t,u)\big) \mathrm{d}t.$$

Hence

$$|I_2(T)| \le \left| \frac{\omega}{n(T)2\pi} - \frac{1}{T} \right| \int_0^{n(T)2\pi/\omega} \left| f\big(\bar{\phi}(t,u)\big) \right| \mathrm{d}t$$

$$+ \frac{1}{T} \left| \int_{n(T)2\pi/\omega}^T \left| f\big(\bar{\phi}(t,u)\big) \right| \mathrm{d}t \right|$$

$$\le \left| \frac{\omega}{n(T)2\pi} - \frac{1}{T} \right| \frac{n(T)2\pi}{\omega} F$$

$$+ \frac{1}{T} \left| T - \frac{n(T)2\pi}{\omega} \right| F$$

$$= \frac{1}{T} \left| T - \frac{n(T)2\pi}{\omega} \right| F + \frac{1}{T} \left| T - \frac{n(T)2\pi}{\omega} \right| F$$

$$= 2 \cdot \frac{1}{T} \left| T - \frac{n(T)2\pi}{\omega} \right| F$$

$$\leq 2 \cdot \frac{\pi}{\omega T} F$$
$$= \frac{2\pi F}{\omega T},$$

where the last inequality holds, because $\left| T - \frac{n(T)2\pi}{\omega} \right| \leq \frac{\pi}{\omega}$ by definition of $n(T)$.

Now, we look for an estimate of $I_1(T)$. By global exponential convergence, it holds that $|\bar{\phi}(t,u) - \phi(t,x,u)| \leq Ce^{-\beta t}|\bar{\phi}(0,u) - \phi(0,x,u)|$ for some $C, \beta > 0$. So

$$I_1(T) \leq \frac{1}{T} \int_0^T \left| f(\bar{\phi}(t,u)) - f(\phi(t,x,u)) \right| \mathrm{d}t$$
$$\leq \frac{L}{T} \int_0^T |\bar{\phi}(t,u) - \phi(t,x,u)|^\alpha \mathrm{d}t$$
$$\leq \frac{L}{T} \int_0^T \left(Ce^{-\beta t}|\bar{\phi}(0,u) - \phi(0,x,u)| \right)^\alpha \mathrm{d}t$$
$$= \frac{LC^\alpha}{T} \int_0^T e^{-\alpha\beta t}\mathrm{d}t \cdot |\bar{\phi}(0,u) - \phi(0,x,u)|^\alpha$$
$$= \frac{LC^\alpha}{T\alpha\beta} \left[1 - e^{-\alpha\beta T} \right] \cdot |\bar{\phi}(0,u) - \phi(0,x,u)|^\alpha.$$

Hence from (3.4.4), it follows that

$$\left| f_\omega^*(x) - \frac{1}{T} \int_0^T e^{\mathrm{i}\omega t} f(\phi(t,x,u))\mathrm{d}t \right| < \frac{2\pi F}{\omega T} + \frac{LC^\alpha}{T\alpha\beta} \left[1 - e^{-\alpha\beta T} \right] \cdot |\bar{\phi}(0,u) - \phi(0,x,u)|^\alpha.$$
$$\qquad\qquad\qquad\qquad\qquad\qquad\qquad\qquad\qquad\qquad\qquad\qquad\qquad\qquad\qquad\qquad\square$$

3.4.3 Global analysis

Now we have a look at the global analysis introduced in 3.3 applied to convergent systems. In particular, we consider the extended solutions $\Psi_t(x,u) = \big(\phi(t,x,u), \theta_t u\big)$.

Proposition 3.4.17. *Assume that System (3.4.1) is convergent in $\mathcal{X}$ for $\mathcal{U}$. For all initial points $x \in \mathcal{X}$ and all controls $u \in \mathcal{U}$, denote by $\bar{\phi}(t,x,u)$ a steady-state solution, to which the solution to u starting in x converges. Then for any continuous $f : X \times \mathcal{U} \to \mathbb{C}$ and all $\omega \in \mathbb{R}$, $x \in \mathcal{X}$, and $u \in \mathcal{U}$, it holds that*

$$f_\omega^*(x,u) = \lim_{T\to\infty} \frac{1}{T} \int_0^T e^{\mathrm{i}\omega t} f \circ \Psi_t\big(\bar{\phi}(0,x,u), u\big)\mathrm{d}t$$
$$= \lim_{T\to\infty} \frac{1}{T} \int_0^T e^{\mathrm{i}\omega t} f\big(\bar{\phi}(t,x,u), \theta_t u\big)\mathrm{d}t, \qquad (3.4.5)$$

provided that the limits exist.

Proof. Let $u \in \mathcal{U}$ and $x \in \mathcal{X}$, and denote by $\phi(t, x, u)$ the solution of (3.4.1) starting in x. By convergence, the steady-state solution $\bar\phi(t, x, u)$ is asymptotically stable, i.e., it holds that $\phi(t, x, u) - \bar\phi(t, x, u) \to 0$ for $t \to \infty$. So continuity of f and Proposition 3.3.11 imply (3.4.5), provided that the limits exist. $\qquad\square$

The following theorem states that, for periodic controls, all possible frequencies can be detected by continuous maps f.

Theorem 3.4.18. *Let $\mathcal{U}$ be a set of $2\pi/\omega$-periodic controls. Assume that the System (3.4.1) is uniformly convergent in $\mathcal{X}$ for $\mathcal{U}$ to the steady-state solution $\bar\phi(t, u)$. Then for every $k \in \mathbb{N}$, there is a continuous map $f : \mathbb{R}^n \times \mathcal{U} \to \mathbb{C}$ such that $f_{k\omega}^*(x, u) \neq 0$ for all $x \in \mathcal{X}$.*

Proof. Let $u \in \mathcal{U}$ and $x \in \mathcal{X}$. Recall from Proposition 3.4.11, that the steady-state solution is uniquely determined, because the system is *uniformly* convergent. By Proposition 3.4.14, the steady-state solution $\bar\phi(t, u)$ is $2\pi/\omega$-periodic. So Proposition 3.4.17 implies that

$$f_{k\omega}^*(x, u) = \frac{\omega}{2\pi} \int_0^{2\pi/\omega} e^{ik\omega t} f\big(\bar\phi(t, u)\big) \mathrm{d}t$$

for every $k \in \mathbb{N}$ and all continuous maps f. So by Corollary 3.3.19, for every $k \in \mathbb{N}$, there is $f : \mathbb{R}^n \times \mathcal{U} \to \mathbb{C}$ continuous such that $f_{k\omega}^*(t, x) \neq 0$. $\qquad\square$

For exponentially convergent systems and Hölder continuous f, we get the following estimate on how fast the harmonic averages converge to the harmonic limit.

Proposition 3.4.19. *Let $\mathcal{U}$ be a set of $2\pi/\omega$-periodic controls. Assume that System (3.4.1) is exponentially convergent in $\mathcal{X}$ for $\mathcal{U}$. Then for Hölder continuous $f : X \times \mathcal{U} \to \mathbb{C}$, and all $x \in \mathcal{X}$, $u \in \mathcal{U}$, it holds that*

$$\left| f_\omega^*(x, u) - \frac{1}{T} \int_0^T e^{i\omega t} f\big(\phi(t, x, u)\big) \mathrm{d}t \right|$$
$$< \frac{1}{T} \left[\frac{2\pi F}{\omega} + \frac{LC^\alpha}{\alpha\beta} \big[1 - e^{-\alpha\beta T}\big] \cdot \big\| \bar\phi(0, u) - \phi(0, x, u) \big\|^\alpha \right],$$

where $F := \max_{t \in [0, 2\pi/\omega)} |f(\bar\phi(t, u))|$, $C, \beta > 0$ are the constants of exponential convergence (see part 2. of Definition 3.4.6), and $L, \alpha > 0$ are the Hölder constants of f (see Definition 2.1.11). In particular, $\left| f_\omega^*(x, u) - 1/T \cdot \int_0^T e^{i\omega t} f\big(\phi(t, x, u)\big) \mathrm{d}t \right| \in \mathcal{O}(1/T)$ for $T \to \infty$.

Proof. The proof is completely analogous to the proof of Proposition 3.4.16. $\qquad\square$

For convergent systems that have the uniformly bounded steady-state property, which will be introduced in the following definition, one can show the existence of a so-called frequency response function α. This function is useful in the analysis of harmonic limits.

Definition 3.4.20. A system that is convergent in a set $\mathcal{X}$ for a class $\mathcal{U}$ of controls has the *uniformly bounded steady-state property*, if for all $\rho > 0$, there is $\mathcal{R} > 0$, such that for all controls $u \in \mathcal{U}$ with $\|u(t)\| \leq \rho$ for all $t \in \mathbb{R}$, it holds that $\|\bar{\phi}(t, u)\| \leq \mathcal{R}$ for all $t \in \mathbb{R}$. ⌐

Remark 3.4.21. This definition is taken from [PWN07, Definition 3]. Note that Pavlov, van de Wouw, and Nijmeijer give an equivalent formulation of the uniformly bounded steady-state property in [PvN06, Definition 2.17], which can be generalized to topological spaces as the state space and the codomain of the controls. ⌐

The next theorem shows the existence of the frequency response function.

Theorem 3.4.22. *Consider system* (3.4.1), *and assume that it is globally uniformly convergent with the uniformly bounded steady-state property for the class of piecewise continuous scalar control functions. Then there is a unique continuous map* $\alpha : \mathbb{R}^3 \to \mathbb{R}^n$ *such that* $\alpha(\mathcal{A}\sin\omega t, \mathcal{A}\cos\omega t, \omega)$ *is the steady-state solution corresponding to the control* $u(t) = \mathcal{A}\sin\omega t$.

Proof. See [PvN06, Theorem 4.23]. □

Remark 3.4.23. This Theorem can be generalized to controls that satisfy some differential equation and a boundedness condition. Compare the proof of Theorem 4.32 in [PvN06]. ⌐

The following results show how the frequency response function determines the harmonic limit.

Proposition 3.4.24. *If system* (3.4.1) *is such that Theorem 3.4.22 applies, then for every continuous* $f : \mathbb{R}^n \to \mathbb{C}$ *and all* $\bar{\omega} \in \mathbb{R}$, *it holds that*

$$f_{\bar{\omega}}^*(x, \mathcal{A}\sin\omega t) = \lim_{T \to \infty} \frac{1}{T} \int_0^T \mathrm{e}^{\mathrm{i}\bar{\omega}t} f\big(\alpha(\mathcal{A}\sin\omega t, \mathcal{A}\cos\omega t, \omega)\big)\mathrm{d}t.$$

Proof. This follows from Theorem 3.4.22 and Proposition 3.2.8. □

Corollary 3.4.25. *If system* (3.4.1) *is such that Theorem 3.4.22 applies, then for every continuous* $f : \mathbb{R}^n \to \mathbb{C}$, *it holds that*

$$f_{k\omega}^*(x, \mathcal{A}\sin\omega t) = \frac{\omega}{2\pi} \int_0^{2\pi/\omega} \mathrm{e}^{\mathrm{i}k\omega t} f\big(\alpha(\mathcal{A}\sin\omega t, \mathcal{A}\cos\omega t, \omega)\big)\mathrm{d}t.$$

In particular, $f_{k\omega}^*(x, \mathcal{A}\sin\omega t)$ *depends continuously on* ω.

Proof. This follows from Proposition 3.4.24. □

Proposition 3.4.26. *If system* (3.4.1) *is such that Theorem 3.4.22 applies, then for every continuous* $f : \mathbb{R}^n \times \mathcal{U} \to \mathbb{C}$ *and all* $\bar{\omega} \in \mathbb{R}$, *it holds that*

$$f_{\bar{\omega}}^*(x, A \sin \omega t) = \lim_{T \to \infty} \frac{1}{T} \int_0^T e^{i\bar{\omega}t} f\big(\alpha(A \sin \omega t, A \cos \omega t, \omega), A \sin \omega t\big) dt.$$

Proof. This follows from Theorem 3.4.22 and Proposition 3.4.17. □

Corollary 3.4.27. *If system* (3.4.1) *is such that Theorem 3.4.22 applies, then for every continuous* $f : \mathbb{R}^n \times \mathcal{U} \to \mathbb{C}$, *it holds that*

$$f_{k\omega}^*(x, A \sin \omega t) = \frac{\omega}{2\pi} \int_0^{2\pi/\omega} e^{ik\omega t} f\big(\alpha(A \sin \omega t, A \cos \omega t, \omega), A \sin \omega t\big) dt.$$

In particular, $f_{k\omega}^*(x, A \sin \omega t)$ *depends continuously on* ω.

Proof. This follows from Proposition 3.4.26. □

Theorem 3.4.28. *Consider system* (3.4.1), *and assume that it is globally uniformly convergent with the uniformly bounded steady-state property for the class of piecewise continuous scalar control functions. Then for any class* $\mathcal{F}$ *of continuous functions* $f : \mathbb{R}^n \times \mathcal{U} \to \mathbb{C}$ *and* $\mathcal{U} := \{u(t) = A \sin \omega t \mid A \in \mathbb{R}, \omega \neq 0\}$, *the control spectrum* Σ_C *does not contain nonzero isolated points, i. e., for every* $\omega \in \Sigma_C \setminus \{0\}$, *there is an open interval around* ω *that is contained in* Σ_C.

Proof. Let $\omega_1 \in \Sigma_C$. Then there are $x \in \mathbb{R}^n$, $f \in \mathcal{F}$, and $u \in \mathcal{U}$ such that $f_{\omega_1}^*(x, u) \neq 0$. Let A, ω_2 be such that $u(t) = A \sin \omega_2 t$. By Proposition 3.4.26,

$$f_{\omega_1}^*(x, u) = \lim_{T \to \infty} \frac{1}{T} \int_0^T e^{i\omega_1 t} f\big(\alpha(A \sin \omega_2 t, A \cos \omega_2 t, \omega_2), A \sin \omega_2 t\big) dt.$$

Assume that $\omega_1 \neq 0$. As $\alpha(A \sin \omega_2 t, A \cos \omega_2 t, \omega_2)$ is $2\pi/\omega_2$-periodic and $f_{\omega_1}^*(x, u) \neq 0$, Proposition 3.3.16 implies $\omega_1/\omega_2 \in \mathbb{Z}$, i. e., $\omega_1 = k\omega_2$ for some $k \in \mathbb{Z}$. As $\omega_1 \neq 0$ it holds that $k \neq 0$. By Corollary 3.4.25, $f_{k\omega}^*(x, A \sin \omega t)$ depends continuously on ω. Hence there is an open interval $\mathcal{I} \subset \mathbb{R}$ that contains ω_2 such that $f_{k\omega}^*(x, A \sin \omega t) \neq 0$ for all $\omega \in \mathcal{I}$. So it holds that $k\mathcal{I} \subset \Sigma_C$, i. e., ω_1 is not isolated. □

3.4.4 Examples

Example 3.4.29. Consider the system

$$\dot{x} = -ax + u$$

in $\mathbb{R}$. Compare Example 3.3.27. By [PvN06, Theorem 2.29], this is a globally exponentially convergent system with the uniformly bounded steady-state property for piecewise continuous controls. The solution of this system is given by

$$\phi(t,x,u) = \mathrm{e}^{-at}x + \int_0^t \mathrm{e}^{-a(t-s)}u(s)\mathrm{d}s.$$

So for $u_1(t) := \rho \sin \omega_0 t$, we have

$$\phi(t,x,u_1) = \mathrm{e}^{-at}\left(x + \frac{\rho\omega_0}{a^2 + \omega_0^2}\right) + \frac{\rho a \sin \omega_0 t - \rho\omega_0 \cos \omega_0 t}{a^2 + \omega_0^2}.$$

Hence, the steady-state solution for u_1 is given by

$$x_1(t) = \frac{\rho a \sin \omega_0 t - \rho\omega_0 \cos \omega_0 t}{a^2 + \omega_0^2}.$$

This solution is $2\pi/\omega_0$-periodic, so for arbitrary f, by Proposition 2.3.33, $f_\omega^* \not\equiv 0$ can only be true if $\omega \in \mathbb{Z}\omega_0$. ⌐

Example 3.4.30. This example for a uniformly convergent system with the uniformly bounded steady-state property is taken from [PWN07, Section VI]. Consider the system

$$\begin{aligned}\dot{x}_1 &= -x_1 + x_2^2, \\ \dot{x}_2 &= -x_2 + u.\end{aligned} \tag{3.4.6}$$

If $u(t) = a \sin \omega_0 t$, then the steady-state solution is given by

$$\begin{aligned}x_1(t) = \frac{a^2}{(1 + 4\omega_0^2)(1 + \omega_0^2)^2}\Big[&(2\omega_0^4 + 1)\sin^2 \omega_0 t \\ &+ 2(\omega_0^3 - 2\omega_0)\sin \omega_0 t \cos \omega_0 t + (2\omega_0^4 + 5\omega_0^2)\cos^2 \omega_0 t\Big],\end{aligned} \tag{3.4.7}$$

$$x_2(t) = \frac{a}{1 + \omega_0^2}\big[\sin \omega_0 t - \omega_0 \cos \omega_0 t\big],$$

compare [PWN07]. This solution is $2\pi/\omega_0$-periodic, so by Proposition 2.3.33, for arbitrary f, $f_\omega^* \not\equiv 0$ can only be true if $\omega \in \mathbb{Z}\omega_0$. If $f(x)$ is independent of the second component of x, then $f_\omega^* \not\equiv 0$ can only be true if $\omega \in 2\mathbb{Z}\omega_0$, because $x_1(t)$ is π/ω_0-periodic.

For $f(x_1,x_2) := x_1$, one gets

$$f_\omega^* \equiv \begin{cases} \frac{\mathrm{i}a^2}{4(2\omega_0 + \mathrm{i})(\omega_0 + \mathrm{i})^2} & \text{if } \omega = 2\omega_0, \\ \frac{\mathrm{i}a^2}{4(2\omega_0 - \mathrm{i})(\omega_0 - \mathrm{i})^2} & \text{if } \omega = -2\omega_0, \\ 0 & \text{otherwise.} \end{cases}$$

And for $f(x_1, x_2) := x_2$, one gets

$$
f_\omega^* \equiv
\begin{cases}
\frac{-a}{2(\omega_0 + i)} & \text{if } \omega = \omega_0, \\
\frac{-a}{2(\omega_0 - i)} & \text{if } \omega = -\omega_0, \\
0 & \text{otherwise.}
\end{cases}
$$

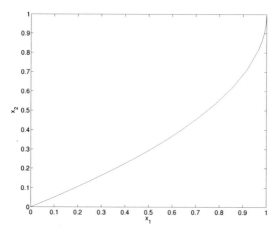

Figure 3.6: Trajectory of system (3.4.6) with $u \equiv 0$ starting in $(1, 1)$

See Figures 3.6, 3.7, 3.8, and 3.9 for some exemplary trajectories. In Figure 3.6, one can see that, without control, the system converges to the origin. In fact, the steady-state solution is constant (namely the origin), as predicted by Proposition 3.4.14. In Figures 3.7 and 3.8, the control $u(t) = \sin t$ was used, and one can see that both trajectories converge to the same periodic steady-state solution, which is given by $x_1(t) = \frac{1}{10}(4\cos^2 t - 2\sin t \cos t + 3)$, $x_2(t) = \frac{1}{2}(\sin t - \cos t)$, see (3.4.7). Figure 3.9 shows a trajectory for the piecewise continuous, periodic control $u(t) = \operatorname{sgn} \sin t$.

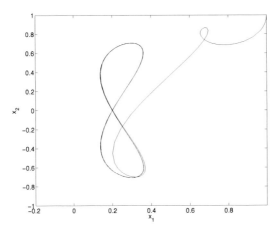

Figure 3.7: Trajectory of system (3.4.6) with $u(t) = \sin t$ starting in $(1, 1)$

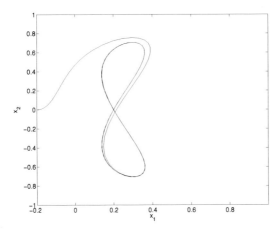

Figure 3.8: Trajectory of system (3.4.6) with $u(t) = \sin t$ starting in $(-0.2, 0)$

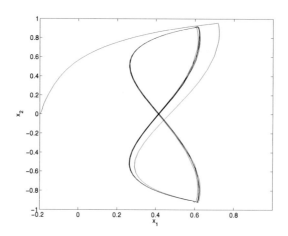

Figure 3.9: Trajectory of system (3.4.6) with $u(t) = \mathrm{sgn}\sin t$ starting in $(1,1)$

List of Figures

Bibliography

The page numbers after each entry refer to the pages where the reference was cited.

[Ama95]   H. Amann. *Gewöhnliche Differentialgleichungen.* 2nd ed. de-Gruyter-Lehrbuch. Berlin, New York: de Gruyter, 1995. See pp. 65, 66, 134–136, 172.

[Ass03]   I. Assani. *Wiener Wintner Ergodic Theorems.* Singapore: World Scientific, 2003. See pp. 25, 76, 79.

[Ass92]   I. Assani. "A Wiener-Wintner property for the helical transform". *Ergodic Theory Dynam. Systems* 12 (1992), pp. 185–194. DOI: 10.1017/S0143385700006672. See p. 25.

[Bau92]   H. Bauer. *Maß- und Integrationstheorie.* 2nd ed. de-Gruyter-Lehrbuch. Berlin, New York: de Gruyter, 1992. See pp. 24, 33, 62, 74, 80, 91, 145, 154, 156.

[BN66]   G. Bachman and L. Narici. *Functional Analysis.* New York, London: Academic Press, 1966. See pp. 71, 93.

[Boh47]   H. Bohr. *Almost periodic functions.* New York: Chelsea Publishing Company, 1947. See pp. 6, 20, 29, 30, 69, 73, 89, 93, 159.

[Bra86]   R. N. Bracewell. *The Fourier transform and its applications.* McGraw-Hill series in electrical engineering. New York: McGraw-Hill, 1986. See p. 12.

[BS98]   A. Bressan and W. Shen. "On Discontinuous Differential Equations". *Differential Inclusions and Optimal Control.* Ed. by J. Andres, L. Górniewicz, and P. Nistri. Vol. 2. Lect. Notes Nonlinear Anal. Nicholas Copernicus University, 1998, pp. 73–87. See p. 174.

[CFJ07]   F. Colonius, R. Fabbri, and R. Johnson. "Chain recurrence, growth rates and ergodic limits". *Ergodic Theory Dynam. Systems* 27 (2007), pp. 1509–1524. DOI: 10.1017/S0143385707000363. See pp. 120, 130, 132.

[CFS82]   I. P. Cornfeld, S. V. Fomin, and Y. G. Sinai. *Ergodic theory.* Grundlehren der Mathematischen Wissenschaften 245. New York: Springer, 1982. See pp. 56, 57, 71, 74, 84, 91, 143.

[CK00]   F. Colonius and W. Kliemann. *The Dynamics of Control.* Systems and
          Control: Foundations & Applications. Boston: Birkhäuser, 2000. See
          pp. 53, 103, 106, 114, 116, 129, 135, 139, 142.

[Dev03]   R. L. Devaney. *An Introduction to Chaotic Dynamical Systems.* 2nd ed.
          Studies in Nonlinearity. Boulder: Westview Press, 2003. See pp. 10, 19.

[DS58]   N. Dunford and J. T. Schwartz. *Linear operators.* General theory. Vol. 1.
          Pure and applied mathematics 7. New York: Interscience Publ., 1958.
          See p. 142.

[Els05]   J. Elstrodt. *Maß- und Integrationstheorie.* Berlin: Springer, 2005. See
          p. 142.

[Far94]   M. Farkas. *Periodic Motions.* Applied Mathematical Sciences 104. New
          York: Springer, 1994. See p. 7.

[Fra03]   J. Franks. "Rotation numbers and instability sets". *Bull. Amer. Math.
          Soc. (N. S.)* 40.3 (2003), pp. 263–279. DOI: 10.1090/S0273-0979-03-
          00983-2. See pp. 1, 8.

[Fra92]   J. Franks. "Geodesics on S^2 and periodic points of annulus homeomor-
          phisms". *Invent. Math.* 108.1 (1992), pp. 403–418. DOI: 10.1007/
          BF02100612. See pp. 1, 8.

[Fur60]   H. Furstenberg. *Stationary processes and prediction theory.* Vol. 44. Ann.
          of Math. Stud. Princeton University Press, 1960. See pp. 25, 76.

[GR07]   I. S. Gradshteyn and I. M. Ryzhik. *Table of Integrals, Series, and Prod-
          ucts.* Amsterdam: Academic Press, 2007. See p. 58.

[Grü00]   L. Grüne. "A Uniform Exponential Spectrum for Linear Flows on Vector
          Bundles". *J. Dynam. Differential Equations* 12.2 (2000), pp. 435–448.
          DOI: 10.1023/A:1009024610394. See pp. 52, 120.

[Hal06]   P. R. Halmos. *Lectures on ergodic theory.* Reprint of original edition
          New York 1956. New York: Chelsea Publishing Company, 2006. See
          p. 43.

[Har82]   P. Hartman. *Ordinary Differential Equations.* 2nd ed. Boston:
          Birkhäuser, 1982. See p. 112.

[Hig08]   N. J. Higham. *Functions of Matrices.* Theory and Computation.
          Philadelphia: SIAM, 2008. See pp. 112, 113.

[HS75]   E. Hewitt and K. Stromberg. *Real and Abstract Analysis.* Graduate
          Texts in Mathematics 25. New York: Springer, 1975. See p. 44.

[IJ90]   G. Iooss and D. D. Joseph. *Elementary Stability and Bifurcation Theory.*
          2nd ed. Undergraduate Texts in Mathematics. New York: Springer, 1990.
          See pp. 107, 117.

[JM82]   R. Johnson and J. Moser. "The rotation number for almost periodic potentials". *Comm. Math. Phys.* 84.3 (1982), pp. 403–438. See pp. 1, 58.

[Kat04]   Y. Katznelson. *An Introduction to Harmonic Analysis.* 3rd ed. Cambridge: Cambridge University Press, 2004. See pp. 81, 84, 85, 93.

[KH06]   A. Katok and B. Hasselblatt. *Introduction to the Modern Theory of Dynamical Systems.* Encyclopedia of Mathematics and its Applications 54. Cambridge: Cambridge University Press, 2006. See pp. 1, 5, 8, 9, 11, 23, 24, 32, 33, 44, 90, 144, 154.

[KM03]   H.-J. Kowalsky and G. O. Michler. *Lineare Algebra.* 12th ed. de-Gruyter-Lehrbuch. Berlin: de Gruyter, 2003. See pp. 98, 99.

[KN74]   L. Kuipers and H. Niederreiter. *Uniform Distribution of Sequences.* New York: John Wiley & Sons, 1974. See p. 76.

[Kön00]   K. Königsberger. *Analysis 2.* 5th ed. Berlin: Springer, 2000. See pp. 40, 41, 66, 94, 95.

[Kön04]   K. Königsberger. *Analysis 1.* 6th ed. Berlin: Springer, 2004. See pp. 65, 121, 149.

[LM08]   Z. Levnajić and I. Mezić. *Ergodic Theory and Visualization II: Visualization of Resonances and Periodic Sets.* 2008. arXiv: 0808.2182v1. See pp. 1, 8, 19, 58.

[LM95]   A. Lasota and M. C. Mackey. *Chaos, Fractals, and Noise.* Applied mathematical sciences 97. New York: Springer, 1995. See pp. 9, 19, 25, 44, 90.

[LY78]   T.-Y. Lie and J. A. Yorke. "Ergodic Transformations from an Interval Into Itself". *Trans. Amer. Math. Soc.* 235 (1978), pp. 183–192. See p. 25.

[LZ82]   B. M. Levitan and V. V. Zhikov. *Almost periodic functions and differential equations.* Cambridge: Cambridge University Press, 1982. See p. 29.

[Maa67]   W. Maak. *Fastperiodische Funktionen.* 2nd ed. Grundlehren der mathematischen Wissenschaften 61. Berlin: Springer, 1967. See p. 7.

[Mañ87]   R. Mañé. *Ergodic Theory and Differentiable Dynamics.* Ergebnisse der Mathematik und ihrer Grenzgebiete. Berlin: Springer, 1987. See pp. 24, 33, 44, 49, 57.

[MB04]   I. Mezić and A. Banaszuk. "Comparison of systems with complex behavior". *Phys. D* 197 (2004), pp. 101–133. DOI: 10.1016/j.physd.2004. 06.015. See pp. 1, 2, 5, 8, 16, 48, 50, 51, 58.

[Mez05]   I. Mezić. "Spectral Properties of Dynamical Systems, Model Reduction and Decompositions". *Nonlinear Dynam.* 41.1–3 (Aug. 2005), pp. 309–325. DOI: 10.1007/s11071-005-2824-x. See p. 50.

[NS89]   V. V. Nemytskii and V. V. Stepanov. *Qualitative Theory of Differential Equations*. Reprint of original edition New Jersey 1960. New York: Dover Publications, 1989. See pp. 75, 158.

[Poi85]   H. Poincaré. "Mémoire sur les courbes définies par les équations différentielles III". *J. Math. Pures Appl.* 4 (1885). See pp. 1, 8.

[PvN05]   A. Pavlov, N. van de Wouw, and H. Nijmeijer. "Convergent Systems: Analysis and Synthesis". *Control and Observer Design for Nonlinear Finite and Infinite Dimensional Systems*. Ed. by T. Meurer, K. Graichen, and E. D. Gilles. Vol. 322. Lecture Notes in Control and Inform. Sci. Springer, 2005, pp. 131–146. See p. 174.

[PvN06]   A. Pavlov, N. van de Wouw, and H. Nijmeijer. *Uniform Output Regulation of Nonlinear Systems*. A Convergent Dynamics Approach. Systems and Control: Foundations & Applications. Boston: Birkhäuser, 2006. See pp. 174, 177, 178, 183, 185.

[PWN07]   A. Pavlov, N. van de Wouw, and H. Nijmeijer. "Frequency Response Functions for Nonlinear Convergent Systems". *IEEE Trans. Automat. Control* 52.6 (2007), pp. 1159–1165. DOI: 10.1109/TAC.2007.899020. See pp. 174, 177, 183, 185.

[Rob99]   C. Robinson. *Dynamical Systems*. Stability, Symbolic Dynamics and Chaos. 2nd ed. Studies in advanced mathematics. Boca Raton: CRC Press, 1999. See pp. 1, 8.

[Row09]   C. W. Rowley et al. "Spectral analysis of nonlinear flows". *J. Fluid Mech.* 641 (2009), pp. 115–127. DOI: 10.1017/S0022112009992059. See pp. 1, 8, 58.

[RS80]   M. Reed and B. Simon. *Functional Analysis*. Methods of modern mathematical physics 1. New York: Academic Press, 1980. See pp. 43, 57.

[Ruf97]   P. R. C. Ruffino. "Rotation numbers for stochastic dynamical systems". *Stochastics* 60.3–4 (1997), pp. 289–318. DOI: 10.1080/1744250970883 4111. See pp. 1, 58.

[Sag94]   H. Sagan. *Space-filling curves*. Universitext. New York: Springer, 1994. See p. 94.

[San88]   L. A. B. San Martin. *Rotation numbers in higher dimensions*. Report 199. Institut für Dynamische Systeme, Universität Bremen, 1988. See pp. 1, 58.

[Sch79]   A. W. Schurle. *Topics in Topology*. North Holland: Elsevier, 1979. See p. 141.

[Ste09]   T. Stender. *Growth Rates for Semiflows with Application to Rotation Numbers for Control Systems*. Vol. 13. Augsburger Schriften zur Mathematik, Physik und Informatik. PhD thesis, Institut für Mathematik, Universität Augsburg. Berlin: Logos Verlag, 2009. See pp. 1, 52, 58, 120–123, 125, 126, 128, 130, 132, 134, 135.

[Str35]   S. Straszewicz. "Über exponierte Punkte abgeschlossener Punktmengen". *Fund. Math.* 24 (1935), pp. 139–143. See p. 132.

[Tak08]   F. Takens. "Orbits with historic behaviour, or non-existence of averages". *Nonlinearity* 21.3 (2008), T33–T36. DOI: 10.1088/0951-7715/21/3/T02. See p. 5.

[Web09]   M. Weber. *Dynamical Systems and Processes*. Vol. 14. IRMA Lectures in Mathematics and Theoretical Physics. Zürich: European Mathematical Society Publishing House, 2009. DOI: 10.4171/046. See p. 25.

[Wer05]   D. Werner. *Funktionalanalysis*. 5th ed. Berlin: Springer, 2005. See pp. 57, 93, 141.

[WW41]   N. Wiener and A. Wintner. "Harmonic Analysis and Ergodic Theory". *Amer. J. Math.* 63.2 (1941), pp. 415–426. See pp. 25, 76.

[Zha03]   C. Zhang. *Almost Periodic Type Functions and Ergodicity*. Beijing: Kluwer Academic Publishers and Science Press, 2003. See pp. 6, 31, 90.

Bibliography

Notation and Symbols

Note that, for Section 2.5, there is an additional list of notation and symbols on page 98.

p' — For a $\mathbb{K}$-differentiable map $p : J \to \mathbb{K}$, where $J \subset \mathbb{R}$ is an interval, and $\mathbb{K} \in \{\mathbb{R}, \mathbb{C}\}$, we denote by p' the derivative of p.

$\subset$ — By $\subset$, we denote set inclusion. Note that we let $A \subset A$, i.e., , we do not exclude set equality.

$p^{(j)}$ — For a j times differentiable map $p : \mathbb{R} \to \mathbb{C}$, we denote by $p^{(j)}$ the j-th derivative of p.

$\mid, \nmid$ — For integers $a, b \in \mathbb{Z}$, we mean by $a \mid b$ that a divides b. Similarly, by $a \nmid b$, we mean that a does not divide b.

$\langle x, y \rangle$ — For two points x, y from an inner product space X, we denote by $\langle x, y \rangle$ their inner product.

$\lfloor x \rfloor$ — For a real number $x \in \mathbb{R}$, we denote by $\lfloor x \rfloor$ the largest integer that is not greater than x, the so-called Gaussian Brackets.

$\#A$ — For a finite set A, we denote by $\#A$ the cardinality of A.

$A^\perp$ — For a subspace A of an inner product space, we denote by $A^\perp$ the orthogonal complement of A.

$\mathbf{1}_B$ — For a set A and a subset $B \subset A$, we denote by $\mathbf{1}_B : A \to \{0, 1\}$ the indicator function given by

$$\mathbf{1}_B(x) := \begin{cases} 0, & \text{if } x \notin B, \\ 1, & \text{if } x \in B. \end{cases}$$

$\arg z$ — For a complex number $z \in \mathbb{C} \setminus \{0\}$, we denote by $\arg z \in [0, 2\pi)$ the argument of z, i.e., the number $\alpha \in [0, 2\pi)$ such that $x = |x| e^{i\alpha}$.

B_1 — By B_1, we denote the unit ball around the origin, either in $\mathbb{C}$ or in $\mathbb{R}^2$, i.e., the set $\{z \in \mathbb{C} \mid |z| < 1\}$ or the set $\{x \in \mathbb{R}^2 \mid \|x\| < 1\}$. It will be clear from the context, which definition we use.

$C(X)$ For a metric space X, we denote by $C(x)$ the set of all continuous functions $X \to \mathbb{C}$.

cl A For a subset A of a topological space, we denote by cl A the closure of A.

D_r By D_r, $r > 0$, we denote the closed disc around the origin with radius r, either in $\mathbb{C}$ or in $\mathbb{R}^2$, i.e., the set $\{z \in \mathbb{C} \mid |z| \leq r\}$ or the set $\{z \in \mathbb{R}^2 \mid |z| \leq r\}$. It will be clear from the context, which definition we use.

$\deg p$ For a polynomial p, we denote by $\deg p$ the degree of p.

denom q For a positive rational number q, we denote by denom q the denominator of q.

diam A For a set $A \subset \mathbb{R}^n$, we denote by diam A the diameter of A.

div f For a map $f : \mathbb{R}^n \to \mathbb{R}^n$, we denote by

$$\operatorname{div} f(x) := \sum_{j=1}^{n} \frac{\partial}{\partial x_j} f_j(x)$$

the divergence of f.

e_j In the linear space $\mathbb{C}^n$, we denote by e_j, $j = 1, \dots, n$, the standard unit vectors.

frac x For a real number $x \in \mathbb{R}$, we denote by

$$\operatorname{frac} x := \operatorname{sgn} x \cdot (|x| - \lfloor |x| \rfloor)$$

the fractional part of x. For example, frac $1.3 = 0.3$, frac$(-1.3) = -0.3$, and frac $2 = 0$.

$\gcd(m, n)$ For two natural numbers $m, n \in \mathbb{N}$, we denote by $\gcd(m, n)$ the greatest common divisor of m and n.

$\Im z$ For a complex number $z \in \mathbb{C}$, we denote by $\Im z$ the imaginary part of z.

lcm N For a set $N \subset \mathbb{N}$, we denote by lcm N the least common multiple of N.

$\log z$ By $\log z$, $z \in \mathbb{C}$, we denote the principal branch of the complex logarithm.

$L^p(\mu)$ For some measure μ on X, we denote by $L^p(\mu)$, $1 \leq p \leq \infty$, the set of complex-valued functions that are p-integrable with respect to μ. If it is clear what measure is used, we omit the argument.

$\mathbb{N}$ By $\mathbb{N}$, we denote the positive integers, i.e., the set $\{1, 2, \dots\}$.

$\operatorname{numer} q$ For a positive rational number q, we denote by $\operatorname{numer} q$ the numerator of q.

$\mathcal{O}\big(g(T)\big)$ For functions $f : \mathbb{R}^+ \to \mathbb{C}$ and $g : \mathbb{R}^+ \to \mathbb{R}^+$, we mean by $f(T) \in \mathcal{O}\big(g(T)\big)$, that f is asymptotically bounded above by g for $T \to \infty$, i.e., that there are $c > 0$ and $T_0 \geq 0$ such that, for all $T \geq T_0$, it holds that $|f(T)| \leq cg(T)$.

$\omega(x)$ For $x \in X$, we denote by $\omega(x) \subset X$ the limit set of x, i.e., $\omega(x) := \{y \in X \mid \exists t_k \to \infty : \Phi_{t_k} x \to y\}$. Note that the limit set ω should not be mistaken for the number $\omega \in \mathbb{R}$.

$\mathbb{R}^+$ By $\mathbb{R}^+$, we denote the nonnegative real numbers, i.e., the set $[0, \infty)$.

$\Re z$ For a complex number $z \in \mathbb{C}$, we denote by $\Re z$ the real part of z.

S^1 By S^1, we denote the unit circle in $\mathbb{C}$, i.e., the set $\big\{z \in \mathbb{C} \mid |z| = 1\big\}$.

$\operatorname{sgn} x$ For a real number $x \in \mathbb{R}$, we denote by $\operatorname{sgn} x$ the sign of x.

S^{n-1} By S^{n-1}, $n \in \mathbb{N}$, we denote the unit sphere in $\mathbb{C}^n$, i.e., the set $\big\{z \in \mathbb{C}^n \mid \|z\| = 1\big\}$.

$\operatorname{spec} A$ For a square matrix A, we denote by $\operatorname{spec} A$ the spectrum of A.

$\theta_t u$ For a function $u : \mathbb{R} \to X$ and $t \in \mathbb{R}$, we denote by $\theta_t u := u(\cdot + t)$ the time-shift u.

Lebenslauf

Persönliche Daten

Name: Tobias Wichtrey

Geburtsort: Augsburg

Geburtsdatum: 27. April 1983

Nationalität: Deutsch

Ausbildung

1990–1994: Besuch der Volksschule Augsburg-Firnhaberau

1994–2003: Besuch des Gymnasiums bei St. Anna, Augsburg; Abitur

2003–2005: Studium der Mathematik an der Universität Augsburg; Vordiplom

2005–2006: Studium der Mathematik im Elitestudiengang TopMath an der Universität Augsburg; Bachelor of Science

2010: Erlangung des Mathematik-Diploms im Rahmen des Elitestudiengangs TopMath an der TU München; Diplom

2006–2010: Promotion am Lehrstuhl für angewandte Analysis mit Schwerpunkt Numerische Mathematik der Universität Augsburg